On Hysteria

On Hysteria

The Invention of a Medical
Category between 1670 and 1820

SABINE ARNAUD

The University of Chicago Press
Chicago and London

Sabine Arnaud is a Max Planck Research Group Director at the Max Planck Institute for the History of Science in Berlin.

The University of Chicago Press, Chicago 60637
The University of Chicago Press, Ltd., London
© 2015 by The University of Chicago
All rights reserved. Published 2015.
Printed in the United States of America

24 23 22 21 20 19 18 17 16 15   1 2 3 4 5

ISBN-13: 978-0-226-27554-3 (cloth)
ISBN-13: 978-0-226-27568-0 (e-book)
DOI: 10.7208/chicago/9780226275680.001.0001

Library of Congress Cataloging-in-Publication Data

Arnaud, Sabine, author.
  [Invention de l'hystérie au temps des Lumières, 1670-1820. English]
  On hysteria : the invention of a medical category between 1670 and 1820 / Sabine Arnaud.
    pages ; cm
  Includes bibliographical references and index.
  ISBN 978-0-226-27554-3 (cloth : alk. paper) — ISBN 978-0-226-27568-0 (e-book)
1. Hysteria—History.   2. Hysteria—France—History—18th century.
3. Hysteria—Early works to 1800.   4. Hysteria—Social aspects.   5. Hysteria in literature.   I. Title.
  RC532.A7613 2015
  616.85′24—dc23

                                                        2014049382

♾ This paper meets the requirements of ANSI/NISO Z39.48–1992 (Permanence of Paper).

For Daniel Carter

Contents

Preface

The words "hysteria" and "hysteric" have become entrenched in today's everyday discourse, immediately calling to mind spectacular manifestations such as convulsions, cries, tears, excessive laughter, and catalepsy. Over the centuries, these words have been used to name an exacerbated body, untimely joy, imminent twitches, the vertigo of desire. Identified with sublime élan or derisory gestures, they alternately provoke fascination and condescension and seem destined to inhabit a world of fiction. In the web of images, disheveled women and artists yearning for spectacle cohabit with mystics in search of transcendence and those distraught by mirage. A whole gamut of oppositions emerges: the private and the public, the intimate and the social, the interior and the codified, the singular and the epidemic, the expressive and the pathological, susceptibility and alienation.

The category of hysteria has a long history in Europe,[1] dating back well before psychoanalysis, to which it later became attached. At the end of the sixteenth century, a number of theorizations emerged to interpret unexplained physiological phenomena in terms of hysteric pathology. A few physicians attempted to rescue women from accusations of demonic possession and from persecution by church and state. Their approach to the body developed over the seventeenth and eighteenth centuries, at a time when a new concern was emerging: the elaboration of diagnoses and their classification into nosologies. There was talk of vapors, suffocation of the womb, fits of the mother, uterine affections, and hysteric and hypochondriac passions. Frequent medical reformulations, the employment of metaphors, and literary and religious references served as opportunities to multiply the significations and connotations of the term "hysteria." They also contributed to the creation of a whole set of genres of medical writing suited to disseminating conceptions about the pathology among the aristocracy and people of letters. Descriptions of hysteric disorders pervaded novels, plays, pamphlets, and correspondence. The diagnosis of hysteria was even used to describe political life, including the crises of the Convulsionaries and the revolutionaries. By 1820, when this book's journey ends, the term "hysteria" had taken firm hold, with two approaches predominating. One determined hysteria as a woman's pathology originating in the genital organs and sexuality; the other, rarer approach situated its origin within the brain. It would take all the skill of psychoanalysis to reorient this first dissemination of explanations and replace them with a model of a "transfer neurosis" provoked by repressed conflicts.

Over the last twenty years, despite impressive developments in the medical humanities, the exchange between medical practitioners and those who attempt to think about medical practices from the outside has remained problematic, with conflicts of authority only aggravating the debate. Indeed, if ever a topic was destined to instigate such quarrels, it is hysteria as it was constructed in the 1970s. The aim of this book is to displace such frameworks for hysteria. Hysteria is not viewed here as a symbol of masculine oppression;[2] neither is the point to posit a historical progression making Mesmer and Puységur forerunners of psychoanalysis. This book does not examine efforts made to cure hysteria, or otherwise attempt to explain hysteria from the viewpoint of the history of medical knowledge. Instead, it examines the history of a category in order to trace the ways that enunciations of medical knowledge operated between 1670 and 1820. In short, hysteria is studied not as an illness, but as an example.

My aim here is to lay out a series of discursive practices that played a major role in the pathology's construction. Because its fits and names were so varied, and because doctors constructed it as a pathology particularly difficult to define, studying hysteria allows us to explore a wide range of issues: (1) It offers a prime opportunity to isolate the role played by language in the definition of any medical category, because of the importance of rhetoric, references, citations, and misunderstandings in its construction. (2) It enables attention to a whole spectrum of written medical genres—dialogue, autobiography, correspondence, narrative, polemic—that have been forgotten by a history of medicine interested in the nineteenth-century pursuit of objectivity. That interest has led historians' readings of the eighteenth century to prioritize what would be retained in later conceptualizations of knowledge: nosologies and the identification of symptoms at the patient's bedside. In contrast, this book reads the eighteenth century in the broad scope of its pursuits, keeping in mind that priorities constantly change with time. (The contemporary renewal of interest in the autobiographical narratives of patients and doctors illustrates this point.) (3) Hysteria became a category in a century marked by the coining of categories (as a result of the ambitious enterprises of Linnaeus and Sauvages, among others). As such, it exemplifies a larger movement, moving from a dozen names to just one and, along with these shifting appellations, between different understandings and definitions of the sick body. (4) The category of hysteria underwent reversals in line with political and religious history that no other category did at the time, opening up our view onto the close connections between epistemology and politics. (5) Hysteria was pictured as a pathology resulting from emerging modernity, so that men of letters, and not only doctors, accorded it great importance— another reason for the category's success. Before and after the French Revolution, many such writers used the pathology to frame an understanding of their time. (6) The conceptualization of hysteria was the scene of a reciprocal construction of sexual difference and a pathology. (7) Finally, the category of hysteria was an ideal means for doctors to draw close to the aristocracy and thereby try to carve out a place for themselves in relation to the higher class. The other important categories at the time, such as plague or venereal disease, lent themselves far less easily to this task. If smallpox was also a subject of literary medical writings, it did not provide a link with sensibility, and thus limited the debate to the value of therapeutic invention and its scientific purpose. Writings on hysteria, in contrast, opened up a constellation of imaginings, caricatures, medical truths, and tools for the building of knowledge that spanned all the discrepancies of observations and references.

Introduction

At the end of the eighteenth century, François Boissier de Sauvages, William Cullen, Philippe Pinel, and Jean-Baptiste Louyer-Villermay assimilated into the single category of hysteria a host of previously known pathologies: uterine furors, suffocation of the womb, fits of the mother, vapors, and hysterical and hypochondriac passions. The very use of the term "hysteria," however, led to radical changes in how the various symptoms and diagnoses were viewed. Physicians secured the authority of the category by inscribing it within a longer tradition of writing. But it was neither the name of the new category, nor its theorization, nor the kinds of patients it treated, nor even its ideological context that enabled such a rereading. What, then, made it possible to view dozens of different diagnoses as variants of a single pathology, hysteria? My hypothesis is that a long process of rewriting and negotiation over the definition of these diagnoses enabled this retrospective assimilation, which was driven by enormously diverse political and epistemological stakes. That process was anything but a linear or untroubled one: the category of hysteria was affirmed through a series of conflicting identifications, equivocations, misinterpretations, and oversights.

This study attempts to untangle the trajectories, associations, and rewritings that developed in the creation of a new medical category. It analyzes how the term "hysteria" spread unsystematically, yet continued to surface as a tool for thought across medical, literary, and political discourses. It also pursues the use of the term in a constellation of different images, commonplaces, roles, and caricatures. The archival sources reveal, for example, that the discrediting of patients and of the diagnosis itself did not occur after its theorization, but rather participated in its construction. They further indicate that the disorders were also constructed as a source of fascination. Medical theorizations espoused, recycled, and displaced these discourses, at times borrowing from their imagery to support their own positions. By studying the construction of hysteria, it will be possible to examine the role played by the forms of its enunciation, which defined the category and in turn transformed the semantic constructions surrounding it.

In reading the texts of physicians, patients, writers, and philosophers about hysteria, my purpose is to examine how, at a certain moment, it became necessary to identify gestures and read them as symptoms of a pathology: what were the stakes that led a number of texts to aggregate around a unique term and thus determine a pathology? Several operations were required: naming, describing a symptomatology, narrating the experience of fits, inscribing symptoms within a tradition of pathologies possibly described by ancient physicians. The questions this book raises are these: How is a pathology created?[1] How is the perception of a diagnosis renewed at particular moments in time, and how does this change who is considered to be afflicted? While interest in the pathology increased throughout the eighteenth century, questionings specific to their times and political contexts prompted the conception of new aspects of the pathology: the crisis of the Convulsionaries in the 1730s, the lettered society of the salon in the 1750s and 1760s, in the late 1780s the French Revolution and the illness of George III in England, and in the 1790s the birth of a new nation in France. In the heart of the eighteenth century, the category of hysteria suddenly became a prominent topic in medical discourse, leading different kinds of patients to be identified and understood through the use of this category.

Eighteenth-century medical knowledge has often been perceived as a series of wavering starts toward the observation of the body. With some noteworthy exceptions, medical texts of the time have been left aside by historians of science in the name of an epistemological break that divides prescientific medicine from the quest for objectivity that has since taken shape.[2] But, in fact, physicians of the eighteenth century saw their role as extending well beyond the cure. Pathologies offered an opportunity to speak about other things: the nature of human being, sexual difference, national differences, social class, modernity, and ways of life. Physicians entered into a dialogue with philosophers in discussing, for example, the relationship between body and mind. During the French Revolution, physicians began to reflect on the role they played in supporting the health of the nation. Thus, eighteenth-century medical knowledge was not an isolated field. Until the publication of Robert James's dictionary in the 1740s, medical terms could be found only in universal dictionaries.[3] Additionally, patients stayed at home rather than entering the hospital space and thus did not fall exclusively under the medical gaze, as they later would. The body's disorders were within the scope of daily life, and this proximity made it a topic to write about beyond therapeutic considerations. It is not unusual to see men and women of letters taking a stand on medical issues or drawing inspiration from pathologies to construct novelistic characters. These texts both invoked and displaced medical knowledge, and together make it possible to observe the constitution of a prescriptive approach to mankind and modernity.

This book addresses the role that writing practices played in shaping medical knowledge. It seeks to demonstrate that different forms of enunciation do not determine just the exterior shape of scientific knowledge, but also its questions and priorities. The purpose is not, therefore, to describe an "evolution" that took place in the definitions or perceptions of hysteria, or indeed to offer a comprehensive catalog of contemporary theorizations. Instead, the book traces the *uses* of the category of hysteria, while considering the functioning and dissemination of texts in which this category appears. This approach makes it possible to consider political and epistemological stakes in conjunction with each other by analyzing the strategies of accreditation and negotiation present in the writings themselves.[4] Another important consideration is that in the eighteenth century, medical knowledge did not start with texts written for colleagues that were later rewritten for a lay audience; rather, it was the outcome of a reciprocal inspiration between texts of different genres, most of which were addressed to the lettered elite. Literary, philosophical, and political uses of the diagnosis continually changed the way that hysteria was perceived and, from the very beginning, the way that the term "hysteria" was used in medicine. A network of references arose that extended beyond the physician's ordinary spheres of competence. The study of medicine thus requires an exploration of the nonmedical corpus in order to measure the echoes and interests that led to future displacements of the category.

The present work considers the development of a medical discourse in its literary, philosophical, anthropological, and political dimensions. It seeks to determine how the construction of the category of hysteria led to the introduction of determining conceptions about men, women, nervous illnesses, sexuality, and modernity, and proceeded to the affirmation of medicine as a discipline. I read medical texts as so many actors participating in epistemological, political, and social history. In the eighteenth century, physicians brought together differing references, representations, myths, and widely circulating ideas. Relationships between pathology and sensibility, genius, social class, sexual difference, and nationality became occasions for readers to identify themselves. Rather than an object of interpretation, the body would appear as an object of symbolic, moral, and political investments.

In tracing the emergence of the category of hysteria the historian faces a complex juxtaposition of histories that crisscross each other and mobilize different aspects of a wide range of pathologies. What was designated and theorized through the various diagnoses assimilated in the term "hysteria" was quite different in 1561, 1667, 1730, 1770, and 1818, with narrations of these afflictions that assigned them different places in society. The years I have named mark key moments in the crystallization of the notions later to become part of hysteria. In 1561, the publication of the works of Ambroise Paré disseminated the definition "suffocation of the womb," locating the origin of convulsions and suffocations in female physiology. In 1667, the publications of Thomas Willis changed the frame: Willis attributed the sporadic and multiple appearances of the convulsions to the circulation of animal spirits through the bodily nerves, and thereby offered a way of unifying the disorders of the entire body and of both sexes. The crises of the French Prophets at the beginning of the eighteenth century in England and of the Convulsionaries around 1730 in France led to another surge of publications, addressing the pathology in terms of religious belief and overexcitable imagination; these works rewrote the existing corpus of medical dissertations on cases of possession in convents. The 1750s and 1760s were characterized by a proliferation of publications on the vapors, this time associating the pathology with class and sensibility and making it one more signifier in a period already dominated by signs, the ancien régime. Breaking the social order, the French Revolution ushered in a complete reinterpretation of the pathology, now considered in terms of gender difference. And the publication of Louyer-Villermay's definition in Panckoucke's 1818 dictionary consolidated that new understanding.

However, this chronological frame by no means equates to a linear progression. Isolating moments of effervescence in the pathology's construction does not do justice to the diversity of surrounding contexts and stakes or to the role played by the cultural imagining of the pathology. The various constructions did not appear in the same kinds of texts, nor were these texts read by the same kinds of readers. From one context to another, different terminology was used and different forms of writing developed, addressing different audiences. The diagnosis first featured in works on women's illnesses and generation in the sixteenth and seventeenth centuries; it reappeared later in treatises devoted to hysteric and vaporous illnesses. In the aftermath of the French Revolution, treatises on the vapors were no longer published, and the terms "hystericism" or "hysteria" were adopted both in nosologies and in works on women, education, and morals.

What justifies gathering together such dissimilar texts? Proceeding from a similarity of symptoms would mean gathering other diagnoses of the time as well, such as tetanus, ergot, and epilepsy. Neither do therapies, theorizations, or approaches to the body provide unifying criteria; and the types of patients varied from one period to another, involving in turn gender, religion, or social class. In fact, the sole element that binds these conceptions into the same history is a network of common references repeated from text to text. The evocation—however brief or unfaithful—of theorizations by previous physicians, quotations, and associations all indicate that their authors thought they were contributing to the development of a shared knowledge.

Clearly, the formation of a new tool for knowledge is anything but a unilateral process. The coexistence of different concepts resulted from the wide range of registers used to write about medical diagnoses that would later be recast as hysteria. These uses were constitutive of so many appropriations and negotiations surrounding the possible significations of the terminology used. A number of distinctive roles for hysteria emerged: an object of knowledge, a metaphor, a word arousing curiosity or complicity, a pretext for stigmatization. From the start, it was impossible to distinguish sharply between literal and figurative meanings. The ambiguity of each diagnosis called for a constant reaffirmation of its meaning, an attempt to circumscribe it, and a struggle to consolidate the category while new uses were displacing it yet again.

Any conceptualization of hysteria must therefore be understood as a collage. Theorizations of the malady operated as a series of grafts, giving hysteria a hybrid character in a body of texts on imagination, sexuality, and generation. Created to gather together physiological phenomena that ranged from convulsion to catalepsy, the category arose in a field saturated with enunciative acts; the naming of convulsive illnesses, women's illnesses, and nervous illnesses proliferated. It was within this sedimentation of conceptions and their diverse and contradictory associations that hysteria was constructed. Rather than recapitulating such interpretations, this study asks how the operation of knowledge configured itself in relation to preexisting constructions of hysteric, uterine, or vaporous pathologies. Both the process of selecting a term and the accompanying theorization were far removed from a progressive selection or exclusion of the causes that provoked the pathology. How, then, did hysteria develop into a singular category; what was the uneven and frequently backtracking movement of its theorization toward assimilation into a single term?

This book thus retraces how different forms of writing on the pathology took shape and shifted across a network of texts. The use of specific references accompanied hysteria, affected its meaning, and made new denominations necessary. It is by analyzing the writing itself that this work examines how political and epistemological stakes participated in the construction of a diagnosis. A history of forms of enunciation in writings gathered retrospectively under a single diagnosis is here privileged over a history of ideas. The term "enunciation" is intended as a way to simultaneously take into account the conceptual choices that make the construction of knowledge possible, the events of writing, political stakes, and the construction of an audience for such writings. I regard the emergence of hysteria as providing the tools for a new narrative of sensibility, ultimately leading to a new distribution of the categories of knowledge.

The construction of hysteria is first and foremost the creation of a discursive space in which a form of knowledge can develop and enter into relations with other forms of knowledge. By examining the rhetoric, narrative, and literary models used in the conceptualization of the pathology, this book follows the path opened by Michel Foucault's works on the role of language in medical knowledge.[5] But it shifts this emphasis slightly. Foucault analyzed the presence of a grammatical and a mathematical model in medicine and saw Philippe Pinel's reference to Condillac as an exemplary case of a medicine that conceived of itself as a semiology. He emphasized how a coherent science developed at the time in a reversibility of the visible and the expressible: the physician can see only what he can say and say only what he can see. To describe is therefore to see and to know at the same time. Foucault's *Birth of the Clinic* powerfully unfolds this chiasmus of seeing and saying at the core of medical thought. Such "measured language" is thought in a mathematical frame, having "the measure of both the things that it describes and the language in which it describes them."[6] However, texts on hysteric affection actually make it possible to identify several other mutations in the role of language. Here, the focus is more the elaboration of a pathology's coherence than the construction of a gaze. At stake is the cohabitation of different systems of writing used to stage different temporalities of discourse. While the need emerged for a language suited to meticulously describing the body—a language born in the very workplace of the hospital and turning all bodily phenomena into a system of signs—there developed, in parallel, various forms of enunciation inlaid with quotations as well as references to political, religious, and moral imaginings.

This book begins by addressing the dissemination of diagnoses later associated with hysteria. Rather than reading the new use of "hysteria" as a sign of changing perceptions, it measures the rupture that the appearance of the term introduced in the enunciation of knowledge, analyzing how specific meanings were created and to what ends. Chapter 1 follows three threads, which would seem at first sight to offer a continuous interpretation: the construction of a female illness; an approach to the pathology that established the role of social class; and a medical interpretation of fits that was proposed in a religious context. Repeated changes emerge in the criteria for identifying the pathology, depending on the political and epistemological priorities of the moment.

To compensate for the stated inadequacy of any definition, physicians agreed on how to characterize the pathology through their selection of metaphors. Chapter 2 analyzes how images such as Proteus, the chameleon, and the Hydra bound together texts from 1575 to 1820 around a similar diagnosis. It also establishes how the use of a reference to Plato's *Timaeus* created continuity among those texts, despite their divergent standpoints. By describing gaps between these usages, this chapter highlights the contradictory nature of the enunciation of medical knowledge.

In chapter 3, the analysis of forms of writing proceeds with a focus on various literary genres adopted by physicians to present the pathology to a lettered and aristocratic audience. Dialogue, autobiography, epistolary treatises, consultation by correspondence, and anecdotes articulated medical knowledge from the beginning of the eighteenth century to the French Revolution. This chapter describes how the development of such formats shaped a new image of the physician as being closer to his patients and presented the pathology as an effect of sensibility and the aristocratic way of life. The focus on genres of writing makes it possible to examine how the development of a new approach to the body, the perception of the readership of medical literature, conceptions of medical activities, and the determination of hysteria were all part of the same process of knowledge.

Men and women of letters used the description of vaporous fits to exemplify the relationship between body and mind irrespective of medical concerns. In chapter 4, works by Paumerelle, Diderot, La Mettrie, Le Camus, and Chassaignon allow us to consider contradictory conceptions of physiological phenomena. Because of their spectacular symptoms, the vapors are repeatedly invoked as an ideal tool for the manipulation either of the sufferer's entourage or of his or her own faculties, pushing these into unexplored areas. The force of the vapors is made to exemplify or stimulate the role of the body in sensibility and creativity.

Chapter 5 examines the increasing role, from the 1750s, of literary and medical narratives in inscribing physiological disorders into a coherent progression. In novels by Lennox, Godwin, and Diderot, the body's manifestations appear as a code, a truth, or a manipulation, depending on the author's specific narrative demands. In narratively structured manuscript reports by correspondents of the Société Royale de Médecine and observations published in medical treatises, hysteria is often presented as a fascinating, and at times a bewildering, pathology. Toward the end of the century, the observations increasingly become a place to see the pathology as a testament to the patient's education, way of life, and emotions.

Chapter 6 analyzes how physicians reframed their social and scientific status in the wake of the French Revolution. A first case study is the writing of Pierre Pomme and his art of mystifying readers by locating his practice within a theological frame. A second case looks at magnetism and the role of somnambulism in envisioning an extraordinary practice. The chapter then investigates how, starting in the 1770s, physicians cast hysteria as a dangerous symptom of modernity. In the aftermath of the French Revolution, this type of discourse served as an opportunity to expand the role of medicine. The new focus on the health of the nation's body shifted the stakes of hysteric illness. It was no longer a diagnosis belonging to the upper classes, but one shaped to facilitate a medical discourse on women.

Initially presented as an isolated and extreme case, hysteria became a common diagnosis in the second half of the eighteenth century, well before it was ever presented as a risk to society. The question this book asks is not how the interpretation of hysteria was transformed during this period, but what was at stake in the writing of the diagnosis.

Names and Uses of a Diagnosis

For it is not by describing that words acquire their power: it is by naming, by calling, by commanding, by intriguing, by seducing that they slice into the naturalness of existences, set humans on their path, separate them and unite them into communities. The word has many other things to imitate besides its meaning or its referent: the power of speech that brings it into existence, the movement of life, the gestures of an oration, the effect it anticipates, the addressee whose listening or reading it mimics beforehand.

JACQUES RANCIÈRE[1]

The Establishment of Hysteria as a Medical Category

In 1818, Jean-Baptiste Louyer-Villermay dedicated nearly fifty pages to hysteria in Charles-Joseph Panckoucke's *Dictionnaire des sciences médicales* (Dictionary of Medical Sciences), thereby establishing its importance as a medical category.[2] The text begins with a list of twelve diagnoses, which are presented as so many synonyms: "*hystéricie*, hystericism, *hysteralgie*, hysteric passion and hysteric affection, uterine affections, suffocation of the womb, strangulation of the uterus, fits of the mother: this illness has also been called 'hysteric vapors,' 'ascension of the womb,' and 'uterine neuroses.'"[3] Louyer-Villermay was attempting to present the category of hysteria as encompassing maladies that had existed for centuries but that could now be approached ahistorically. His contribution was not so much a recognition of specific symptoms as a gathering together of divergent and conflicting interpretations. Louyer-Villermay's act of definition served as an opportunity to eliminate certain features of hysteria and force through others, with the aim of facilitating the physician's use of the diagnosis. Through a series of equivalences, he reduced a broad spectrum of approaches to the body to a unique pathology.

After placing these theories under the aegis of a single category, Louyer-Villermay proceeded to assign them a single point of origin: the female uterus. He justified his selection of the term "hysteria" as follows: "It expresses the idea attached to it quite well, and . . . use has consecrated it."[4] The noun was read as metonym, functioning as proof that the origin of the illness was in the womb. Louyer-Villermay continued:

Were it not so, one would at least have to change the denomination; for the word "hysteria" implies the nonexistence of this illness in the man. The impropriety of terms being, in science, the first twist given to reason, this word could not have been kept if it did not represent an exact idea, that of an illness proper to woman.[5]

The choice of the word "hysteria" was thus confirmed in the name of upholding the validity of contemporary science, along with an explicit concern for the exacting use of terminology. An equivalence was postulated: a perfect reciprocity between idea and terminology would lead the way toward a complete and exact grasp of the pathology. In 1816, Louyer-Villermay summarized in a single sentence the transformation of the suffocation of the womb, the vapors, and various hysteric illnesses into what would come to be seen as hysteria: "A man cannot be hysterical, he has no uterus."[6] He presented this assertion as evidence that was firmly grounded in centuries of medical history. The creation and use of the category hysteria had thus become an accomplished fact, and the lack of any direct opposition to the use of the term allowed Louyer-Villermay to claim a consensus—even making this consensus the crux of his explanation. Still, in 1818 it was a recent and fragile consensus, and occurrences of the term "hysteria" had been rare until its appearance in Panckoucke's dictionary.

As it happens, it was the use of the term by two eminent physicians during the first two decades of the nineteenth century that allowed Louyer-Villermay to assert the category in such an authoritative manner. The physicians represented rival institutions: Joseph-Marie-Joachim Vigarous was professor at the Faculty of Medicine in Montpellier, while Philippe Pinel held the equivalent post in Paris.[7] In 1801, Vigarous assimilated seven diagnoses under the banner of hysteria in his *Cours élémentaire de maladies des femmes* (Elementary Course on the Illnesses of Women).[8] Meanwhile Pinel defined hysteria as a kind of neurosis in his nosographical works. Pinel, the chief physician at the Salpétrière, did not merely localize the origin of the pathology in the female organ; he further identified the dysfunction of the body as resulting from unassuaged sexual needs. To confirm the origins of the pathology, he referred to an ancient medical practice: "The primitive seat of hysteria, as its name indicates, is the womb, and very often too an austere continence is one of its determining causes; this has given rise to a method known to all matrons, and which Ambroise Paré describes with his usual naïveté, prescribing then the use of frictions."[9] Citing a therapeutic practice in this manner gave Pinel an expedient means to establish the origin of the pathology.[10] On the one hand, Pinel attempted to remove all doubts by appealing to the authority of Ambroise Paré, Henri II's adviser and surgeon, famous for having revolutionized surgery. On the other, the reference to Paré allowed Pinel to present himself

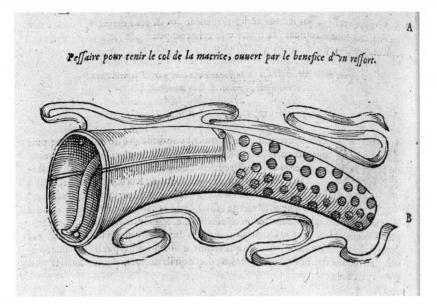

FIGURE 1. Pessary (1) (Ambroise Paré, *De la Génération*, in *Œuvres* [Paris: Gabriel Buon IX, 1585], 978). Image courtesy of the Bibliothèque interuniversitaire de santé.

and medicine from a relatively decorous perspective, separating the more abstract world of physicians from that of surgeons and midwives. The "method known to all matrons" was the use of masturbation and the insertion of a pessary into the vagina (fig. 1). This instrument was a perforated metallic cylinder through which the fumes of various herbs would pass (fig. 2). These fumes were meant to attract the womb toward the lower part of the belly, thereby combating the suffocation of the womb thought to result from its ascension. By using the crude authenticity of Paré's words as additional proof of his assertion, Pinel was able to present hysteria as an exclusively female affliction, with its origin localized in the uterus.

For Pinel, Vigarous, and Louyer-Villermay, the very name "hysteria" operated as proof of the origin of the pathology, which was now attached to a specific gender. Further, for lack of a better argument, the etymology of "hysteria" made it possible to explain the causes of its symptoms. A strict equivalence was established between the noun, the idea, and the origin of the pathology, exhibiting a theoretician's desire for methodological rigor. Here again, the etymological root was used as a means to ennoble science, and the category of hysteria was in part validated through an appeal to the authority of classical antiquity.

When justifying their terminology, Louyer-Villermay and Pinel did not mention that "hysteria" was a relatively new term, one that was absent from

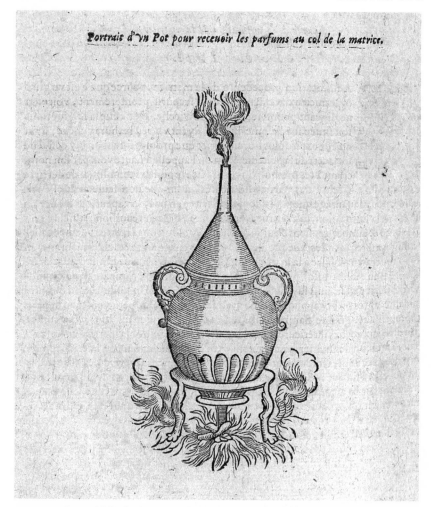

Portrait d'vn Pot pour receuoir les parfums au col de la matrice.

FIGURE 2. Pessary (2) (Ambroise Paré, *De la Génération*, in *Œuvres* [Paris: Gabriel Buon IX, 1585], 978). Image courtesy of the Bibliothèque interuniversitaire de santé.

the treatises they themselves considered most relevant.[11] Admittedly, the evocation of a hysteric pathology with an origin in the womb went back to ancient times. Ilza Veith identifies a first description in the Kahun Papyrus (1900 BC), though she adds that no equivalent of the term "hysteria" itself can be found in any of these texts.[12] When examining the long genealogy of "hysteria," it is easy to neglect the fact that in French and English the adjective "hysteric" does not appear until the second half of the sixteenth century.[13] King Lear's mention of *hysterica passio*[14] was one of its first appearances in the literary space,[15] but the noun does not appear in French until 1703 or in English until 1764.[16]

As Étienne Trillat and Helen King have shown, in the Hippocratic corpus the treatise *On the Illnesses of Women* expounded on illnesses "of the uterus" or those "localized in the uterus."[17] One of these illnesses, referred to as *suffocatio hysterica*, was characterized by convulsions, breathlessness, and suffocation. An aphorism from the Hippocratic corpus proposed marriage as its most appropriate treatment. Because the description was so vague, authors would be able to cite these texts for centuries to come. Until the twentieth century, for example, the necessity for marriage would continually appear as a polite means of speaking about the womb's inopportune desires. An ancient aphorism, the similarity of a few symptoms—these were the seeds of a lengthy tradition, the origin of a palimpsest, a series of references repeatedly rewritten and erased, constituting a sedimentation of references for thinking about hysteria. Cross-references to hysteric illnesses would give shape to future medical treatises despite the absence of the term itself. By reading hysteric affection as Hippocrates's discovery, physicians of the sixteenth, seventeenth, and eighteenth centuries became invested in a highly specific genealogy of texts.

What happened, then, at the beginning of the nineteenth century, for the etymology of the word to make the cause of this pathology a statement of the obvious? What specific symptoms were implied by the diagnoses that Louyer-Villermay compiled in Panckoucke's dictionary? And was the uterus singled out as the origin of these symptoms? How did hysteria become a "women's illness"? Around 1800, interest in hysteria fell within a movement to reorganize medicine into a unified and closed body of knowledge. Traditions at times parallel, at times opposed, were associated with each other through echoes and quotations that were repeated or at times transformed. Vigarous, Pinel, and Louyer-Villermay conceived of their work as an attempt to achieve convergence, a retrospective attempt to recognize and clearly identify that which had previously been known only indistinctly. Charles E. Rosenberg has argued that "in some ways disease does not exist until we have agreed that it does, by perceiving, naming and responding to it,"[18] and the reinscription of symptoms and diagnoses in new categories also results in a repeated refashioning of diseases. Looking back to the late sixteenth century and following the terms and interpretations attached to each of these illnesses will demonstrate just how shifting and variable these interpretations had been. However, considering the use of the terminology rather than its origins also makes it possible to ask how a term prompted a way of thinking and erased the gap between naming an illness and situating its origin. The word is examined here in its capacity to anchor and transform an accompanying discourse, as a point of departure for a series of utterances within the field of knowledge. The use of the term produced an epistemological shift as different afflictions converged in the diagnosis "hysteria." Between 1575 and 1820, perceptions of the pathology's association with gender, with class, and with religion developed along different trajectories and through different forms of narrating, explaining, and citation. How were these categories used, on what occasions, and with what agendas? What kind of displacements and gaps appear in the use of terms that were later gathered into the category of "hysteria"?

An Intermingling of Terms

Until the end of the seventeenth century, writings on the suffocation of the womb,[19] uterine suffocation,[20] uterine furor,[21] *mal di matrice*,[22] melancholic furors,[23] loving rage,[24] erotic melancholia,[25] and fits of the mother[26] indicated a variety of symptoms, notably paleness, stifling, stiffness, swoons, and catalepsy. The theorizations and references contained in these texts closely followed a long tradition of Greek, Latin, and Arabic works, retraced by Helen King.[27] The predominant category of the era was "suffocation of the womb," for which the first description in vernacular appeared in the work *De la Génération* (On Generation) by Ambroise Paré, published in 1561.[28] From then on, this type of diagnosis became a subject of medical inquiry in works addressing medical practitioners and devoted to women's illnesses and generation.[29] Despite the variety of names and contexts in which the description of such symptoms took place, most works indicated that hysteric illnesses were caused by vapors, which would eventually become a diagnostic term in its own right. Vapors circulating throughout the body became the explanation for pathologies exhibiting all manner of symptoms, attacking different parts of the body, without an immediately assignable cause or progression of symptoms. In the sixteenth and seventeenth centuries, some physicians also described a "wandering womb" and ascribed the cause of suffocation to the womb's movements.[30]

Jean Liébault, the sixteenth-century medical practitioner at the Faculté de Paris, wrote that vapors were a consequence of retaining menstrual blood.[31] In his *Trois Livres des maladies et infirmitez des femmes* (Three Books on Women's Maladies and Infirmities), published in 1598 and reprinted and expanded in 1651,[32] Liébault situated the origin of the pathology in the lower belly. According to him, the womb could move upward and along the side of the woman's body—even if the rare dissections of the time did not make it possible to identify where the womb would migrate or the vapors circulate.[33] Such movements provoked contractions of the diaphragm, resulting in the womb's suffocation as it sought to expel vapors and excess air. But in which directions did these vapors migrate? Liébault described the vapors as moving through veins, arteries, and occult openings inside the body.[34]

Similarly, André du Laurens, Marie de Médicis's physician (and later First Physician of Henri the Great), found a source for the suffocation of the womb in the "malignant & venomous vapor coming & rising from the suppression of menstrual blood & the semen, leading to suffocation & at times interception of breathing."[35] According to him, the womb moved based on its attraction to or repulsion from the vapors of specific odors. In his *Le septiesme Livre des œuvres anatomiques* (The Seventh Book of Anatomic Works), Du Laurens questioned masculine desire and the drive to insert the male organ so close to "the shameful part of women, soiled by so many infections & set for this reason in the lowest parts, like a sewer & sink of the body."[36] The physician thus resuscitated a common trope from the Renaissance, also present in a text attributed to Aristotle on generation.[37] For Du Laurens, the body consisted of noble parts and lower parts, reflecting the hierarchy between the celestial order and the human one, which was marked by sin. Generation appeared as a mystery, tied as it was to the natural human attraction for the body's nether regions.

Du Laurens was not the only one to speak of the womb as a sewer, at a time when no physiological definition of a sewer could be found in dictionaries.[38] It is difficult to gauge today the symbolic intent of this seemingly degrading term. In writings on hysteric affection, a "sewer" generally referred to all liquids that could cause harm if they were not eliminated in time from the body, or to a vessel through which such liquids passed. Men were able to disperse such waste thanks to their active way of life, which encouraged perspiration. Women, on the other hand, converted unexpelled liquids into white flowers and menstrual bleeding, though, as Michael Stolberg has shown,[39] doctors often regarded men's periodical hemorrhoidal bleeding or other blood loss as parallel to the female menses.[40] Cathy McClive notes the variety of terms used to talk about women's blood loss, dating the first written occurrences of the word "menstruation" to Astruc's *Traité des maladies des femmes* (Treatise on Women's Illnesses), published in 1761.[41] According to physicians, menstrual blood was a noxious substance that could have various different effects on bystanders, or even on objects such as mirrors. It could sour milk and wine, make plants infertile, sicken dogs, and kill flies.[42] The womb was seen not only as playing a role in generation, but also as protecting the body. All noxious substances coalesced with semen, which was thought to spoil if the woman was not impregnated. The belief that this reservoir could hinder as well as facilitate the evacuation of waste was only a step away. In either case, the dysfunctions of the womb became the primary means to explain hysteric fits. Seeing the residue of the womb as the source of vapors also led to a connection between the disruption of menstrual discharges and vaporous affections. When menstrual bleeding was irregular, the physician had to artificially recreate it so that the body would not be invaded by harmful vapors. A woman's health was at the womb's mercy. Other maladies involving stiffness and occasional trembling, such as melancholia,[43] "perturbations of the Mynde,"[44] and hypochondriac affection, were also explained by vapors that came from the spleen or remnants of semen (although the latter case was considered rare).[45]

Publications on the suffocation of the womb began to proliferate at the beginning of the seventeenth century, and similar symptoms were gathered under the heading of hysteric illness, *hysterica passio*, hysteric passion, hysteric affection, or fits of the mother (where "mother" referred to the womb), without these pathologies being distinguished from each other. In a discourse dedicated to Du Laurens, the apothecary Henry Joly presented the first more detailed observation on hysteric passion. He intermingled hypochondriac and hysteric diagnoses, associating vapors from the spleen, the mesentery, and the liver with those from the womb.[46] In 1613, Nicolas Abraham de la Framboisière, physician at the Faculté de Paris, likewise saw the suffocation of the womb as resulting from rotten semen,[47] as did Helkiah Crooke, King James I's physician, who two years later explained the "fitte of the mother"[48] that was caused by such vapors. Physicians collected accounts of extraordinary cases, such as one about Vesalius saving a woman believed to be dead as he was about to dissect her body. One stroke of his scalpel awakened her from a hysteric catalepsy. Similarly, Jacques Ferrand published a book in the same period devoted to erotic melancholy, in which he insisted on its frequent occurrence in women and cited famous cases such as the young women of Lyon who threw themselves into a well to calm their uterine furors.[49] The same conception seemed to predominate throughout Europe. In 1641, the Danish doctor Thomas Bartholin spoke about "nasty vapors" that stayed hidden and "excite[d] the womb," thus provoking a suffocation.[50] Fifteen years later, a translation of Jean Varandée's work appeared in print. This professor at the Faculty of Medicine in Montpellier blamed contents of the womb, which he named "excrements," in older women and in victims of a "furor of the womb."[51] François Mauriceau, chief surgeon at the Hôtel-Dieu in Paris, echoed this theory in his famous *Traité des maladies des femmes grosses et de celles qui sont accouchées* (Treatise on the Illnesses of Pregnant Women and

Those Who Have Given Birth).[52] At the time, the German physician Michael Ettmüller enumerated a series of "vices" (a term that referred to physiological as well as moral flaws) associated with menstruation. In his chapter on hysteric affection, Ettmüller also describes white flowers as a "sewer created by the natural parts, consisting in a gastrorrohea, a thick & mucilaginous humor, or aqueous & serous humor, more or less abundant, at times acrid, saline, & pungent, at times benign & sweet."[53]

Over the final thirty years of the seventeenth century, there was a significant transformation in the way hysteric illnesses were viewed and in the terms used to describe them. In his *Pathologicae cerebri, et nervosi generis specimen,* published in 1667 and translated into English in 1681 under the title *An Essay of the Pathology of the Brain and the Nervous Stock in which Convulsive Diseases are Treated of,* Thomas Willis, a professor of natural philosophy at Oxford with a medical practice in London, took the similarity of symptoms experienced by men and women as evidence of a widespread interpretive error.[54] Willis, who was a founding member of the Royal Society, concluded his description of male symptoms by stating: "If this strange distemper had happened to a woman, it would presently have been said, that it was the mother, or histerical, and the Cause of it would have been laid on the fault of the womb; especially, for that the accent of something, like a bulk, began the fit, from the bottom of the belly."[55] For Willis, such symptoms had their origin in the brain and the nervous system, thus attesting to the agitation of animal spirits and vital spirits.[56] As George Rousseau puts it, "This *animus,* or *anima,* was destined to infiltrate much eighteenth-century physiological thought and experiment, and was transformed as the base building-block of later theoretical systems: Haller's sensibility and irritability, Condorcet's sensory physiology, William Cullen's basic disease theory as nervous and many others."[57] Writing a few years after Willis, Thomas Sydenham went even further.[58] Known as the father of English medicine or the English Hippocrates, Sydenham had studied in Montpellier before qualifying in Westminster. His 1682 *Dissertation* was devoted to hysteric and hypochondriac affections, which he considered as indistinguishable as two eggs.[59] Sydenham estimated the number of sufferers at more than the half the population, noting that most women were afflicted, as were sedentary men. Many physicians repeated this assertion (as Mark S. Micale has pointed out[60]), eventually making hysteric and hypochondriac passions the most frequent of all diagnoses and fueling the need to study these affections further.

Willis's and Sydenham's authority was such that in the following decades it became rare for physicians to localize the origin of hysteric illnesses in the womb, and those who did not abide by the new theory were accused of ignorance. The use of such a wide spectrum of terms, including "suffocation of the womb," "uterine furors," and "fits of the mother," began to fade. In France, the physician M. Lange claimed to radically revise the approach to such fits when he published the first treatise devoted to the vapors in 1689. In the preface, he asserted that his doctrine was entirely new, and requested the support of the Faculty of Medicine, which he obtained from Daquin, Louis XIV's First Physician. Although Lange borrowed some conceptions from Raymond de Vieussens, who had just dedicated a chapter of his *Histoire des maladies internes* (History of Internal Illnesses) to vapors,[61] he abandoned Vieussens's use of the theory of temperaments, as well as their role in affecting the blood.[62]

In his own treatise, Lange railed against the ignorance surrounding vapors, stating it was "one of these vague words to which no distinct idea has been attached, & used by physicians to cover their ignorance, or to amuse the People with words."[63] If Lange was the first to title a medical treatise with the word "vapors," he was the last to invoke excrements to explain them. He used the language of the newly developing field of chemistry[64] to draw an analogy between bodily vapors and fermentations that emerged from the earth. He also compared vapors to the air we breathe, which consists of both beneficial and pernicious vapors. The body also produced vapors, some of which could result in vaporous disorders.[65] Although Lange did not break with interpretations that positioned convulsive pathology as a female malady, he did downplay the role of the womb, making it just one of many possible explanations. In fact, he distinguished between four types of vapors, located according to their origins in different parts of the body. He spoke of "epileptic vapors" that came from volatile ferments, "melancholic vapors" from the spleen, "glandular vapors" from a lack of ferments in the glands, and "hysteric vapors" from the fermentation of salts, itself provoked by sensations and the imagination, which directed the animal spirits toward the organs of reproduction. In keeping with Willis, the movement responsible for the fit did not progress from the lower body parts to the higher, but from the brain to the rest of the body. Lange compared vapors to an effervescence, a perspiration, or a tickle produced by the agitation of animal spirits.[66] His words were without derogatory intent and merely aimed to describe the extraordinary character of the pathology. In this, he joined Marin Cureau de la Chambre, another physician and adviser of the king, who saw vapors as a way to explain the physiological effects of the emotions in his *Caractères des passions* (Characters of Passions).[67]

Following Lange's publication, explanations of the vapors assumed many forms. Most pursued the idea of a pathology that affected males and females alike, regarding the distinction between the hysteric and hypochondriac as a mere convention of discourse. Physicians recorded cases that attested to the presence of similar symptoms in men and women. Stating that women were not the only ones afflicted with vapors was more than a way of speaking about patients: extending the diagnosis meant that vapors were now an affection of all mankind and should be treated as a pathology in their own right.

Soon, treatises on the vapors were being published with great regularity. In 1691 Jean Chastelain, professor at the Faculty in Montpellier, conceived of vapors as a vicious contraction of the "mainspring fibers."[68] Pierre Brisseau's publication followed a year later, describing hysteric affections and "sympathetic movements" that activated the reflux of animal spirits. In 1702, John Purcell interpreted such movements as the consequence of repeated indigestion, and Dr. Dumoulin, in 1703, as that of an overabundance or lack of blood. For Noel Falconet in 1723, the presence of insects or worms in the body caused the vapors, while for William Stukeley it was nerves situated in the spleen. Richard Blackmore pursued this hypothesis, blaming the nervous system and the poor circulation of animal spirits in 1725, soon followed by Jean Viridet in 1726 and Nicholas Robinson in 1729, who each developed a theory on the effects of liquids and solids on the nerves. Describing all the interpretations of vapors by eighteenth-century physicians would require a book in itself, especially given the numerous other variations presented in the writings of Richard Browne in 1729, Charles de Barbeyrac in 1731, David Kinneir in 1739, William Cullen in 1742, Richard Manningham and John Andree in 1746, Jean Baptiste Joseph Lallemant in 1751, Charles Perry in 1755, and Pierre Hunauld in 1756. In 1758, Joseph Raulin identified seven general causes of vapors as well as seven proximate ones. Despite all these theoretical divergences, however, "vapors" remained the preferred term, and a steady stream of medical treatises appeared until 1789, each bearing its own nuances.

While sexual difference did not disappear from the medical discourse, over the eighteenth century vapors were thought of in terms of "sensibility,"[69] or of something that acted on the fibers and limbs of the body.[70] The noun "vapors," sufficiently vague to accommodate any theory, operated beyond the realm of sexual difference, which perhaps contributed to its frequency in medical and literary works, outnumbering the uses of the expression "hysteric affection." On the other hand, this very repetition led some physicians to criticize its usage as rash. In his 1758 *Traité des affections vaporeuses du sexe, avec l'exposition de leurs symptômes, de leurs différentes causes, & la méthode de les guérir* (Treatise on Vaporous Affections of the Fair Sex Including the Exposition of Their Symptoms . . . & the Method to Cure Them)—reprinted the following year with a meaningful addition to the title, *On y trouve aussi*

des connoissances rélatives aux affections vaporeuses des hommes (One Also
Finds Knowledge about Male Vaporous Affections)—Joseph Raulin, adviser
and physician to the king, echoed Willis in opposing any attempt to assign
the pathology to a specific gender.[71] In his preliminary discourse, he presented the association of the pathology with the womb as an idea stemming
from classical antiquity, one that had since been refuted by the presence of
the same symptoms in men.[72] Like Willis, he noted the sensation of a ball in
the belly in men and radicalized it with the assertion that "one would have
sworn they were women, if one had not been certain of their sex."[73] Raulin
conscientiously refused to employ the adjective "hysteric," and also used the
noun "vapors" parsimoniously. Instead, he favored the somewhat less well-
worn expression "vaporous affection," which also left open the patient's gender. Raulin went to great lengths to challenge the associations that had been
suggested by the word "vapors":

> If a woman has restlessness, yawnings, hiccups, spasms, irregular moves in the
> nerves, she complains bitterly; her parents, her friends, her neighbors reply
> with indifference, those are vapors. These light vapors progress imperceptibly,
> the sick woman becomes sad, she sheds tears, or seems cheerful, she articu-
> lates words one does not hear, or she says pretty things, she laughs, she sings,
> or she cries & laughs alternately, still without knowing herself; one laughs like
> her, jokes about her state, saying that these are vapors. These vapors eventu-
> ally become more violent, that is their ordinary course; one is first agitated
> with slight convulsive movements, one falls into convulsions, one loses the
> use of the senses, the limbs stiffen; occasionally they become inflexible with-
> out one noticing it by any external sign. . . . After the attack, one continues to
> say in the same tone, these are vapors. Nevertheless these incidents multiply,
> they become very frequent; the functions are injured; dangerous obstructions
> occur in the viscera.[74]

Raulin's concern came at a time when the term "vapors" found broad use
outside the medical sphere, which itself brought into question the credibility
of the diagnosis and lessened its scientific authority. He lamented this usage
and denounced an approach that neglected the violence and dangers linked
to these affections. By reevaluating these perils, the physician supported the
patients affected with such fits. The treatise not only determined causes and
appropriate treatments, but also tried to combat the ironic perception of the
pathology, insisting instead on the sufferings of patients.

Raulin emphasized the parallels between masculine and feminine symptoms.[75] He viewed the frequency of hysteric affections in women as a function
of each person's sensibility or fragility and hence of physiological inequality.
Raulin's conception lay within an Aristotelian tradition, opposing man and

woman, warm and cold, strong and weak.[76] The text presents female vapors as the result of a feminine nature, and male vapors as the result of a man's feminization. Women are seen here as double victims: of a frail constitution and of a domestic education that, seeking to protect them, weakens them even more.[77] As for men, they "become vaporous just as women do when their nerves lose their natural firmness; sometimes they are weak as a result of the temperament, but they become so most often as a result of debauchery, exhaustion, idleness, strain of the mind, &c."[78] If a man's way of life led him away from his nature, he could be stricken by this pathology. Reading and staying indoors, for example, could draw his nature toward the feminine. Thus, for Raulin the universalization of vapors was not the result of the pathology's indifference to gender, but rather of a feminization of manners. He adds later in the text: "We will soon see that by imitating women & their way of life, men will eventually contract all women's illnesses."[79] Cultural similarities reduced physiological differences and mitigated the natural inequality inscribed within the fibers and nerves.[80]

In the seventeenth and eighteenth centuries, aristocrats displayed feminine traits as a matter of prestige.[81] Feminine manners functioned as a sign of education, and of a sensibility that, because of the sympathetic relation between mind and body, at times weakened the latter.[82] According to the image of vaporous affections he wanted to present, Raulin counterbalanced the words "debauchery" and "idleness" with "vulnerability," "delicacy," and "the passions." He was not interested in emphasizing the differences between the sexes; physiological inequality was a starting assumption, but was never fully articulated. Through the fourteen chapters listing as many causes for the vapors, he set out all the obstructions of organs that could result in vapors, including the role of air, nutrition, menstrual bleeding and "white losses,"[83] and the passions. The great number of possible causes for vaporous affections cleared the physician of any suspicion of dogmatism. Even more, it created a tactful context for treating patients, who could receive such a diagnosis without experiencing any guilt at having provoked it. The author asserted: "All passions can cause vapors."[84] More than providing an exhaustive account of the pathology, Raulin aimed to describe the context within which it was likely to develop. Although the physician did present the bodies of aristocrats in opposition to those of peasants, noting that peasants' bodies were constantly active and hence resistant to vapors, he did not accuse or stigmatize any of his patients. That he was Montesquieu's physician—and had recently become one of the king's physicians—doubtlessly contributed to the care he took not to attack the aristocratic way of life too strongly.

A few years after Raulin, Robert Whytt, a renowned physician who trained and practiced in Edinburgh, also distinguished between general (or "predis-posing") and proximate (or "occasional") causes. In contrast to Albrecht von Haller's experiments on irritation, he returned to the concept of "sym-pathy" and adapted it to a new framework for understanding the body, one that was based on the nerves.[85] As Christopher Lawrence argues, "Whytt's principal contention was that the body's responses are purposeful and not the result of blind mechanism," and Whytt thus contributed to the disap-pearance of Boerhaavean physiology.[86] Certainly, in the seventeenth century the concept of sympathy had already allowed physicians to posit an infinite set of relationships between the organs and the sympathetic power of the nerves. Because of a constant intermingling between the moral and the physi-ological, this power affected the entire body, although, as Julius Rocca notes, for Whytt sympathetic action was "always centrally mediated through the nervous system,"[87] and all illnesses could thus be said to have some nervous origin. Within this framework, Whytt conceived of strictly nervous illnesses, which he also named hysteric and hypochondriac disorders, resulting from the "delicacy or unnatural sensibility of the nerves."[88] He understood hysteric and hypochondriac disorders with respect to their physiological and moral dimensions, noting the obstruction of the nerves and passions of the mind as possible causes. Thus, Whytt too came to regard the difference between male and female as secondary, at least with regard to hysteric, hypochondriac, and nervous disorders, all of which he classed together. His discussion of these pathologies followed a lengthy chapter on the nerves, which—in the foot-steps of Willis—he defined as "those small cords, which rising from the brain and spinal marrow, are distributed to every part of the body."[89]

First Occurrences of the Term "Hysteria"

It was not until the 1760s that the term "hysteria" finally came into use by a number of physicians in France. The term was not defined in the first appear-ance I have been able to trace, in 1703; rather, it was used as a chapter heading in an anonymous work published as part of Dumoulin's *Nouveau Traité du rhumatisme et des vapeurs* (New Treatise on Rheumatism and Vapors).[90] This was followed by only scattered appearances:[91] a second use can be dated to 1760, in an article published in the *Journal de Médecine, Chirurgie, Pharma-cie* (Journal of Medicine, Surgery, Pharmacy), but without the term gaining wide acceptance in subsequent decades.[92] The article's author, Dr. Dufau, was commenting on Pierre Pomme's treatise, a treatise that would later become the most famous one in France on these pathologies, but was in fact devoted to vaporous affections and did not use the term "hysteria." In 1771, the noun "hysteria" reappeared, this time in a pocket medical dictionary by Dr. Hélian. Addressing "priests & . . . persons living in the countryside who take care of the sick far away,"[93] the dictionary was intended for a broad readership not confined to the ranks of the privileged. Instead of a definition, however, the reader was referred to the expression "hysteric vapors"; Hélian's dictionary thus contributed to the process of accrediting the term only as a synonym, not on its own account. Evidently, the new word "hysteria" was seen as a duplicate of a previously existing diagnosis. In the same year, Nicolas used the word "hystérie" in his French translation of François Boissier de Sauvages's *Nosologia*.[94] Here the noun appeared as a chapter heading, followed by a series of terms identified with it: "in the vulgar tongue 'hysteric passion,' 'amary' for the Languedocians, 'hystero-hypochondriac malady' for Sthalians [i.e., Stahlians], 'nervous melancholy' for Lorry, 'spleen' and 'vapors' for the English, '*isterismo de Cochi Bagni*' in Pisa."[95] This edition would be supplanted in the following years by a new French translation by Gouvion, who preferred to retain the Latin form used by Sauvages: *hysteria*.[96]

The first appearances of the term "hysteria" in French were not, however, associated with a radical change in the perception of any of its related pathologies. In the term's first occurrences, there is no mention of a pathology specifically coming from the female organ. Neither Dumoulin, Dufau, Hélian, nor Sauvages situated hysteria's origins in the uterus.[97] Each of these physicians offered a choice of interpretations, associating hysteric symptoms with a host of pathologies. At most, Sauvages saw hysteria as the consequence of a "flabby and effeminate constitution,"[98] thus following the contemporary medical literature on vaporous affections.

Sauvages's nosology represented a rupture with previous attempts to classify illnesses, such as that of Vieussens, his senior at Montpellier, or his own earlier publication of 1731, or the classifications proposed by Daniel Sennert,[99] John Jonston,[100] Friedrich Hoffmann, Willis, or Sydenham (all of whom he occasionally referenced).[101] What made Sauvages's attempt so novel was his claim to approach pathologies through a methodology grounded in and nourished by the study of mathematics, geometry, and physics. In his preface to the *Nosologia*, Sauvages insisted on the relevance of history, philosophy, and mathematics to the theory and practice of medicine. By history, he meant the knowledge of facts; by philosophy, the knowledge of causes, principles, and experience; and by mathematics, the need to measure and quantify strength, speed, and intensity. He presented his nosology as a "science of the pathologies, or the art of demonstrating everything concerning them in a positive or negative manner."[102] He stressed the importance of words as "signs of our ideas"[103] and each word having a "fixed, constant, and known value,"[104] and established a systematic nosological method based on the need to understand each pathology in relation to a given species, with each species considered within a series of genuses, which in turn fell into a series of classes. As Hubert Steinke notes, in Sauvages's works "diseases were classified by their appearances and not according to some causal relationship."[105]

Sauvages stands out as a pioneer in the classification of diseases. He published his first nosology in 1731, and maintained a correspondence with Carl Linnaeus, to whom he paid homage on several occasions.[106] Indeed, Sauvages's selection of the term "hysteria" in his *Nosologia methodica* was consistent with the approach to knowledge defined by the famous botanist.[107] Linnaeus postulated in his *Philosophia Botanica* that naming "demand[ed] immediately names,"[108] adding: "We must banish from our Republic generic Names, compounds of two complete and separate words," "Generic words, whose roots are neither Greek nor Latin, must be rejected," and "Adjectives are less good as nouns for generic Names." By naming hysteric and vaporous affections as "hysteria," the Montpellier professor of medicine selected a noun with a classical Greek etymology, thus eradicating any vague terms that might lend themselves to endless speculation. His choice aimed to ground a recently identified pathology through the use of an acknowledged methodology.

But Sauvages's description of hysteria raised a number of new methodological problems with regard to taxonomy. The category of hysteria came up repeatedly in his work, at times as a spasmodic illness, at others as a moral one. Spasmodic illnesses constituted a distinct species, characterized by a constant or interrupted contraction of certain muscles. Sauvages distinguished between four types of spasmodic illnesses, with the fourth type including hysteria.[109] In contrast to his colleagues, who since Willis had tended to view hysteric and hypochondriac affection as a single illness, Sauvages radically separated hysteria from hypochondria, which he classified as one of the "extravagant illnesses," alongside mania and melancholia. He then proceeded to subdivide hysteria according to its different origins or localizations: white flowers,[110] menorrhea, libidinous hysteria, stomachic hysteria, and febrile hysteria. In his definition he pointed out the connection, specific to hysteria, between the pathological and the moral: "a clonic & tonic spasm of the limbs & even of the inner organs, whose transient symptoms vary slightly & are accompanied by a very intense fear of death."[111] This fear of death distinguished hysteria from other spasmodic illnesses. Hysteria thus required an entirely different treatment approach: the physician could no longer prescribe antihysteric therapies alone, but also had to soothe the patient with appropriate words.

Sauvages complicated his classification even further when, at the end of his *Nosologia*, he contrasted both hysteric and hypochondriac affections with spasmodic ones. He placed the first two pathologies within the class of moral illnesses, associating them with melancholic illnesses, epilepsy, dissimulation, and the vapors. Spasms, on the other hand, were seen as a form of bodily error, taking shape "by mistake, whim, bad habits, & without reason."[112] They were simply depicted as unexplained convulsions. Hysteric and hypochondriac affections, on the other hand, could arise from the emotions and the unsettling of the imagination—hence the equivocal relationship between hysteria, lies, and dissimulation. The class of moral illnesses under which "several types of hysteria" were mentioned itself included forty-four diagnoses.

But Sauvages had a very singular way of constructing his nosology, one that would not find future advocates. He created numerous bridges between diagnoses, linking them in a crisscrossing structure of pathologies. This led him to use a great variety of adjectives as well as noun forms. For example, he employed the adjective "hysteric" to qualify a number of other illnesses: hysteric palpitations, hysteric tetanus, hysteric exhaustion, hysteric syncope, hysteric asphyxia, hysteric catalepsy, hysteric apoplexy, hysteric hypochondria, hysteric vertigo, hysteric panphobia, and hysteric insomnia.[113] Characteristics and symptoms were repeated from one category to another. How, then, is the use of the adjective form "hysteric" to be understood? Did it signify the intensity of the symptoms, the difficulty in eradicating them, the moral nature of their origins, or the gender of the patient? Diagnoses were not exclusive, ruling each other out, but inclusive, allowing the creation of hybrid pathologies. For Sauvages, classification did not necessarily mean drawing up unique, fixed categories; rather, it meant putting into play differences and variations while maintaining the possibility of constant comparison. Sauvages's system did not function strictly as a table, or if it did, it was as a table whose data could be infinitely recombined. The inclusion of numerous case studies could shed new light on qualities that were beyond the definitions of medical categories. As such, the system was continually open to further divisions. This does not imply a flaw in Sauvages's "classificatory reason,"[114] but a system of classification that strongly differed from those that came afterward, for example in William Cullen,[115] Jean-Louis Alibert,[116] David Hosack,[117] or François-Joseph-Victor Broussais,[118] for whom a nosological table implied a strict use of delimited and predefined categories.

Despite these numerous ambiguities and ambivalences, the inclusion of hysteria in a work listing 2,400 other pathologies was a radical step toward the recognition of the pathology. Inserting this noun alongside so many other pathologies, and describing it based on the same methodological approach, implied that speaking about hysteria as a unique pathology was not the main

priority. While the inclusion of such a nebulous pathology may have threatened to undermine the clarity of the nosological edifice, simply belonging to
the nosology implied that hysteria had to be recognized according to universal criteria. What interested the physician more than individual patients was
the clear identification of a set of symptoms. The patient was viewed as a body
to be analyzed according to manifold criteria. The vertiginous structure of
the nosology was meant to overcome the indescribable character of hysteria.
No matter how indeterminate hysteria's symptoms might be, they could be
understood by means of rules.

After Sauvages, nosologists expected names to function as a clear means of
differentiation. By the very fact of its name, hysteria was neither hypochondria nor epilepsy nor a convulsive spasm. The recognition of this pathology
functioned as a reduction, a statement of nonequivalence, creating distinctions and exclusions as much as it asserted a belonging to a given typology.
This is not to say that the borders of the pathology were clear-cut, but that it
was meant to function as an autonomous and irreplaceable medical category.

The same tendency can be observed in texts in English, where the term
"hysteria" did not appear until the 1760s.[119] Following Sauvages, William Cullen used the term "hysteria" in his lectures of 1765,[120] although he rejected
the creation of subdivisions for medical categories and chastised Sauvages for
including an excessive number of pathologies in his nosology and thus blur-
ring distinctions.[121]

For Cullen, all pathologies were of nervous origin. He created a new term,
"the neuroses," to bring together those pathologies emanating directly from
the nervous system. By casting hysteria as a neurosis, Cullen undid the link
between pathology and dissimulation, and posited instead a strictly physiological origin: the mind and the body suffered in parallel from a nervous
imbalance. Cullen's description of hysteria is very short, as are all the other
diagnoses in his nosology. He insisted on the difficulty of distinguishing between hysteria, dyspepsia, and hypochondria. His choice of the word "hysteria" might seem paradoxical; according to him, men just as well as women
could be afflicted by it. The selection of this term can therefore be seen as
part of the nosologist's desire to reinforce categories that were already in
use by colleagues, with the additional goal of creating parallels between the
vernacular languages of French and English. The use of scientific terms in
this way is characteristic of Félix Vicq d'Azyr's thesis. The secretary of the
Société Royale de Médecine, and later a co-opted member of the Académie
Française, Vicq d'Azyr asserted that "while sciences progress every day, their
idioms expand, and with them the art of thinking is perfected. Recognizable technical expressions, being, so to speak, the same thing in every coun-

try, shape a universal language, equally written, understood, and spoken by everyone."[122] Such was the willingness at the end of the eighteenth century to anchor a new approach to diagnostics by using denominations that were universally recognizable.

The term "hysteria" was still missing from the *Dictionnaire encyclopédique* published in 1790, but appeared in Nysten's dictionary of medicine in 1814,[123] as well as in the 1818 dictionary edited by Panckoucke, where the category was finally defined in an authoritative manner. That the term had been used so little and so recently had by this time been forgotten. Now recognized and cataloged, hysteria would henceforth be regarded as a diagnosis that stood in relation to all others, one that no longer required a completely unique treatment approach. While hysteria's alleged irreducibility disappeared, the diagnosis had also gained a new validity. The difficulty of its identification had become a mere detail, and uncertainties attending its explanation a mere characteristic.

The transition from the word "vapors" to "hysteria" also led physicians to construct an exclusively female diagnosis that could be contrasted with hypochondria.[124] The latter would become a nervous affection affecting both genders, but applied more specifically to men, and even more specifically to men of letters.[125] The link to poetry lent hypochondria an aura of prestige. Writing in 1818, Louyer-Villermay revived this tradition and radicalized its theory:

> Hypochondria is an illness of all times and countries, arising in all seasons and temperatures. It is common to both sexes, but does not affect all ages or social classes alike. Its frequency is, up to a certain point, the effect of the development of human understanding and of civilization's progress. The illness prefers to choose its victims from men of letters, well-read citizens who assiduously frequent their libraries, artists, poets, the most distinguished writers; and especially those endowed with the most ardent imaginations or the most vivid sensibility.[126]

Hypochondria is here seen as a noble pathology, linked to custom more than nature and afflicting both the sexes.[127] Such a conception allows Louyer-Villermay to challenge the equation of hypochondria with hysteria and cast the latter as a female illness with no equivalent in men. The physician specifies from the start that women are little predisposed to hypochondria. In his texts on hysteria, case studies present the uterus more as a sexual organ than a reproductive one, and the investigation of the pathology's causes gives way to an examination of his patients' sexual practices. Writing about one of his patients, Louyer-Villermay relinquishes modesty in the name of medical knowledge: "During the last three months of her pregnancy, she experiences new attacks, which seem to her to be caused by the slowness with which her lover proceeds, since they do not happen when he comes quickly to the goal."[128] Here the narrative of physiological disorders becomes an opportunity to stigmatize women and cultivate a mythology of the lustful and insatiable female. Sexual difference is the means by which to assert an inequality with respect to pathology. Louyer-Villermay's theorization finds its justification in etymology and exemplifies the emergence of a new way of perceiving woman, who is no longer considered in terms of her entire body, way of life, sensibility, or intellect, but from the perspective of her sexual organs. Louyer-Villermay sees her not as the possible victim of a pathology, but as a hot-blooded firebrand. He conflates sexual desire and pathology, regarding hysteric pathology as revealing the "greed" of the uterus, and hysteria as a natural effect of the organ. As we will see in chapters 2 and 6, this move to equate pathology with the greedy womb made all women potential hysterics. Sander Gilman recounts that when, some sixty years later, Charcot, and later Freud, spoke of hysteria as a masculine illness, they would meet with much disapproval. Despite continued conflicting interpretations, hysteria as a women's illness had already become a cliché.[129]

Étienne-Jean Georget, one of the rare physicians to directly oppose Louyer-Villermay, offered the term *cérébropathie* to replace "hysteria," further indicating that one should add "the epithet 'spasmodic' or 'convulsive' to *cérébropathie* to distinguish it from hypochondria, which should simply be called *cérébropathie.*"[130] Aiming to lend authority to this diagnosis through linguistic rigor, he assimilated a list of previously used categories: "Perfecting ideas must lead to perfecting language. It is by changing nomenclature, by grounding it in solid bases, that chemists have made their science able to make huge progress, and classify without confusion a huge quantity of bodies."[131] To further ground his thesis, Georget recommended reading Le Pois[132] and Willis, physicians of the early and late seventeenth century, respectively, who identified the brain as the source of hysteria. After contemptuously rejecting hysteria's localization in the womb, Georget instead located its origins in the moral affections and nervous fragility. In a chapter devoted to sexual difference, he clearly opposed men and women: men's "strength of reason" was "made to subjugate, without appeal, one half of mankind to the other."[133] He explained the larger number of convulsive illnesses and "so-called hysteric illnesses" among women by citing the importance of moral affections in women. In his chapter on hysteria, Georget indicated masturbation as a frequent cause,[134] thereby reinscribing sexual desire at the core of the pathology and identifying as a cause what others, two centuries earlier, had offered as a cure. He went so far as to say he had never seen "spasmodic *cérébropathie* as a result of intellectual tasks."[135] Georget considered spasmodic *cérébropathie* to afflict only virgins and young women and therefore advised marriage as the appropriate therapy.

It was by using examples, referring to specific observations, and contrasting hysteria to other pathologies such as hypochondria and epilepsy that the interpretations of symptoms became increasingly associated with gender at the beginning of the nineteenth century. The works of Louyer-Villermay and Georget thus exemplify a trend toward consensus despite the theoretical conflicts at play.

Around 1800, interest in hysteria fell within a broader movement to reorganize medicine into a unified and closed body of knowledge. Physicians selected categories of reference in such a way as to further the establishment of their discipline: they delimited the pathologies, species, and sexes that would be used as diagnostic reference points. Hysteria had lost its mystery; its history had been rewritten. Asserting the validity of scientific discourse became just as important as asserting the differences between male and female pathologies. The literal succeeded the vague, the variable, and the contradictory. Doctors broke out of an infinite regress of symptoms to arrive at a narrower set of meanings defined by etymology. Naming organized the relationship to afflictions, and the enigma of hysteria vanished, to be replaced by a semantic law. The diagnosis now acted as the delimitation of the cause. Whereas in the eighteenth century physicians varied the terms used to talk about similar affections, for their successors the use of the category of hysteria operated as a verdict. The authority of medical discourse was constructed based on prior reflections on language that allowed semantic considerations to serve as the foundation for a diagnosis.

The increase in acknowledgment of the term "hysteria" changed the approach to hysteria even more than the term's first appearance did. Although the term had already existed since the beginning of the eighteenth century, it was only at the turn of the nineteenth century that it began to operate in earnest—it became the key that enabled the identification of a whole gamut of different troubles. The question is what inscribed the term as a solution to so many problems: to the dismay aroused by troubles that cannot be categorized or explained; to the need to replace religious explanations of fits; to the questions posed by political upheavals. Both the centuries of neglect and the term's sudden success tell us much that is crucial about the construction of a medical category.

Vaporous Affection and Social Class

In the sixteenth and the first part of the seventeenth century, diagnoses of hysteric passion, suffocation of the womb, and fits of the mother were not strictly associated with a specific status in society, as can be seen in the writings of the physicians Ambroise Paré, Liébault, Du Laurens, Crooke, Varandée, Ferrand, and Bartholin, to name but a few. Nonetheless, the expression "fits of the mother" (*mal de mère*) was most commonly used to refer to women of the lower classes or those from the countryside. La Framboisière writes that this expression is used by the commoner (*vulgaire*), distinguishing it from his own preferred term, "suffocation of the womb."[136] Judging from Henry Joly's dissertation on hysteric and hypochondriac illnesses, these pathologies were also interpreted as being chiefly the unfortunate lot of the people:

> It is something that leaves no doubt among Naturalists that while the little people grovels for the great, it is conducted, governed, and moved by the latter; it is made healthy or sick, joyful or melancholic, liberal or thrifty, pusillanimous or brave, rich or poor, vicious or virtuous, & in all subjects to receive its influences according to what the stars beat down and send, whether good or bad.[137]

In other works, they were feminine affections, irrespective of social class or even linked to the higher ranks. From the 1650s on, however, the term "vapors" entered the language as a diagnostic term to describe a pathology specific to the aristocratic classes. This is probably what most clearly distinguishes the corpus on vapors from that on madness. For "unreason, and with it insanity," as George Rosen has powerfully argued, "were related primarily to the quality of volition and not to the integrity of the rational mind. Endowed with reason, man was expected to behave rationally, that is, according to accepted social standards."[138] Madness was the name given to manifestations disrupting social constructs of *bienséance*; the vapors was the name for extravagances that were still espoused by the higher social class of the time and that never overturned the social boundaries of society—as if in this class such physical displays negotiated ways to disrupt yet simultaneously abide by the customs of high society.[139] From 1658, Antoine Daquin, First Physician to Louis XIV,[140] regularly wrote on the king's vapors in his *Journal de santé de Louis XIV* (Journal of the Health of Louis XIV). In 1675, his remarks extended across a number of pages, an excerpt from which follows:

> The king has been a victim of vapors for seven or eight years, but much less than before, vapors rising from the spleen and the melancholic humor, whose marks they bear in the sorrow they transmit, and the solitude they make one

long for. They slip by the arteries to the heart and lungs, where they excite palpitations, restlessness, exhaustion, considerable breathlessness; from there, rising up to the brain, they cause vertigo and whirling of the head by agitating spirits in the optic nerves, and elsewhere striking the principle of the nerves; they weaken the legs, in such a way that one needs assistance to hold oneself up and to walk, an unfortunate trouble for anyone, but especially for the king, who needs his head to apply to all his affairs. His temperament has a propensity to melancholy, his sedentary life most of the time, his time spent in committees, his natural voraciousness which leads him to eat large quantities, have given this pathology opportunities to develop. Strong and deep-rooted obstructions that crudenesses have aroused in the veins hold back the melancholic secretion, prevent it from oozing out by the natural channels, and by their presence prompt overheating, fermenting, and the arousal of all this tempest. In fact, there is no reason to be surprised that bleeding so strongly reawakens all this disorder, since it is certain that, by the movement it creates in the mass of blood and in all the veins, it agitates this secretion in its seat without evacuating it, and stirs up foaming and evaporations in it.[141]

Louis Marin, the first scholar to draw attention to this text, stressed that the health journal provides an opportunity to extol the king's greatness even while discussing his pathology.[142] The unique character of Louis XIV lies in the excesses of his physiology, which the physician describes in hyperbolic terms. Whirling, vertigo, and breathlessness are not markers of feminine fragility, but those of a body entrusted with a unique vocation. The king's way of life, rich diet, and frequent evening gatherings only contribute to these symptoms, and his incomparable worries and responsibilities also take their toll. With the theme of melancholic secretion, Daquin also raises that of "genius," as described by Aristotle in *Problem XXX*. Vapors are presented as the price of the ultimate privilege, royalty, and exemplify the status of the king, his activities, and, in the ancient sense of the word, his genius. Based on the correspondence and anecdotes of the time, one can surmise that the physician's writings were not addressed to Louis XIV alone, but circulated in the court's inner circle in an attempt to inspire the emulation of his subjects. The body-emblem of the reign of representation, Louis XIV also served as a symbol of the pathological body, to be reproduced and perpetuated by his courtiers.

In this period, the affliction became an object of correspondence, and "vapors" a word that served all purposes. "You don't want us to say 'vapors,' but what will we do if you take this word away from us? For we put it everywhere. Waiting for you Cartesians to find another one, I beg to be allowed to use it,"[143] wrote Madame de Sévigné in a 1689 letter to her daughter. She occasionally mentioned the king's vapors, the convulsions of the queen, and quite frequently her own vapors. A few letters later, she continued:

> You seem to be reconciled to the word "vapors," which you did not want to utter any longer until it was explained to you. You relented for the sake of social intercourse, which would have been entirely broken up if you had banished the word; it helps to explain a thousand things that have no name. Our ignorance puts up with it, and uses it as a *quinola* to win a premium.[144]

According to Madame de Sévigné, it was simply for lack of a better word that the term "vapors" was needed to explain physiological phenomena. "The vapors" operated as a flexible and changing term, like the symptoms it described, and lent itself to use in any phrase. The term appeared repeatedly in intimate journals and letters and was appropriated by epistolary writers to give voice to the turmoil of the soul. Slowness, torpor, idleness, and excessive crying and laughter found themselves renamed as "the vapors." The correspondence of Madame du Deffand, Mademoiselle de Lespinasse, Diderot, and Sophie Volland delighted in evoking them. In these writings, the vapors had lost their physiological dimension; their origins and effects were more nebulous than ever. The term was applied more to an aesthetic of existence than to a pathology, and it found its representative image in the mists of the eighteenth-century paintings described by Diderot in his *Salons*.[145]

Ephemeral fits according to some, the vapors were often considered from a distance in order to diminish their frightening characteristics:

> I write to you, Madame, with half, if that, of the common sense I brought into the world. You will ask me what made me so foolish: nobody, thank God! But the fact is that I am, prodigiously. The difficulty is to tell you clearly what I have, for it is not migraine; the ache I feel is not so painful, but it is blacker, heavier, and I don't know how to name it if not "vapors." But what is it to have the vapors, you are going to ask! I don't know anything about it, Madame: you need to have felt this nasty little ache to know it. All that remains for me to tell you is that, along with many properties that are all very disagreeable, it has one that is very singular: it dulls the senses, extinguishes the imagination, goes even as far as weakening reason, and do not think that it does so in the way the passions do. Reason falls silent before the passions only because the passions speak louder. But here the passions don't say a word, and reason is not the better for it. Here is roughly my affliction, Madame, it is not possible for me to define it to you better. What would be much more useful to me would be to recover, but the whole of medicine can do nothing for it. One of these gentlemen the Physicians whom I consulted lately told me, "Keep yourself occupied." Here is the advice of a simpleton; for what is an occupation that one gives to oneself? Monsieur the physician would have to give me one, and

it would also have to be an agreeable one. Another told me to take a lover; isn't that another lovely prescription, and wouldn't one say that one finds a ready-made lover at the apothecary's? All well examined, Madame, I believe I will hold on to my vapors. May I bring them to you tomorrow? I don't know any gentler way to have them wear off; accept in the meantime the note I am sending you. You will find it a little black, but don't scold me about it, it is our color: we have broken up, we vaporous women, with the colors pink and pale blue, and thus we make rather bad company. Fortunately for the public, we are aware of it, which makes us a little reclusive. But why do they have to tell us, and isn't it more respectable for us to flee people than to annoy them?[146]

This undated letter, signed by Madame Aimond de Saint Marc, is today held in the manuscript archives of Avignon. Taking the form of an overview, the letter avoids a moral approach, which would normally present the tumultuous character of the passions and imagination in opposition to a reasonable person's quest for health.[147] Instead, the letter writer describes her muffled faculties, an invading torpor that saps her strength. She gives voice to shifting viewpoints on the pathology held by physicians, family members, and personal attendants, and varies her characterizations of the vapors. Her style is by turns familiar, ironic, descriptive, and even philosophical when she speaks about the impossibility of defining the term. These changes in register allow her to maintain a certain detachment. For her, "vapors" stands as a word to be played with: she can use it without being condemned by it.

The perception of vapors as an aristocratic pathology was supported in the eighteenth century through the publication of treatises insisting on the link between a pathology, a sensibility, and a way of life.[148] Samuel-Auguste Tissot, long a physician to the poor before being feted by the upper classes,[149] does not even mention hysteric or hypochondriac illnesses in his *Avis au peuple sur sa santé* (Advice to the People with Regard to Their Health), even though his other publications attest to his interest in the matter. When this work was reprinted in 1763, a second title was added, *Traité des maladies les plus fréquentes* (A Treatise on the Most Frequent Illnesses),[150] and the edition was expanded, including a new paragraph devoted to swoons. Tissot described in particular "swoons depending upon nervous illnesses,"[151] which he said afflicted those who lived in town or only partly in the countryside. He proposed two denominations: "vapors" for people living in towns and "mother"[152] for those in the countryside. In Diderot and d'Alembert's *Encyclopédie des arts et des lettres*, the colics of artisans and painters were based on the same symptoms (such as spasms, convulsions, and fainting) but were never associated with vapors, as if the social class of those afflicted was the key to interpreting its symptoms. In Vandermonde's *Dictionnaire*, the article on a colic found in

Poitou, attributed to repeated exposure to sulfur, describes the same symptoms as belonging to a strictly physiological realm.[153] Vandermonde lists a whole series of artisans' maladies characterized by tremblings and swoons: the launderers' affliction, the bakers' malady, the distillers' malady, the chemists' malady, the foundry artisans' malady, and so on.[154] According to the dictionary, all the artisans mentioned here (launderers excepted) were exposed to noxious vapors. But these vapors were viewed as so different from hysteric and hypochondriac ones that no one took the trouble to distinguish between them, despite the similarity of the symptoms they exhibited. Hysteric and hypochondriac vapors were now thought to be a pathology of the idle rich. George Sebastian Rousseau has identified the same association between nervous illnesses and class in England: "If consumption was . . . the disease of poverty and deprivation, nerves was the condition of class and standing."[155]

The opposition drawn notably by Tissot between people living in towns and those living in the country developed further over the course of the century. Joseph Raulin asserted: "For more than a century, vapors have been endemic to large towns. Most women who enjoy life's conveniences are vaporous; one can say that they buy the agreeableness of riches with languor."[156] It was not only the cities' foul air, so eloquently deplored by Louis-Sébastien Mercier, that physicians blamed, but even more so their inhabitants' way of life.[157] By opposing a pure, idyllic world to the one they aimed to condemn, physicians could further strengthen and legitimize their advice.

The attribution of the vapors to the aristocratic or urban way of life differed, however, from the manner in which doctors used the term on a daily basis. Guenter Risse's study of the treatment of hysteria in Edinburgh's hospital shows that the majority of patients came from the peasant classes.[158] If inactivity and high social status were often presented as the source of the vapors in order to make the pathology more fashionable, a number of treatises and manuscripts assigned the same diagnosis to aristocrats and artisans alike.[159] Manuscripts sent to the Société Royale de Médecine certainly attest to the use of the diagnosis in the countryside as well. A correspondent shared his surprise when faced with the development of the pathology in a peasant milieu: "Another illness that one sees every day in our countryside, and which might be as common as fever, are the fits of the mother, hystericism."[160] Questioning the link between idleness, wealth, and hystericism, he exclaimed: "A thousand village women are obsessed with vapors: I have seen terrible crises."[161] By identifying in villagers a pathology long thought to be the price of luxury and idleness, such comments established its diffusion across the entire population. A few years earlier, Jean-Baptiste Pressavin had written: "Few people can flatter themselves never to have been afflicted with a fit of vapors."[162] In 1786, the Société Royale de Médecine offered this type of pathology as a topic in its annual dissertation contest, indicating that although it had been largely redefined as a nervous illness, its determination had still not been fully resolved and was now considered a priority.[163]

Publications on the invasion of such phenomena into the working class emerged at around the same time in England. In 1787, an article published in *The Gentleman's Magazine* described an epidemic of convulsions starting at a cotton factory in a small town in Lancashire.[164] The first fit occurred when a young female worker threw a mouse at the breast of her neighbor, who then experienced convulsions over the next twenty-four hours. Three other female workers were afflicted the next day, and six more the day after that. One day later, the number of convulsive workers had increased to twenty-four, including three males. The phenomenon had already spread to another factory five miles away, afflicting those who had not even seen the incident. According to the author of the article, the mouse had by then been forgotten, and a rumor arose that the cotton bags used in the factory contained a noxious substance. The very idea that the cotton could be infected was enough, he wrote, to provoke new fits. On the fourth day, Doctor St. Clare finally arrived with his electrical machine and treated the workers with electric shocks.[165] As soon as the doctor confirmed the nervous nature of the symptoms, there were no further victims of convulsions. A glass of wine was served to all, and the eradication of symptoms celebrated with a dance.

Yet regardless of how or where the diagnosis of vapors was made, its associations with the aristocracy persisted. In this regard, Franz Anton Mesmer (of whom more will be said in chapter 6) provides another example.[166] Even though he saw patients from all social classes and offered a free treatment tub to be used by the poor, caricatures of Mesmer consistently depicted him in the company of aristocrats. Clearly, hysteric affections and vapors were still seen as a pathology of the nobility. As a consequence of repeated enunciations addressing this specific audience, the vapors remained associated with aspirations to luxury. They also became a frequent theme in song, indicating how commonplace the pathology had become by the 1770s. A song entitled "The Vapors," published in a collection of *chansons choisies* (selected songs), went as follows:[167]

> Ah! You will perfume with amber
> My bedroom;
> Leave, leave.
> I have such a keen sense of smell!
> Unceasingly
> You go to my head.

Does my taste have to accommodate
These odors,
This insipidness?
Abbot, you're awkward!
I have the vapors![168]

Soon after, the diagnosis began to attract revolutionary verve, setting the stakes far beyond the realm of irony. Pamphlet writers looked for images that could illustrate the excesses of the great. They played with notions of vaporous affections as a characteristic feature of the aristocracy. The pathology became a mark of social dereliction and a call to revolutionary action. Vapors, nymphomania, and uterine furors were used in particular to stigmatize the queen and Madame de Lamballe. The spread of this terminology outside the medical world displaced its existing connotations. Antoine de Baecque has studied the role of sexual imagery in the development of the Revolution,[169] following the use of such imagery in a series of caricatures, anecdotes, and pamphlets that were distributed from the late 1780s to discredit the king and the monarchy. They repeated a series of images: the queen[170] with her uterine furors,[171] Princess de Lamballe with her hysteric swoons,[172] the king with his impotence,[173] and the king's son with his practices of masturbation and incest. Maurice Lever and Chantal Thomas have also examined the political use of these metaphors. *Les Fureurs utérines de Marie Antoinette, femme de Louis XVI* (The Uterine Furors of Marie Antoinette, Louis XVI's Woman), *Le Bordel patriotique* (The Patriotic Brothel), and *L'Autrichienne en Goguette* (The Austrian Woman Out for a Good Time) are some of the titles of pamphlets that delineated and reproduced sexual and pornographic imaginings. The creation of such stereotypes strengthened the association between hysteric affections and women.

In an irony of history, according to some accounts the Revolution presented the perfect opportunity to cure the aristocracy of its vaporous fits. Madame Campan, the queen's first lady-in-waiting, spoke in her memoirs of the transformation of the queen's body during the most virulent periods of the Revolution:

> A torrent of tears followed these painful exclamations. I wanted to give her an antispasmodic potion; she turned it down, saying that nervous pains were the illnesses of happy women, that the cruel state to which she was reduced made this assistance useless. Indeed, the queen, who often had spasmodic crises during the time of her happiness, had the most equable health since all the faculties of her soul had been supporting her physical forces.[174]

Physicians themselves noticed this newfound resistance of the body within aristocratic milieus. According to some, fear of the guillotine appreciably reduced the number of vaporous fits that were reported by the upper classes. In an anonymous manuscript from the Paris School of Medicine, a report similar to that about the queen assumed an epistemological tone:

> A long-sustained impression of active passions gives the nervous system a sufficient degree of force to have it ward off vaporous dispositions; in fact we have seen in France, in those times when revolutionary terror was the order of the day, vaporous subjects who inhabited towns where the government deployed the most rigor, where the revolutionary calamities exploded most markedly, who were delivered of their fits for as long as moral causes acted upon them.[175]

Such transformations became a matter of legend, appearing regularly in anecdotes and reinforcing the image of a capricious nobility that only became worthy when forced to gather its strength.[176] Hysteric and hypochondriac vapors were cast as an affectation used by the privileged to draw attention to themselves or otherwise pursue their desires.

Another imagining, that of the nervous temperament, was opposed to that of hysteric nobility. In this context, nerves illustrated the capacity to react with sanity and dexterity to events. The journal *La mère Duchesne* praised the political value of a nervous disposition in a dialogue published in January 1792: "In a kingdom composed of twenty-five million men, most of whom have neither religion nor morals, a simple and very nervous government is absolutely necessary to maintain the right order, and with it security and freedom."[177] Here, nerves are no longer associated with a pathology, but with masculine vigor, representing the valorous force of the Revolution. In 1806, Jacques-Antoine Guibert resorted to the same terms in his *Eloge du Maréchal de Catinat* (Praise of Maréchal de Catinat): "France presented the spectacle of a land rejuvenated by the very storm that had seemed destined to destroy it. The torrent of dissent had set in its breast new seeds of vigor. Character grew more nervous, courage more male, the soul higher."[178] Nerves were by then associated with a combative virtue, ensuring the renewal of a society that had previously been consumed by vapors.

At the same time, the Revolution was presented in newspapers and pamphlets as a form of hysteria, an epidemic of convulsions or of political vapors. The character Taconnet in a parody by Jean-Baptiste Artaud says mockingly: "I will dissipate the political vapors that disturb your brains; a light dose of this attic salt, which you liked in me when I was on stage, will suffice to cure you."[179] In *Le Cri de l'humanité pour les victimes égorgées sous Robespierre* (Humanity's Cry for the Victims Whose Throats Were Slit under Robespierre), Dartaize exclaims: "Good God! In what chaos, in what convulsion, would you plunge the universe again with such monstrous principles!"[180] He worries about the "foolishness of public officials" becoming "imprudent chemists" and "swept along by this fermentation and this universal delirium," their "minds bubbling to the point of dementia."[181] Descriptions of convulsion and fermentation were used to qualify revolutionary movements in police reports before they resurfaced in novelistic writing after the Revolution.[182] At the time when the region of Vendée was in uprising, others spoke of the "convulsions experienced by the Republic."[183] In his *Reflections on the Revolution in France*, Edmund Burke describes the anarchy of the convulsions and uprisings taking over France. He presents France as a feminine body whose suffocation is tearing it apart. The therapy designed to save France, revolution, has revealed itself as the worst poison that could be employed.[184] Later in the text, Burke touches on the notion of the nervous body, diagnosing an absence of the nerves that would be required to hold a measured and responsible discussion.[185] In Burke we find the same glossary of terms as is used in the pamphlets and journals of the time. Referring to the privileged nobility who met under the king's aegis at Versailles in 1789, he writes: "They breathe the spirit of liberty as warmly, and they recommend reformation as strongly, as any other order. . . . [The monarchy] breathed its last, without a groan, without struggle, without convulsion."[186] Burke constructs his analysis by opposing the convulsions of fanatical and superstitious people to a moderate and sincere aristocracy.

It was during this constant reframing of the associations between hysteria and the kind of people best defined by it in the context of the Revolution that George III of the United Kingdom received his diagnosis of "hysteric" illness. The use of this adjective did not spring from mockery, but from the infinite caution of a physician. The stakes were significant: George III was declared unable to reign in 1788, and would remain so until mid-February 1789, when Francis Willis, appointed to take care of him from the time of his confinement at his Kew residence,[187] declared him able to exercise sovereign power again.[188] For Willis, an appreciation of the right words to use was crucial. The first episode in 1788 would be followed by four others that entailed the king's dismissal, the last of them ending with his death.

The power relationship between doctor and royal patient was most extreme in the first phase of revocation. At this point the sovereign was reduced to a state of absolute dependency, with those meant to serve him now holding the authority to take coercive measures in response to his fits, or even to pronounce the permanent revocation of his reign.[189] Ida Macalpine and Richard Hunter have described the "system of government of the King by intimidation, coercion and restraint," which meant each account of his illness arose from the "repressive and punitive methods by which he was ruled."[190] The clergyman and physician Willis, chosen for his experience as the keeper of a madhouse in Lincolnshire, was assisted by other physicians, with whom he wrote a daily health report. These physicians were present as much to monitor the king as to provide alternate information and viewpoints. The language of the reports had to account for the king's absence while providing diagnoses that would not necessarily exclude him from ruling. Each physician was subjected to a thorough, independent interrogation by the House of Lords,[191] in which the same questions were repeated: each doctor was asked to justify the therapies prescribed, variations in the king's daily activities, and any exceptions granted. These interrogations, conducted in George's absence and published regularly, allowed the Lords to compare the information they received from different physicians.[192] The physicians' responses revealed a variety of tensions around the exact choice of words used in the king's health journal. As Dr. Warren related,

> The Report proposed to be sent was written thus: "His Majesty passed Yesterday quietly, has had a very good Night and is calm this Morning." Dr. Willis desired that some Expression might be made use of, indicating that His Majesty was advanced since the Day before, in His Cure; I objected to this, because I had ample Reason, from my Conversation with His Majesty, and from the Information which I had received from Mr. Charles Hawkins, to think the contrary true. Dr. Willis then said, "a certain *great person* will not suffer it to go so, and it will fall upon you."
>
> Chamber of Lords: Are you sure you are correct in those Words?
>
> Dr. Warren: I believe I am; I took the Words down as soon as I came Home.[193]

It is no surprise that political stresses may have played a role in the portrayal of the king's health, but the extreme attention given to the choice of each word is striking. Physicians debated the appropriateness of every word and later recorded their negotiations, including the expressions that were discussed, the terms selected or set aside, and the reasons for adding or withdrawing them. Language offered a space for both dispute and agreement, a space where agreement was reached only by skirting around what was actually said.

At the level of the interrogation itself, it is possible to see Willis's answers as a succession of detours. The Lords did not ask Willis or the other physicians to name a pathology, even though the careful selection of a diagnosis was becoming the hallmark of famous physicians, at a time when they presented nosologies as an important new medical tool that distinguished them from early modern physicians.[194] Instead, there was talk of the "melancholy state" of the king, or at times just his "state." The interlocutors tacitly agreed to the use of the expression "melancholia," though subsequent discussions indicate that this was a mere euphemism. Pointed as the Lords' questions were, they also relied on a series of circumlocutions. This circumspect approach enabled Willis to avoid the use of the one category, madness, that could disqualify the king from his reign.

Willis thus displayed great skill in providing vague answers to precise questions. He spoke of "progress towards a cure," even though his descriptions repeatedly mentioned the king's inability to maintain concentration. He tried to justify the difficulty of establishing a prognosis in order to insist on the possibility of remission. Required to provide an explanation for the king's symptoms, Willis seems to have invoked "hysterical symptoms" in an attempt to portray their extreme nature while also alleviating the fright they provoked. He used a locution that allowed him to avoid attributing any pathology to the king, reporting that the symptoms were similar to "what appears in Hysterical Cases."[195] It was not the prevailing association of hysteria with women that necessitated Willis's reservations. After all, in Cullen's recently published nosology hysteria was conceived of as both a male and a female illness. Rather, Willis's detour ("what appears in") allowed him to qualify the king's symptoms without actually identifying an illness. Put another way, Willis could suggest a possible reading of symptoms while avoiding having to assign the king to a specific illness.

Within the narrow frame of the interrogation, Willis attempted to establish a space of doubt in order to preserve the possibility of recovery. The variable qualities of language Willis used to describe the king's pathology allowed the committee to envision a time after the fits. Terminology was negotiated not only in relation to the choice and duration of treatments, but in relation to the identification of an illness. Expectations existed not only at a therapeutic level, but at the level of discourse itself. If the physician's responsibility resided in speaking—the physician's primary tool for the enumeration of symptoms, diagnosis, and prognosis—Willis's conjectures were a means to delay all prognoses and even the implied ultimatum for the king to regain his health. Here the physician appears as one who appreciates how to manipulate language in order to divert the need for hasty conclusions, decisions, and verdicts. He weighs what could be said and avoids the use of specific words, or applies them in a way that lessens their power of attribution, placing them in a context that could undermine them. The physician's ability to coin, phrase, and carefully position words here asserts his power to capture the state of the body and, as such, helps to establish his authority.

Encounters between Medical and Religious Spheres

If this chapter turns last to the use of medical categories in a religious context and the encounter between physicians and the clergy, that is in order to downplay the significance usually assigned to spirit possession in the history of hysteria. In fact, physicians began to reread cases of possession in terms of hysteric illness very early on. With the work of Jean-Martin Charcot,[196] writings on possession became a central locus for the conceptualization of hysteric illnesses by physicians, historians, and philosophers. Over the past forty years numerous studies have analyzed how, in the sixteenth century, convulsions and other unpredictable bodily phenomena once likened to possession or ecstasy came to be seen as medical phenomena.[197] Here, however, texts written in the encounter between the church and physicians are considered only insofar as they used medical categories that later fell under the aegis of hysteria. Because texts on possession represent only a small part of the corpus involving terms that would later be assimilated to the category of hysteria, only a few historical moments will be presented in relation to possession.

From the late fifteenth century on, convulsions and swoons were the subject of intense interest among the church, the state, and physicians.[198] Religious dissertations, court documents, and medical observations all give the historian the sense of a broad interpretive debate taking place across the elite and public spheres. The main question addressed by these texts was whether such bodily phenomena fell under the province of medicine or theology. With his *De Praestigiis daemonum* (Witches, Devils, and Doctors in the Renaissance),[199] Johann Weyer was one of the first physicians to attempt to counterbalance the influence of the theological work *Malleus maleficarum.* Weyer discussed how melancholic humors could affect the mind and body, provoking visions and any number of physiological disorders, most notably convulsions. According to him, such humors could affect the imagination by creating fancies and visions that people interpreted as the presence of the devil. He recast these visions as illusions orchestrated by "infamous magicians" who convinced melancholics they had been the subject of a supernatural visitation. Even so, Weyer remained open to the possible influence of the supernatural in such convulsions, as did so many physicians of the sixteenth and early seventeenth centuries. In fact, it was quite common in the sixteenth century for medical readings to be juxtaposed with theological readings of the body, thus allowing for the possibility of demonic influence. In Pierre Le Loyer's *Discours des spectres et des visions apparitions d'esprits* (*A Treatise of Specters or Straunge Sights, Visions and Apparitions*), for example, epistemological and religious registers are combined with no sense of contradiction:

> If the devil sees that the brain is insulted by illnesses that are specific to it, such as epilepsy, falling sickness, mania, melancholia, lunatic furors, and other similar passions, he takes the opportunity to torment it further, grabbing the brain with the permission of God, and blurs the humors, dissipates the senses, captivates the intellect, occupies the fancy, and offends the soul. And speaking about the organs of mankind, [the devil] flows within them, reveals himself & and shows what he is, utters different languages, recounts what happens in the world, prophesies the future, and all this so much that he happens to lie most times and make marvels that one cannot believe to come naturally from a human being.[200]

To Le Loyer, the evocation of a pathology does not as such revoke the influence of the devil, nor does a medical understanding of the body imply the abandonment of a theological approach to the body.

From the very end of the sixteenth century, however, the term "hysteric passion" was used to promote a different understanding of possession. As Michel Foucault and Michel de Certeau have noted, clergy members themselves invited physicians to give rulings on convulsions. Far from prejudicing the authority of the church, the judgment of physicians was welcomed as a way to assert the competence of both the medical and the religious establishments. Given that exorcisms were not particularly successful in appeasing demonic bodies, recasting convulsive phenomena as physiological in nature freed the church of its responsibility in these matters.

In 1599, for example, the physician Michel Marescot published one of the first vernacular case studies in his *Discours véritable sur le faict de Marthe Brossier de Romorantin, pretendue démoniaque* (True Discourse on the Fact of Marthe Brossier de Romorantin, So-Called Demoniac). Cardinal Guondi sent Marescot to visit Marthe Brossier de Romorantin, as members of the clergy claimed she was in need of exorcism. Instead, Marescot argued that this was a case of "simulation," meaning that she displayed signs of witchcraft without in fact being possessed. Henri IV asked Marescot to publish a case study on Marthe Brossier de Romorantin after the *parlement* sentenced her to house arrest, a move designed to protect her and avoid the need to proclaim her a witch. To support his hypothesis of simulation, Marescot cited the 1583 Synod of Jerusalem's edict on the "rule approved by the Apostolic See to exorcise demoniacs."[201]

By the late sixteenth century, the association of such phenomena with the devil was no longer considered self-evident, even for the church. But it was still rare for a physician to be able convince the *parlement* of the pathological origin of convulsions. Shortly after Marescot, the English physician Edward Jorden unsuccessfully pleaded the case of Elizabeth Jackson, who had been taken with convulsions and condemned to prison and the pillory (a sentence she would never serve) for witchcraft and possession.[202] Jorden published his account of the Jackson trial and reframed the fits of Mary Glover—Jackson's alleged victim—as deriving from pathological disorders of a physiological nature: he employed the diagnosis of "suffocation of the womb."[203] Like many physicians of his time, Jorden tried to account for a wide-ranging set of symptoms by invoking a lengthy stream of diagnoses, including uterine suffocation, fits of the mother, uterine epilepsy, melancholy, hypochondriac illness, and hysterical passion.

The true purpose of Marescot's and Jorden's texts, then, was not to assert a specific diagnosis, but to validate a pathological approach to bodily phenomena. Wide-ranging medical terms were used as a means to lend credibility to a physiological interpretation of bodily fits. In fact, physicians and their followers needed not only to oppose clergy members and a superstitious population frightened by the spectacle of convulsions, but also to rephrase and redirect the statements of the afflicted, which tended to assign a metaphysical origin to their fits and interpret them in terms of possession.[204] Even before prescribing a treatment, the physician's role was to reinterpret signs. When a physician could not heal, he devoted himself to transforming the general approach to the body. The very possibility of a cure depended on a physiological rereading of the turmoil's origin.

In both these texts, the goal was to reinterpret bodily manifestations as a sign of excess or bodily dysfunction, thus establishing a different view from those of exorcists and other clergy members. When summoned to contribute their opinion in a trial, physicians were able to assert their social and political status; on the other hand, when attending to the victim as patient, the physician could play both a therapeutic and a moral role by discussing the melancholy and frights associated with the pathology.

These new spaces for the dissemination of knowledge would open up in ensuing years, leading to an increasingly clear definition of the roles of the church, the state, and physicians with regard to convulsive phenomena. By appointing physicians to observe cases of possession, the state contributed to a reimagining of the respective roles of the church and physicians. In 1680, a royal edict forbade the execution of witches in France,[205] and witchcraft ceased to be a punishable act under British law with the Witchcraft Act of 1735.

Nonetheless, in medical texts physicians were careful to acknowledge the authority of the church. In England, the physician Thomas Willis made his piety clear by dedicating his 1684 "Essay of the Pathology of the Brain and Nervous Stock" to the archbishop of Canterbury, "Primate and Metropolitan of all England." Willis went on to describe his attempts to stave off the influence of superstitions and noted the importance of his book for religious congregations. Though localizing the origin of convulsions in the brain, Willis mentioned God's help for the physician in making a diagnosis.[206] Many other physicians took care to cite the Bible in the reports on convulsions that they drew up at the request of the church or court. François Bayle and Henri Grangeron, appointed by the court to examine cases of convulsions in a Toulouse convent in 1682, reaffirmed their submission to ecclesiastical orders throughout their report. They insisted that the devil's existence should not be called into question and cited the Old Testament numerous times. Interpreting the nuns' symptoms variously as the result of weakness of the mind, an excess of bilious or melancholic humors, or an effect of the imagination, they underlined that such physiological disorders were in accord with "the laws God established in nature and with the structure and economy of specific bodies."[207] Through such submissive statements, the physicians carefully indicated that they did not aim to modify anyone's religious beliefs or question the power of the church; rather, their aim was simply to identify pathological manifestations in a God-given world.

In the eighteenth century, the church and the police began to appropriate medical discourse, and the right to describe convulsive bodies in pathological terms no longer belonged to doctors alone. Reverend Pierre Le Brun, priest of the Oratory, extensively cited the dissertation of Dr. Rhodes in his 1702 *Histoire critique des pratiques superstitieuses qui ont séduit les peuples et embarassé les savants* (Critical History of Superstitious Practices Which Have Seduced the People and Embarrassed Savants).[208] In this large volume, Le Brun welcomed Rhodes's many references to Hippocrates, Descartes, and Willis,[209] and detailed a few cases of convulsions himself, adopting the format of a medical observation for his purposes. Le Brun wrote of "black vapors,"[210] which he presented as an effect of the imagination aroused by fasting and abstinence. In his text, a pathology could result from an intermingling of faith and superstition and could be exacerbated by too much discipline or unwarranted fears.[211] Convulsions, in particular, were described as the result of frights of the imagination or physiological "vices" (a term used to refer to any weakness of the body). Le Brun's pastoral voice aimed to subdue cries and trembling by appealing to a faith that was circumscribed by reason.

With the Camisards in France, the French Prophets in England, and the

Quakers in the United States, pamphlets and dissertations again served as platforms for adopting medical conceptions within religious and political discourse and carrying them forward in public debate. Court cases were another space where such encounters flourished. In France's famous trial of 1731, which opposed the nun Cadière and her Jesuit confessor, Père Girard, the charge of physical ecstasy was reframed as an instance of vaporous fits.[212] This time, it was the Jesuits who adopted a medical interpretation in an attempt to prove the nun was conscious during her encounter with Père Girard. The nun claimed not to know what the priest had done nor that she had become pregnant. Père Girard and his supporters, for their part, attempted to prove the nun's consent, her role in seducing him, and her guilt in taking a potion to kill her unborn child. In this encounter between religious, medical, and political voices, the medical interpretation of the body became a political tool.

A much broader debate arose during these same years, when people began to gather around the tomb of the Jansenist deacon de Pâris. Following the excommunication of the Jansenists at the beginning of the eighteenth century, the religious movement had lost a great number of followers. When de Pâris was buried in the Saint Médard cemetery in 1728, following a life of extreme deprivation, those who knew him perceived the event as the inhumation of a saint (fig. 3). The recounting of several miracles that took place by the tomb was enough to strengthen de Pâris's fame far beyond the sphere of the Jansenists. Men and women of all ages and social conditions soon gathered around the tomb, experienced fits, and thought themselves cured from disabilities as various as lameness, paralysis, and blindness. The reverse also happened: one woman reported that she was suddenly paralyzed after traveling to the tomb, even though her original aim had been to ridicule the crowds. All those gathering by the tomb, Jansenist or not, were named "Convulsionaries," who together caused such a stir that they became the subject of fierce public debate as well as literary and philosophical inquiry.[213] By 1731, the uproar was such that King Louis XV decided to close the cemetery, but Convulsionaries continued meeting in private until the Revolution. Their physiological disorders only increased as their public gatherings were prosecuted by the state. Some asked to be beaten by other Convulsionaries; others spent hours nailed to a cross; yet others explained that submission to violence was the only way they could find relief from suffering.[214]

The church refused to accept the hypothesis of divine intervention, an interpretation that would have conflicted with its grounds for excommunicating the Jansenists. The Convulsionaries were so stigmatized that many Jansenists themselves, fearing their beliefs would be discredited, took sides

FIGURE 3. "Le Tombeau du B. François de Pâris" (Louis Basile Carré de Montgeron, *La Verité des miracles opérés par l'intercession de M. de Paris* [Cologne: Chez les libraires de la Compagnie, 1727], 9). Image courtesy of the Bibliothèque interuniversitaire de santé.

against the Convulsionaries.[215] Physicians were again brought in to explain the convulsions, and those who opposed a religious interpretation of the fits resorted to the medical glossary to exclude all possibility of God's intervention. Diagnoses such as the vapors and hysteric affection began to appear in texts such as *Consultation sur les Convulsions*,[216] and physicians insisted on the importance of "sympathy," a phenomenon thought to link the imagination and the emotions to the body.

Strikingly, the instrumentalization of hysteric and vaporous pathologies also served as an opportunity to portray these illnesses in a new light. While medical treatises published until that time had insisted on the mysterious and inexplicable character of vaporous and hysteric illnesses, they could be presented in this new context as "known illnesses," with one author noting that they were "quite ordinary to the female sex."[217] Previously considered among the most difficult illnesses to cure, these pathologies were now described as "most often cured by nature, by art, or by the assistance of both together."[218] Of the famous case of Anne le Franc,[219] two physicians peremptorily declared: "She could have been cured much sooner, had she had the constancy to observe an appropriate diet following the use of remedies proper to the hysteric affliction."[220] What had been presented as a miracle by some was here reread as a simple failure to maintain a prescribed treatment course.

Partisans of the Convulsionaries took up this challenge and opposed the diagnosis of hysteric affection. The second letter of Jean-Baptiste Desessarts, *Recherche de la vérité*, published anonymously in 1733, is devoted to undoing the association between the Convulsionaries' bodily manifestations and a hysteric pathology.[221] While the author admits the similarity between convulsive symptoms and the vapors, he points out that Convulsionaries never experienced vaporous fits before meeting with other believers and that the vapors had never before been known to cure an infirmity. The opposition between the weak constitution of many of the Convulsionaries and the violence of their disorders served as additional proof of the convulsions' extraordinary origin. According to Desessarts, physicians and their partisans had lost all credibility by invoking the involvement of an increasingly wide array of pathologies, including vapors, hypochondriac vapors, epilepsy, hysteric affection, melancholia, and religious enthusiasm. Desessarts explained that if

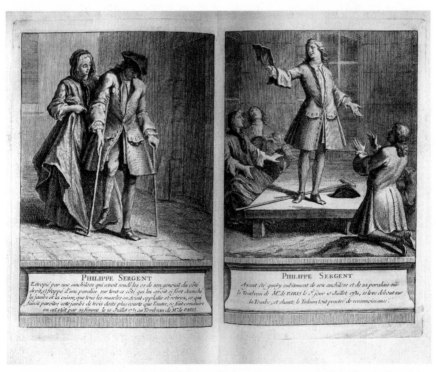

FIGURE 4A & 4B. "Philippe Sergent" (Louis Basile Carré de Montgeron, *La Verité des miracles opérés par l'intercession de M. de Paris* [Cologne: Chez les libraires de la Compagnie, 1727], 18). Image courtesy of the Bibliothèque interuniversitaire de santé.

no diagnosis was sufficient to account for a physiological manifestation, then none could be considered originary.[222] As a result, the convulsions were seen as miraculous in nature. From the mid-1730s, many writers favorable to Jansenism sought to challenge any association between a person who had been miraculously cured and those suffering from hysteric illnesses. They cited the complexity, etiology, and length of the disease. Even so, the need in contemporary publications to distinguish the fits of the Convulsionaries from hysteric illness attests to the importance attributed to this diagnosis—and to making a diagnosis in general—at the time.

In fact, the right to make a diagnosis no longer belonged exclusively to physicians. Convulsionary women arrested by the police were released if they admitted that their convulsions had in fact been the vapors.[223] Those who refused were imprisoned in the Salpétrière and accused of imposture. That the accused could free themselves by invoking this pathology once again indicates the generalized currency of the diagnosis of hysteric vapors. But as Lindsay Wilson has discussed, some women refused to rephrase their case as a pathology or instance of "simulation," and wrote to *parlement* requesting a reinterpretation that would do justice to the true and metaphysical origins of their bodily transformation.[224] Such reinterpretive gestures were meant to put an end to public disruption; the motivation of the police was not, of course, to cure but to affirm a certain nexus of physiological disorders. Since the Convulsionaries had been excommunicated, punishment was administered on the basis of both their religious interpretive divergence and their disturbance of public order. It was to last only as long as individuals persisted in their beliefs. The use of the diagnosis had thus become a social fact.

A testament to the overlap between the religious, medical, and political investments in the convulsive body of the eighteenth century can be found in Joseph-Antoine Cérutti's *Épître en Vers Irréguliers* (The Village Paper. Epistle in Irregular Verses) of 1790, which assumes a historical perspective to summarize revolutionary movements, convulsionary violence, and Mesmer's practices in a single sardonic passage:

> The same faces—I remember seeing them among the convulsionaries of Saint Médard and later in the mesmeric tub. Among other resemblances, the Man with a large beard, nicknamed the Headhunter, struck me. I remembered having seen among the first Jansenists a Wild man, just like him, at times distributing blows of a log to convulsionaries, at times nailing them down on the cross, according to the practice of this miraculous and tumultuous sect.

It would be quite singular that the same man who crucified the martyrs of Jansenism today decapitates those of the aristocracy. That sometimes makes me believe that the great machine of our present-day plots is not led by democrats, nor by the aristocrats we suspect, but by savant hands, which directed these plots and financed Jansenist quarrels. Let us follow the route I indicate; we will see marked traces: some leave from Saint Médard; others from the Tub; they gather in the Palais-Royal; these are the same actors. Who are the directors of the troupe? I do not know, but they are not new to this work.[225]

Cérutti quickly evokes the staging of bodies in the throes of fits, then compresses the Convulsionaries, the healer Franz-Anton Mesmer and his attendants, and political revolutionaries and their opponents into a single category. Over the course of these major crises, bodies repeatedly serve as receptacles for eighteenth-century causes célèbres. According to Cérutti, religious, medical, and political causes are able to take possession of bodies through convulsions and extreme behavior. Convulsive bodies reveal the extent to which these causes rend the social fabric and overturn the church's foundations, the traditional respect accorded to the Faculty of Medicine, and the authority of the state. To Cérutti, bodies had become mere objects of violence, devoid of any sense of origin or belonging, that could be manipulated in predictable ways according to prevailing ideologies. To conflate the Revolution with both Jansenism and Mesmerism was to gloss over the motivations at play in each of these movements and present them as a kind of puppet theater in which cries and convulsions were enacted. For Cérutti, the rich variety of physiological phenomena he observed was simply the result of rhetorical manipulations. Drawing a parallel between such different political and religious contestations was meant to hollow out their stakes, leaving nothing but a recurring feast of cruelty. Discounting ideological causes, Cérutti instead presents hysteria as the form of articulation for violence in an eighteenth century that he considers fundamentally hysteric.

Medicine, social class, religion: Cérutti gathered together the three issues that writings on vapors, hysteric affections, and similar medical categories all dealt with. Later, these same themes would provide spaces for the theorization of diagnoses that would be reread as hysteria. Yet, as this chapter has shown, none of these themes formed a continuous thread throughout time. Sometimes the types of knowledge involved in these approaches overlapped or mirrored one another, but they did not necessarily belong to the same inquiry; rather, each form of knowledge evolved at its own pace. The use of different hysteric and vaporous diagnoses was the result of different political and epistemological stakes that effected different modes of understanding.

In 1789, the last treatise to use the word "vapors" in its title was published. The author, Joseph Bressy, did not expound on any new theories, instead suggesting novel therapies such as the use of leather or water-filled mattresses.[226] The years 1689 and 1789, the publication dates of treatises by Lange and Bressy, respectively, thus marked the beginning and end of the ascendancy of the vapors, concluding with the French Revolution.

In Search of Metaphors: Figuring
What Cannot Be Defined

Here is the effect of the unevenness of language: replacing the meaning of a word with
the meaning of the repetition of that word.

ANTOINE COMPAGNON[1]

Throughout the centuries, treatises on the vapors, nosologies, and manuscripts all joined in stating the difficulty of defining the hysteric and vaporous
illnesses. Physicians equated their characterization with an exercise of style,
and their identification with a brilliant coup. They pinpointed two main
problems. On the one hand, proving the coherence of hysteric (or vaporous)
symptoms was no easy task. On the other, the hysteric affection hid itself; it
simulated other pathologies by displaying the same symptoms, which made
its diagnosis extremely difficult. Such a conception raised two corresponding,
major epistemological problems.

First, the establishment of any pathology depended on the identification
of a series of symptoms happening simultaneously or in a certain order. In
the texts on hysteric and vaporous illnesses, the theorization began by insisting on the impossibility of circumscribing the pathology in the diversity of
its bodily manifestations and of the circumstances of the fits. The difficulty
of predicting when the crises would happen and when their frequency would
decelerate or resume, and of explaining their cause, only increased the obscurity. According to Laurence Brockliss and Colin Jones's masterly overview of
early modern French medicine, in the seventeenth century

> the fact that an individual set of symptoms had been given a pathological la-
> bel, such as dysentery or pleurisy, did not mean that each individual display-
> ing all or most of those symptoms had exactly the same internal qualitative
> changes. Diseases in reality were individual and plastic states which could be
> difficult to assign within contemporary taxonomy and whose prognosis was
> consequently difficult to predict.[2]

Yet even if, in the early modern period, doctors did not have the strict expectations regarding the theorization of a diagnosis that they would embrace from the late eighteenth century, they themselves spoke of hysteria as the illness that defied all theory. Treatises also frequently echoed Sydenham's assertion that it was impossible to list all the symptoms.[3] Worse still, there were no exclusive or constant symptoms, and no symptoms whose presence would rule out the possibility of a hysteric or vaporous affection. If a common denominator is to be found across the diagnoses later recast as hysteria, it is in their construction of these diagnoses as paradoxes resisting all attempts at determination.

The second problem was even more serious, as it cast into question the very project of reading the body. For all pathologies, symptoms operated as signs; they were seen as the voice of the body, forming a language. Symptoms were constructed as the more or less direct expression of a physiological disorder, and the diagnosis was grounded in the body's innocence: the doctor postulated the truth of the body, in the name of which he described symptoms and discovered an illness. The color of the skin, the urine, the temperature, the pulse, were so many testimonies to be interpreted. But in the case of the vapors and hysteric afflictions, the possibility of translating bodily signs according to a symptomatology disappeared. Because hysteric and vaporous illnesses were said to display all symptoms, they were seen as borrowing signs from other pathologies, and indeed often as mimicking other pathologies. The physician therefore had to remain on his guard against a possible simulation by what was nothing but a hysteric illness.

With this danger in mind, Jean-Baptiste Pressavin, a member of the Royal College of Surgery in Lyon, urged the greatest caution:

> However ancient this malady [the vapors] may be, it has not yet exhausted all of the forms it is likely to take. Every day it appears in a new disguise; there is no malady whose characteristics it does not affect, in such a manner that it commands respect among the most skillful of the Art. Is it thus surprising that it has given rise to so many systems?[4]

Pressavin was not alone in this view: physicians in the eighteenth century agreed in asserting that hysteric and vaporous affections "disguised" themselves. Any symptoms could be suspected of miming hysteric and vaporous illnesses rather than signaling the pathology with which they had traditionally been associated. The vapors and hysteric illness could, it was argued, take possession of the body and divert the signs. Such a conception destroyed the equivalence between signifier and signified. The hysteric body lied because its signs were dispossessed of their original meaning and invested by the hysteric affection. These signs now served as a costume, or more precisely as a mask. The result was the paradox of a pathology that had no character of its own, one resembling all others and, by resembling all, resembling none.

If the difficulty of distinguishing a specific pathology was the only thing that could serve as a sign of that pathology, a new danger arose: hesitating in the face of versatile and confused signs, a doctor might too hastily diagnose a hysteric illness or the vapors. A lack of knowledge of the features of hysteric or vaporous illnesses meant that other affections might be overlooked, their symptoms attributed to the hysteric cause. Indeed, in 1759 Joseph Lieutaud, Louis XV's First Physician, accused his colleagues of using such diagnoses to mask their ignorance, and Robert Whytt reproached his contemporaries in similar terms a few years later.[5] This tendency might explain why most physicians of the eighteenth century declared such diagnoses to be the most widespread ones. In any case, for anyone who still doubted the existence of such pathologies, it provided an irrefutable justification for their study.[6]

To characterize the pathologies that would come to be called hysteria, physicians had to resort to particular types of writing. Specifically, the abandonment of definition occasioned discussion about the role and forms of medical writing. In the foreword to the second edition of his *Traité des affections vaporeuses du sexe*, for example, Joseph Raulin defended his work against a *Journal de Médecine* review that had objected to the vagueness of his theorization and set the value of definition against Raulin's focus on description. Raulin insisted on the particularity of vaporous affections and of the forms of writing suited to giving an account of them. He distinguished two types of definition, one intended to be exact, the other taking the shape of description, and stressed that even geometers did not disdain this second kind of writing but used it to project figures in space.[7] The objection voiced by the *Journal de Médecine*'s reviewer and countered by Raulin indicates the extent to which physicians were compelled to reflect on the writing of knowledge itself. Here, description was a way of avoiding having to circumscribe the pathology. In the 1750s, physicians saw their role as building a pathology as a series of cases, without claiming to abstract an entity from those cases. They mainly positioned themselves not as savants but as observers, thereby displacing the role of the savant from systematization into observation. Certainly, their approach redefined the activity of medical knowledge,[8] for renouncing the claim to fix a pathology permitted a better grasp of its diversity, its variability, and its contingency. Their descriptions focused on types of sensations and physiological reactions, and aimed to draw up a larger inventory of physiological phenomena so as to facilitate the practitioner's identification of symptoms, whether ones already specified or not.

The reluctance to define hysteric and vaporous illnesses was not restricted to treatises, but also appeared in courses on medicine, in the shape of hesitations, incertitude, and ellipses. Even at the end of the eighteenth century, François Broussonet, professor at the Montpellier Faculty of Medicine, admitted the confusion created by the diagnosis. His lecture manuscript notes: "We must be far more attentive and circumspect to look for the signs which seem to characterize this illness, undetermined as it is."[9] He adds:

> It is quite difficult, not to say impossible, to define hysteric passion so that this definition gives a fair and true idea of this illness. We see nevertheless that we can define for better or worse the hysteric passion as a convulsive illness accompanied by a despair of curing it, by fright and an almost continuous fear of dying. We don't guarantee the accuracy of this definition; we give it for what it's worth without worrying too much whether it pleases or not.[10]

The professor undertakes to carry on regardless of the obscurity of hysteric passion. Brisk and capacious, the definition is undermined for his students even before it has accomplished its role: determining a pathology, enumerating its aspects, delimiting its form. Broussonet renounces the responsibility for clarifying his words, or more precisely does not seem to feel obliged to do so when facing an affection that does not submit to his inquisitive gaze. His reticence discredits a pathology that did not lend itself to the exercise of knowledge. The physician here transmits not knowledge, but impatience.

A Catalog of Images: Proteus, the Chameleon, and the Hydra

Unable to outline the pathology with precision, physicians' writings insisted on their desire to give the reader an image of it. These images were mostly situated at a strategic point, at the beginning of the book or as a chapter heading. Far from being the result of chance, or invented for the occasion, they were repeated from text to text during the seventeenth and eighteenth centuries. It soon becomes clear that these rhetorical figures had more than a merely ornamental function. Significantly, they were less crucial in writing on epilepsy, melancholia, or Saint Vitus' dance, all regarded at the time as pathologies similar to the hysteric and vaporous case. In texts on hysteric illness, their role was partly to please and to create a mystery, but even more to the point, their inclusion became a means of compensating for the absence of a key symptom that could be recognized in the sight or in the suffering of patients.[11] This section examines the operation of these metaphorical figures.

The role of words in constituting visibility at the turn of the nineteenth century is magisterially described by Foucault in *The Birth of the Clinic*:

"A hearing gaze and a speaking gaze: clinical experience represents a moment of balance, for it rests on a formidable postulate: that all that is *visible* is *expressible*, and that it is *wholly visible* because it is *wholly expressible*."[12] In contrast to this descriptive language, the use of metaphorical language for many centuries opened the text to another dimension, that of evocation. Here description, as the language proper to evocation, appeared inadequate to express the singularity of the illness's manifestations. Rhetorical figures indicated that the text remained at the level of the unsaid, the suggested, the indirect. The insertion of terms extracted from other contexts operated as a substitute for a discourse that was not to take place. When talking about hysteric pathology, different types of images were called on in line with the different kinds of interpretation developed. Rural animals, mythological animals, and Greek gods personified in turn the pathology, the struggle of physicians, and the suffering of patients. The relation to the extraordinary, the unexpected, and the troubling was transformed and given a space within the qualification of the pathology. As a result of this process, the operation of connotation became essential to the functioning of knowledge. It signaled that not everything could be described, indicating interplays of power, value, and affect that constituted symbolic investments in the production of knowledge.

A first set of metaphors privileged rural animals. Physicians used these to describe "premonitory symptoms" and sensations experienced by patients during their fits. "Swarms of insects,"[13] "reptilian movements,"[14] and the "swarmings of ants"[15] were the recurring images in the description of premonitory symptoms.[16] These comparisons reinforced the opposition between body and mind, making the hysteric affection a pathology in which the individual suddenly discovered the unknown and disturbing presence of a body that might veer out of control. The choice of insects and reptiles as metaphors gave prominence to despicable animals—the low, the horizontal, as Georges Bataille would say. These are creatures that flee humans, and hide to attack them by surprise. The repetition of such metaphors in the description of cases and in theorization is intriguing: it is unlikely that the patient would have formulated the sensations with such uniformity. The figures may have been suggested by the physician and accepted by the patient in the search for a language to communicate sensations.[17]

The sight of hysteric fits led to another range of metaphors, where images of dogs and wolves were used in reference to patients, both by physicians and by the afflicted themselves. Such comparisons place hysteria in another tradition, that of lycanthropy, a form of melancholia diagnosed in antiquity.[18] John Freind, a member of the College of Physicians who was later appointed physician to Queen Caroline, described such a case in an observation that became a reference text in the discourse on the vapors. The observation was first published in English in the late seventeenth century, then in French in *Journal de Trévoux*, republished in Dumoulin's *Nouveau Traité du rhumatisme et des vapeurs* (New Treatise on Rheumatism and the Vapors), and published again in its entirety in French in François Planque's *Bibliothèque* some fifty years later.[19] In the observation, Freind recounts the experience of the physician Thomas Willis when visiting an Oxfordshire family whose girls were afflicted with cries and convulsions, and goes on to relate a similar case he encountered himself in Blackthorn. The doctors' curiosity about an uncommon phenomenon is marked by the care they take with their eyewitness accounts of the potentials of the pathology. Freind's "observation" strikes the reader with its precision. His narration reconstitutes the scene in the way Willis might have seen it, giving the reader, in turn, the impression of being able to imagine the scene and to see the symptoms in the order that they appeared. His account sustains a delicate balance between contradictory features: the repetition of the notion of ordinariness[20] underlines the frequency of the crises for the patients, but this contrasts sharply with the particularity of symptoms that is described. To vindicate that particularity, the physician insists on the trouble he has taken to describe what he saw and heard, adding that one had to have seen it oneself to have an accurate idea of it.[21] The extraordinary violence of the fits astounded the physician all the more because it occurred daily, producing bodily movements that were repeated identically each time. Faced with the difficulty of conveying such a pathology, the doctor resorts to an image, that of a dog. The comparison gives a visual force to the narration, further heightened by the absence of any action on the physician's part. Freind describes Willis and himself as witnesses, who admit their ignorance and do not submit the patients to the risk of a therapy undertaken merely as an experiment. Nor do they claim to explain the fits: they abstain from any hypothesis about the causes. The fits are recounted precisely for their incomprehensible and irreducible character. Freind takes no pains to distinguish the phenomenon from a case of possession, despite the similarity of the description[22]—an omission indicating that this frame was now dismantled fully enough for the physician not to have to reference it. The humility of Freind's discourse is both that of a practitioner and that of a theoretician reduced to the status of a spectator. Symptoms and their circumstances take up the entirety of his report. The observation's impact was to further circulate the image of the body-become-dog.

Such comparisons with the barking of dogs or the maws of wolves to account for convulsive fits reappear in the late eighteenth century in manuscripts sent to the Société Royale de Médecine.[23] From insects to wolves, comparisons illustrate the increased violence of fits, the savagery and unpredictability of crises. The experience of the body is read as that of a metamorphosis, which alone can account for its uncontrollable character. At the same time, the reference to animals commonly found in the countryside inscribes the pathology in daily life. In the texts using images of this kind, physicians avoid technical terms and theorization of the pathology. They put forward a medical innocence, the admission of being surprised and the aim of sharing that surprise, so as to draw their colleagues' attention to the incompetence of medicine in the face of such phenomena.

Some physicians focused not so much on the denomination of specific animals as on the description of a state that characterized them. In the late seventeenth century, Ettmüller compared uterine furors to the rut of the female.[24] Varandée compared hysteric catalepsy to hibernation, later referring to the venom that linked hysteric pathology to the torpedo and the scorpion's venom, and also drawing a parallel with rabies.[25] Cureau de la Chambre[26] and Kenelm Digby,[27] explaining the diffusion of the vapors in relation to different emotions, also referred to rabies, and compared it to the excitation produced by the spleen and melancholic humors.[28] These images presented hysteric and uterine pathology as a transitory phase of the body, and thus as part of its balance. Such comparisons attenuated the extraordinary character of symptoms, casting them instead as a reaction of the body to a specific context. Illustrating their points with references taken from popular imagery or natural sciences was a way for physicians to make an idea of the hysteric pathology accessible. Lacking explanations, physicians offered analogies that allowed them to subdue the pathology's mysterious character.

For as long as an approach to the body in terms of temperament dominated, humors were the criteria of definition of the body, but in the late seventeenth and early eighteenth centuries this frame progressively disappeared; physicians focused on symptoms and sought to identify them all. Mentioning, as Freind did, that the patients and their parents were free of any intent to attract curiosity or profit from the convulsions reinforced the idea of the patients' innocence. No connection was made between the patients' affliction and their education, their way of life, or their sensibility. The affection, even if characterized by acute physiological sensations, remained foreign to the body, and the image of the animal served to mark this duality of the body at the moment of suffering. The convulsions implacably took the body over, inhabited it, but without owning it; the physicians' concern was to detect a pathology, not a type of patient. From the 1770s, however, the doctor-spectator would become a doctor-interpreter. It would become equally important for him to recognize a pathology and a type of patient for hysteric vapors. The affection would then lodge in the patients' bodies, the bodies of those considered at high risk.

The use of each image underwent periods of great success and others of neglect. However, a motion to graft together imaginaries, generate identifications, and write in echoes led to a great regularity in the use of three particular images: the chameleon, Proteus, and later the Hydra. All three contributed to shaping the discourse on hysteric and vaporous illness, yet what was said with the same image differed widely, for in the process of repetition, the orientation of the images began to diverge according to the medical writings in which they were inserted. What changed was not the motifs themselves, not the words, but their value—the manner in which those comparisons and metaphors were used and their tone. To trace this transformation, it is necessary to examine how the metaphors of the chameleon, Proteus, and the Hydra were cited and displaced in various texts over time.

François Mauriceau's *Traité des maladies des femmes grosses et de celles qui sont accouchées* first published in 1668, is the first widely circulating text to compare hysteric passion to the sea god Proteus:

> This malady is ordinarily accompanied by such a great number of different accidents, according to the various dispositions of persons afflicted with it, & causes so many different changes & such great alteration in the functions of the body & spirit of women, that it can well be compared to the power that Proteus, the sea God of fable, had, the power to change into all sorts of different forms.[29]

In Greek mythology, Proteus was the son of Oceanus and Tethys, and was characterized by his capacity to assume any form, animal or vegetable. He was invested with a gift of prophecy, but refused to make predictions to mortals who questioned him. Through this comparison with a privileged image of the baroque era, Mauriceau envelops hysteric affection with a magical aura. Shortly afterward, the Countess of Winchilsea deployed it to characterize the spleen in a poem that would be cited in the first pages of William Stukeley's medical treatise on the same subject:

> What art thou Spleen, which every thing do'st ape?
> thou Proteus to abus'd mankind,
> who never yet thy real cause could find,
> or fix thee to remain in one continu'd shape![30]

Proteus quickly became a synonym for hysteric and vaporous affections in medical works.[31] Thomas Sydenham, John Purcell,[32] Pierre Hunauld,[33] Pierre Pomme,[34] Charles Perry,[35] Jean-Jacques Menuret de Chambaud,[36] William Perfect,[37] and Joseph Capuron[38] all repeated this rhetorical figure, which became almost a commonplace in eighteenth-century medical discourse. George Cheyne[39] and Noël Retz[40] invoked it to justify their resignation to the lack of defining features for the pathology, affirming the importance, instead, of devoting oneself to the cure. Pierre Pomme, in contrast, announced in his preface an ambition to "fix this Proteus,"[41] and later calls the pathology a "proteiform monster."[42] In 1763, the association was so widespread that François Boissier de Sauvages, who usually rejected literariness and equated truth with precision and rigor, took it up in his *Nosologie méthodique*.[43]

A further metaphor soon accompanied the Proteus notion: the chameleon, another subject of fascination in baroque literature and salons.[44] Sydenham used the combination in his dissertation on hysteric affections: "To simply enumerate all the symptoms of hysteria would be the work of a long day; so numerous are they. Yet not less numerous than varied, proteiform and chameleon-like."[45] Pierre Pomme and Claude Révillon took up the image in Sydenham's wake.

How should the choice of these two images and their success be explained? To begin with, the mediation of these figures enabled physicians to enroll hysteric affections into an existing tradition of texts, for they were not the first to use the images in a discourse on bodily afflictions. Proteus and the chameleon had already appeared in Johann Weyer's *De Praestigiis daemonum*, first published in 1563, and André du Laurens's *Discours de la conservation de la veue, des maladies mélancoliques, des catharres et de la vieillesse* (Discourse on the Conservation of Sight, on Melancholic Illnesses, on Catarrhs and on Old Age), published in 1594. In both texts, the metaphors are associated not with a pathology, but with the imagination. In fact, these writings themselves borrowed the figures from one of the philosophical authorities of their time, Marsilius Ficinus, who adopted them in his commentary on a text by Priscian that interprets a chapter of Theophrastus on "phantasia & intellectu" (fantasy and intelligence).[46] According to Ficinus, the fantasy or imagination was the representation of what is perceived, the faculty to imagine, and the faculty to compare different perceptions. It was also the faculty to extract from mere sensations something that could be analyzed by reason, and that could appropriate all perceptions by committing them to memory.[47] In Ficinus's commentary, it is the versatility of the imagination itself that is compared to a Proteus figure and a chameleon. The two images underline the imagination's capacity to take different shapes, as well as its impulsivity: Theophrastus's text seems to have insisted on the imagination's characteristic of reacting to stimulation and responding by transforming itself. Theophrastus explains that the imagination changes according to what happens and what it has previously assimilated. It is therefore both passive and active, receiving images and, in turn, producing them. It has the power to figure things that are not there and to modify itself not only in relation to the present, but in relation to memory perceptions that it can summon up itself. In evoking Proteus and the chameleon, then, Ficinus stressed the transports of the imagination as it solicits reactions and metamorphizes accordingly.

When Weyer borrowed the comparison offered by Ficinus, his aim was not limited to a description of the faculties of the imagination. His treatise insists on the magnitude of the movements made by the imagination with the objective of explaining phenomena that had previously been perceived in terms of demonic possession and presenting them, instead, as a pathology. The comparison of the imagination with a chameleon and with Proteus thus served to stress the imagination's capacity to react to absent things and to make them present. Weyer indicated that the production of inner images was followed by motions of appetite, desire, fear, and pain.[48] Du Laurens's work, in contrast to this conception of struggle and pain, evoked the imagination to praise mankind. He associated it with memory and reason, making those the features that distinguished humans from animals: "Man having in his soul the graven image of God & representing in his body the model of the universe, can in an instant transform it into anything like a Proteus, or receive in a moment the impression of a thousand colors like a chameleon."[49] The two comparisons underline the searing intensity of the imagination's metamorphoses, and magnify a faculty seen as "the prime power of the soul,"[50] serving both daily affairs and contemplation.

As part of this tradition, the use of Proteus and of the chameleon articulated the physician's fascination with the hysteric fit. The metaphors figured the versatility and virulence of symptoms. Medical books here lined up with the works of morals and of letters, in which the chameleon meant great skill and an absence of moral rules of conduct. In these works, the image designates those who know how to manipulate the people around them by presenting themselves under the most expedient aspect at any one moment. Medical texts adopted the existing connotations of Proteus and the chameleon, but also displaced them: the attributes remained the same, but their understanding was altered by the stakes of therapeutic practice, and marveling yielded to mistrust. When Menuret de Chambaud, for example, refers to Proteus in a discussion of stupor, it is in the form of a warning that hysteric affection cannot be approached like other maladies: "A true Proteus, it assumes all forms, masks itself, & hides itself in a hundred ways, glides and is found everywhere & and it is quite essential not to lose sight of it."[51] By the late eighteenth century, when the figure of Proteus appeared in the courses given by Victor Broussonet, the imperative to distrust had been inscribed at the heart of the perception of the hysteric passion:

> Having no definite character, it takes that of others in turn, under the mask of which it always shows itself, beyond the frequent errors into which these various forms that are foreign to it make it fall or can nevertheless make the doctors fall, so to speak, if they are not employed in unmasking it with as much address as they are able. These errors have consequences for the fault and the life of the sick as well as for the reputation of doctors. For if he does not devote special attention to unveiling this new Proteus, guided by deceitful appearances, he will be led despite himself to resort to a method of treatment entirely opposed to the one required by hysterical passion.[52]

Here, Proteus no longer suggests a fascination with the extraordinary, no longer connects the diversity of symptoms with the marvelous. "Mask," "unmask," "deceitful appearances," and "despite himself" resonate to stress the guileful character of the malady. The image warns of risks associated with a pathology that cannot be localized, whose character cannot be fixed, and whose description is always already obsolete.[53] Not only does the hysteric affection elude definition; it also mystifies those who attempt to treat it. The use of the image of Proteus marks a dual danger emanating from the affliction as lie, which risks both the doctor's reputation and the patient's life. The marvelous has lost its baroque glow; it is invoked in order to be demystified in a medical discourse that aims to expose everything.

The very presence of a hysteric affection ruins the innocence and truth of the body in the name of which the physician previously decrypted symptoms and identified pathologies.[54] If, like Proteus, the hysteric passion can appear only in disguise, the body can no longer be approached in terms of transparency. According to late eighteenth-century physicians, the hysteric pathology diverted the signs usually associated with other pathologies; it hid and lied about its own nature. Investing symptoms in new ways, physicians cast hysteric illness as an illness become *lie*. The metaphors of Proteus and chameleon did not signify the physician's wonder at physiological capacities, but rather his awareness of their dangers. Fluctuations became the mark of inconstancy. Metaphors drew a profile of duplicity and deceitfulness.

From the 1750s, a new figure was enlisted that further amplified this imaginary nexus. In the course of an observation published in his treatise, Pomme offers another metaphor originating in Greek mythology, the Hydra: "To combat a Hydra whose heads were reborn every day, it was necessary to use the same weapons but even more powerful, in order to insure the defeat of this monster."[55] Théodore Tronchin cited this figure in his correspondence of the same period, writing to a victim of the vapors:

> As I have told you, the weakening they cause to the nerves is what you call the vapors, for this word—to which the one who utters it and the one who hears it rarely attaches a fixed idea—expresses a pain one feels but cannot explain. It is a mysterious word which expresses quite clearly that no one understands anything about it. What one would want to express is a Hydra with a hundred heads. It is an incomprehensible monster. Do you not recognize it, Mademoiselle? I wager you do.[56]

Pierre Sue also took up the image in his *Anecdotes historiques, littéraires et critiques* (Historical, Literary, and Critical Anecdotes) to absolve the physician from responsibility for the failure to cure hysteric illness; following the myth of the Hydra, he asserted that only a hero could vanquish it.[57] Guillaume Daignan, in turn, cited it in a chapter on spasmodic and convulsive illness: "Science, ignorance, art, artifice, trickery, illusion, deceit, imposture; everything has taken up arms furiously to combat this Hydra."[58]

The Hydra of Lerna, a serpent with five or six heads that regrew as soon as they were cut off, was created as a trial for Hercules. The reference is again mythological, yet the register evoked is a different one: Proteus referred to the marvelous, Hydra to the violence of a monster. With the Hydra, the accent was no longer on the description of symptoms, but on the resistance of the hysteric affection. What characterized the pathology in this representation was the difficulty not only of identifying it, but also of eradicating it. The struggle overturned the classical vision of therapy as formulated by Galen: it was no longer a hand-to-hand battle between the patient and nature, observed by the doctor.[59] The struggle now pitted the doctor against the affliction, while the patient was effaced behind the hysteric affection. Changing the adversaries meant changing the position of the patient, who was now associated with the pathology at a time when the latter seemed always to be victorious. Hysteria was now what summoned the doctor, challenged him to a duel, and threatened his career.

In the seventeenth century, metaphors operated as a chiaroscuro effect to dazzle the reader. At the very moment when he enunciated his knowledge, the doctor indicated its limits; considering the pathology from a distance marked his modesty and reserve. Whether it aroused fascination or fright, the metaphor signaled an admission of inability to describe, grasp, and reconstruct. But if the use of metaphors initially avowed the incapacity to account for hysteric vapors, in the mid-eighteenth century it gradually assumed another role. The image was not passive and did not simply illustrate the text; it led a life of its own. It was represented as being a pragmatic tool that allowed physicians to present figurally what they could not define. However, its function was not hermeneutic alone, but also ideological: metaphor allowed the anticipation of a whole set of ideas that could not be grounded in a logical process, but were gradually attributed to the pathology. Its repetition from one text to another brought about a transfer, as metaphors lent the pathology their traits and sparked off a chain of presuppositions. Physicians personified the hysteric pathology, thus suggesting what they could not demonstrate. Since nobody knew how to delimit it, the pathology was identified with a will to refusal, simulation, and trickery. Not mastered as an object of knowledge, it was imagined as an irreducible and treacherous subject whose motivations were contrary to medicine. The pathology's extravagance was interpreted as a tactic in which the patient was an accomplice or a victim.

In providing a figure for the reader to interpret, the physician compelled the reader to be active and led him toward a dynamic relation with the text, thus conveying an image not only of the pathology and its risks, but also of the stakes involved in its identification. The process by which the reader was invited to make meaning from the image evoked the power of his imagination, while the images themselves might permanently imprint the imagination with the extravagance, savagery, or indomitable force of the affliction. The text gave voice to the emotional force at play in the encounter with the patient, whether as a source of fascination or terror. The surplus generated by the metaphor was therefore not solely epistemological. It also created a fissure in the writing that could transmit fright. It assigned the pathology to the unknown, the mysterious, the frightening, more vibrantly than the absence of definition alone would have done. The physician's inability to define was erased by the force and permanence of the chosen images. Medical discourse drew on them as inherent parts of a thinking process that aimed to elucidate the pathology.

Repeated Quotations, Divergent Readings

Even in the late eighteenth century, when physicians saw medicine as a discipline in revolution, they operated with the repetition of figures and the citation of models, of true founders. The aim of this section is not to contest that a radical renewal took place,[60] but to argue that its novelty consisted in the rewriting and reinstrumentalization of texts that had gone before. Legitimating references, fading in the first half of the eighteenth century, returned in force at the century's end. And in the early nineteenth century, far from limiting themselves to denominations and recent theories, physicians invoked a whole genealogy of texts—a rereading of the ancients was undertaken in the service of radical newness. For example, Philippe Pinel wrote in his unpublished notes that in order to be a good pathologist, one must be "very discriminating in the choice of observed facts," yet equally have "deeply studied Greek medicine."[61]

At the very moment that Pinel presented observation as an essential tool for reflection, he reasserted the importance of Greek medicine as a model for comparison and the identification of diagnoses. His recourse to the past was not limited to the space of theorization. It also took place in the exposition of what he called "observations." The case studies in Pinel's nosology, *La Médecine clinique*, presented as the fruit of his experience at the Salpêtrière, were by no means restricted to patients he had met himself, but included a large number of cases related by others, notably by Friedrich Hoffmann, a German physician who died in 1742. The nosology referenced the ancestors at precisely the point when it claimed to set out a new way of seeing and appreciating patients.

Even in works that saw themselves as radically new, establishing medicine's credibility still meant inscribing it into a tradition. Quotations placed as epigraphs on the title page presented the author as a man of knowledge. They might be taken from a philosopher (usually Plato, Aristotle, and, in the late eighteenth century, Rousseau), a literary authority (most often Virgil), or a physician—usually Hippocrates, Galen, or an allusion to Hermes, Asklepios, or Aesculapius. More recent forebears were Sydenham, Willis, and Descartes. Other references were located inside the exposition, situating the author's point of view and establishing the credibility of his assertions. They exempted the author from providing further details or even having to justify his claims, representing these as an accumulation of ideas whose extreme longevity meant they need not be specially discussed. As well as reiterating particular, well-established allusions and quotations, these physicians were careful to refer to their ancient (or more rarely modern) predecessors even if only to mark their contradistinction. The claim to continuity was not only a way to gain acknowledgment from a traditional structure, the Faculté de Médecine or the Royal College of Physicians, but also a way to formulate knowledge and fashion meaning.

In writing on hysteric illness, quotations became instrumental in asserting the authority of a text that was undermined by the difficult object it attempted to grasp. At times, they were provided en passant, without even naming the authority to which they referred. Such was the case in Ambroise Paré's *On Generation* (1573). In his short chapter devoted to the suffocation of the womb, Paré makes reference to a description of the womb that would later become a favorite topos:

> In the suffocation of the womb, the putrid vapors sometimes rise up to the diaphragm, lungs, and heart, making it so that a woman can neither inhale nor exhale. Such vapors are carried not only by the veins and arteries, but also by the hidden spiracles within the body. If the said vapors rise to the brain they cause epilepsy, catalepsy, . . . lethargy, apoplexy, and often death. To put it simply, the womb has its own sentiments beyond the woman's will, such that it is said to be an animal, since it dilates and rushes about more or less, in response to a variety of causes. And when it desires, it wriggles and shakes, making the poor weakling female lose patience and all reason, and causing a great racket.[62]

As mentioned in chapter 1, the suffocation of the womb was seen as the ascent of vapors provoked by the movement and the rising of the womb itself. With the word "sentiment," Paré employs a concept that contained a wide range of meanings during his era: it signified the senses (as in the five senses), the capacity for perception, and sensation, along with emotion, awareness, presentiment, consciousness, intuition, and instinct. By asserting that the womb carries sentiments beyond a woman's will, Paré makes of it a totally independent entity. As for *tintamarre* or "racket," sixteenth-century dictionaries describe such noise as blaring, muddled, and disagreeable, consisting of a prolonged echo or a drone. The component *marre* comes from a Dutch word meaning "nightmare." Describing a disruptive violence that shakes and disturbs, Paré's text associates animality with disorder, trouble, and the unknown. The animal serves to emphasize how one region of the body works against another, and becomes another name for this physiological pandemonium. Pleasure and its demands resound loudly. Paré emancipates the womb, while regarding the woman as a weakling or *femmelette*. He offers a vision of extremely violent crisis; woman's desire is not conceived as a continuous state, but as a sudden and complete unsettling. For Paré, unsatisfied desires destabilize natural order and physiological balance.[63] When the animal is presented as a model of the womb, the use of the locution "it is said" refers the reader to one of Plato's works. This strategic comparison is taken from *Timaeus*, where the philosopher compares the womb to a hotheaded animal that occasionally takes control of women:

> With the so-called womb or matrix of women; the animal within them is desirous of procreating children, and when remaining unfruitful long beyond its proper time, gets discontented and angry, and wandering in every direction through the body, closes up the passages of the breath, and, by obstructing

respiration, drives them to extremity, causing all varieties of disease, until at
length the desire and love of the man and the woman bringing them together
and as it were plucking the fruit from the tree.[64]

To grasp the full scope of this passage, which was presented across the
centuries as one of the finest descriptions of the pathology, it is important
to look at earlier parts of the text as well. For Plato, the female animal has a
masculine equivalent, though this was seldom mentioned in citations:

> And this was the reason why at that time the gods created in us the desire for
> sexual intercourse, contriving in man one animated substance, and in woman
> another, which they formed respectively in the following manner. . . . Where-
> fore also in men the organ of generation becoming rebellious and masterful,
> like an animal disobedient to reason, and maddened with the sting of lust,
> seeks to gain absolute sway.[65]

Plato's text constructs women in reciprocity with men. The phrase "like an
animal" introduces the animal as an object of comparison, not as a determining feature. Some paragraphs earlier, too, the human body is also discussed in
terms of being an animal; when Plato speaks of a physical attribute of woman
as "animal" he is employing an image, not a categorical notion.

Paré, having summoned the authority of citation, proceeds to substan-
tially distort Plato's imagining of the womb, an appropriative move that ren-
ders the citation from *Timaeus* almost unrecognizable. Here, the text veils
the operation of selection, while carrying the original concepts far off course.
It freezes the metaphor, locking it into a singular meaning that is, in turn,
repeated until it becomes standard. The womb is made foreign to a woman's
body, independent of her will, and beyond the rule of reason. Across such citations, the feminine body, its needs, its function, and its effects are reframed.
The desire for reproduction becomes a desire for copulation in a rewriting of
Plato's text made possible by truncation. This erasure facilitated a new distinction between the sexes; it also led to an association of womanhood with
the sexual organs. In this consideration of the feminine body, woman was
woman-animal, remote in her alterity.

In medical texts published from the late sixteenth century on, the use
of the animal metaphor invited an understanding of the malady, an assess-
ment of the sick body, a specification of the condition in women as compared
to men, and an interpretation of sexuality. But the metaphor's durability in
medical discourse did not translate into consistency of usage. Texts published
over a span of 250 years directly or indirectly cited this same passage, yet their
accounts of the malady's symptoms and their interpretations of its causes
shifted considerably. Doctors proved skillful at harnessing the metaphor for a

FIGURE 5. *The Nightmare*, Henri Fuseli, 1781. Image courtesy of the Detroit Institute of the Arts.

range of evocative purposes, and it was instrumentalized to specific political and epistemological ends. The role of the excerpt in the original text disappeared as it took on a new scope in the texts that cited it. The process recalls Antoine Compagnon's discussion of the practice of quotation, as developed in *La seconde Main; ou, Le Travail de la citation* (The Second-Hand; or, The Practice of Quotation). Compagnon proposes a definition of the citation that encapsulates what I retrace in this chapter, as "a repeated enunciation and a repeating enunciation."[66] The citation signifies on two levels: on the one hand the content of the quotation itself, and on the other its enunciation, that is, its meaning within its new context. This second level is affected by the way the citation is deployed, and its acquired value as a repeated sign—what Compagnon calls the "interdiscursivity of the citation."

Studying the changing use of Plato's words reveals how these two levels can come into play, and adds an epistemological dimension to the history of gender as defined by Joan Wallach Scott in her study of social contexts.[67] At different times and with different textual uses, the same citation favored different associations, inscribed different meanings. Looking at a selection of medical, literary, religious, and philosophical texts, three main kinds of use emerge, which mark three main periods in approaches to the illness. The usage starts in the late sixteenth century and shifts in the late seventeenth century, then again in the eighteenth century, with Denis Diderot's 1772 intervention.

Montaigne's *Essais* (1588) is one of the rare sixteenth-century texts to mention reciprocity between the sexes in the animality of desire.[68] Overwhelmingly, Montaigne's contemporaries identified the desire that Plato discussed in both men and women as a characteristic uniquely of the female condition. Doctors introduced Plato's characterization exclusively in the framework of the female malady of suffocation, neglecting its masculine parallel, which was not thought to develop into a pathology.

The omission of the masculine animal would shape visions of woman for centuries to come. At the end of the sixteenth and the beginning of the seventeenth century, references to the animal reappeared frequently.[69] The suffocation of the womb was seen as a form of impatience driving the organ in all directions, as described by Plato. But the womb also affected regions beyond the scope of its actual movement. Although its stirring was restricted to the lower part of the body, it produced vapors that could reach the head and organs. Depending on where it reached, the suffocation of the womb could provoke heaviness of the head, palpitations, coughing, nausea, epileptic convulsions, suffocation, or fainting. It was characterized most thoroughly in the work of Jean Liébault. In 1582, Liébault revisited the ancient conception of the body and wrote of a "vagabond womb."[70] He explained that the womb "becomes indignant like an animal"[71] when its desire goes unsatisfied. This conception of animality led to the therapy mentioned in chapter 1: as in the Renaissance, doctors used the pessary to spread unpleasant odors so that the womb would breathe them in and be drawn back down into the body. Varandée expanded on the principles of this therapy,[72] describing the womb as an animal that rushed toward the scents it preferred and fled from those it abhorred. The womb's preferences were different from those of the nose. When odors that pleased the womb, such as asafetida and sulfur, were placed near the mouth, the womb moved up to reach them. If they were spread near the abdomen, however, the womb would return to its habitual place.

The womb was thus personified. It sensed, reacted, enjoyed, and fled. Liébault and Varandée followed Paré in considering the womb completely independently of the rest of woman's physiology, constantly at risk of creating an imbalance. The womb threatened to disturb woman's physiological equilibrium. Woman could not resist her womb, but neither was she one with her own organ. On the contrary, she was its victim: to an extent, pathology indicated the woman's refusal to abide by her physiological and sexual needs and listen to animal desires, which was a proof of her morality, since she had not indulged in copulation just to fulfill her needs. As such, this vision did not lend itself to moral judgment against individual patients.

Around 1670, when physicians abandoned conceptions of the womb as an origin of the suffocation and convulsions, they began to focus increasingly on the brain and nerves to explain the pathology. For a time, references to *Timaeus* and animality disappeared.

The Convulsionaries crisis in the 1730s, however, marked a change. Writing on sexual desire as a source of hysteria took on new significance. In this period, explanations based on hysteric affection dominated understandings of the malady, and returned the question of sexual difference to the heart of its theorization. It is only at this point that writings connecting hysteric illness with sexual desire emerged again. A simple diagnosis would have been unable to combat the aura around the Convulsionaries' manifestations, and powerful images were required. In fighting the imagining of miracles and the growing fervor among the movement's followers, few portraits would prove as effective as that of a lustful animal. A second mode of citing Plato's *Timaeus* was thus inaugurated.

Philippe Hecquet, a Jansenist doctor opposed to the Convulsionaries, seized on the metaphor and viewed the Convulsionaries' manifestations as a form of animality.[73] *Le Naturalisme des convulsions* (The Convulsions' Naturalism), published in 1733, makes reference to Hippocratic texts,[74] and then identifies the symptoms of the malady:

> This malady is neither obscure nor unknown among the ancient authors of Medicine, . . . a beast whose excited movements, as *Plato* said, from a base of *concupiscence* (*animal concupiscencia*) rose from low to high, or toward parts in all regions of the body, but particularly toward the throat, where stranglings arise; toward the chest, where hysteric suffocations arise, and the enormous swelling of all parts set into commotion, *uteri strangulatus, suffocatio, inflatio.*[75]

Hecquet returns to this image some pages later: "These characteristic signs of hysteric passions are the movements of this supposed beast, imagined in antiquity to be within the bodies of girls, where it causes, as it rises, this ball that swells so enormously in the lower abdomen of hysteric persons."[76] The vacillation of Hecquet's framing statement is significative in itself. He first associates Plato with the metaphor, elaborating the image with an explanation and bolstering it with Latin terms. Once the reference has fulfilled its role, the author returns to it in a more measured manner, and moderates the evocative power of the image by adding the adjective "supposed." One sees a Janus-like staging of the authority of antiquity and the authority of modernity in quick succession, the latter bringing nuance to the discourse. The metaphor is not invoked as truth, but rather as a way of creating meaning; to inspire an understanding of the malady—as an imagining rather than a theory. The metaphor is mentioned not for its capacity of denotation but for its value of connotation. It invokes the possibly nonexistent animal to suggest a more general beastliness[77] and to personify the powerful hold of the senses. It directs attention away from possible manifestations of a higher power in the eruptions of the body, localizing them instead within a well-defined parameter: the belly. Insisting on the movement of rising and swelling, the figure presents the body as a spectacle. Far from regarding these ecstatic moments as a transcendence of the soul over the body that transports the subject away from its physical needs, the doctor establishes that the vapors lower the soul down into the body, its flesh inflamed by eroticism.

However, Hecquet departed from the interpretation of hysteric illness as a malady of the womb. And he did recognize the presence of male Convulsionaries, without drawing on the Platonic animal to explain their troubles. Rather than identifying sexual pathology among these men, he cited a vulnerability of tissues and sensitivity exacerbated by the sight of other Convulsionaries. A truncated citation of Plato's text thus allowed the absence of difference between symptoms to be reconfigured as pathological difference, and inscribed a physiological distinction. The symbolic significance affected women alone.

In *Examen critique, physique et théologique des convulsions et des caracteres divins qu'on croit voir dans les accidens des Convulsionnaires* (Critical, Physical, and Theological Examination of the Convulsions and the Divine Character That One Believes Can Be Seen in Convulsionaries' Fits), also published in 1733, Louis de Bonnaire refers to the Hippocratic text *On the Sacred Malady*, recalls the case of the nuns of Loudun as a recognized example of the vapors' effect on the female body, and revisits the metaphor of the animal. De Bonnaire's first citation promises his readers some astonishment:

> One of the symptoms that surprised most greatly him from whom I borrow these instances in the study of convulsions was a weight that would come from the lower abdomen of a man and rise to his chest, from there up to his head, in such a manner that it seemed there was a living animal that slid under his skin: the same thing happened to the girl spoken of in the first example; this resemblance of fits led him to conclude that vapors rising from the lower regions were not so much a proper feature of one of the sexes that the other would be exempt from them.[78]

The idea of a ball rising from the lower back was common in texts on hysteric symptoms, but with de Bonnaire it became an animal. In this passage, the doctor alludes to the example offered by Thomas Willis, describing a man taken by convulsions. By identifying a movement up from the abdomen toward the heights of the body, de Bonnaire aims to illustrate the error of doctors who see the malady as an effect of the womb. He rewrites Willis's interpretation, adding a living animal sliding under the skin that is nowhere to be found in Willis's text.

De Bonnaire's second mention of Willis is just as visually powerful. It comes in his criticism of a machine used to restrain Convulsionaries and control their thrashing movements:

> Doctors order similar things: they find that it is very good to strap the lower abdomen up to the armpits, as was done to Mademoiselle le G. They would lay one of Willis's convulsionaries in the bosom of a servant who hugged her with all her might while others pressed her belly and lower abdomen; and the English Doctor believes that if by such violence one can prevent the living animal, which departs from the lower abdomen, from reaching the head, the convulsions will be fewer in the other outer regions.[79]

Louis de Bonnaire references Willis while leaving his theories ambiguous, even though Willis distinctly rejected the view of convulsions as resulting from the womb or desire. It is true that the English doctor mentions several servants. In Willis's texts, these servants are themselves attacked by convulsions, with pains in the chest and abdomen, and at no point does he draw a connection with sexuality. In fact, de Bonnaire himself appears to identify the origin of convulsions in a variety of causes other than the womb—namely, fasting, study, work, emotion, and the suggestive sight of other convulsionaries.[80] Violent desires are just one among many causes he offers. The living animal is thus only included for the sake of the connotations it allows. This second-order citation shows the doctor concerned less with explaining the body than with fashioning the way it is perceived. The sick body becomes dangerous for spectators and for the exercise of religion, and invoking an animal makes it possible to attribute independent intention to that body.

In de Bonnaire's text, the metaphor is not presented to illustrate a theory. On the contrary, it seems that the perception of the body has come to illustrate the metaphor. The metaphor organizes visibility, offering an opportunity to directly envision and circumscribe what is seen. And the convulsive body is set in equivalence to the desiring body.

Many other doctors, too, incorporated into their writing the animal as described by doctors citing Plato; they did not directly source their citation in

Plato himself, but in Paré or other physicians. To reference Willis rather than Plato brought into play a modern approach to medicine, seeking to explain all the body's manifestations in physiological terms. These interpretations faded toward the end of the 1730s, as the movement of the Convulsionaries became less strident—but they had left an indelible mark on medical discourse, and the question of sexual difference would henceforth remain at the center of writing on hysteric vapors.

It was Denis Diderot's transformation of the reading of Plato's passage and, with it, conceptions of the womb, that signaled a third period in the function of the citation. Diderot's text "Sur les Femmes" (On Women) grounds itself in a discussion on hysteric affection to craft an economy of woman's emotions. First published in 1772 in *Correspondance Littéraire* (Literary Correspondence), as a response to Antoine Léon Thomas's *Essai sur le caractère, les mœurs et l'esprit des femmes des différents siècles* (Essay on the Character, Morals, and Mind of Women of Different Centuries), the text was republished twice and revised by Diderot just before his death. In it, women are carried away by the very force of their physiology, taken beyond themselves in a constant movement of transformation: "The woman bears within herself an organ given to terrible spasms, governing her and arousing in her imagination phantoms of every species. . . . It is from the organ proper to her sex that all her extraordinary ideas begin."[81] The womb is at the root of woman's unlimited sensitivity, in ecstasy as in fury. Diderot offers a series of examples to illustrate the overwhelming force that affects the hysteric woman. He is particularly interested in the Convulsionaries, whose bodies seem to struggle against themselves to attain visibility, heeding their physiology only in tearing it apart. "The woman dominated by hystericism experiences something of the infernal and the celestial. At times she makes me tremble. I have seen and heard her in the fury of a ferocious beast that makes up part of herself. How strongly she felt! How she expressed herself! What she said had none of the qualities of a mortal."[82] Rather than an organ apart, existing only in relation to desire, Diderot sees the womb as woman's center and as characterizing her whole being. A division is drawn between the brain, the organ of reason, and the womb, the organ of instinct. Physiology becomes the origin of woman's sensibility. Diderot's argument is built on a reversal: the womb does not belong to woman; it is woman that belongs to her womb. Ironically, while Diderot sought to defend women's valor by stressing their sensibility, he inaugurated a reading of women from their organs that would subsequently be reproduced without such a positive slant and would become a means of stigmatization.

From this point on, the notion of animality, now cast as a physiological reality, gained even more ground. Nicolas Chambon de Montaux, author of a volume entitled *Des Maladies des femmes,* quotes Aretaeus of Cappadocia's reference to an animal in an animal, giving the second-century physician an authoritative role in the definition of hystericism Chambon de Montaux prepared for the 1798 *Encyclopédie méthodique, médecine par une société de médecins* (Methodological Encyclopedia, Medicine by a Society of Physicians):

> In the hypogastric region is found the womb, an organ which almost has the character of a special animal, since it moves out from itself in all directions, rising up to the chest in the ensiform cartilage, throwing itself to each side, now to the right, now to the left, towards the liver or other viscera.
>
> Nevertheless it has more of a tendency to descend to the vulva. In brief, it is an errant being that loves agreeable odors, goes near to the place from which they emanate, is saddened by the sensations exhaled by fetid bodies, and moves away from them. It absolutely resembles an animal locked up in another animal.[83]

While Chambon de Montaux denies the actual existence of the animal, he chooses to cite a text in which it is mentioned three times. He could have cited another equally celebrated ancient text like that of Soranus of Ephesus, who invoked the image of the animal only to negate it,[84] or Galen, who, a century later, also distanced himself from this theory.[85] Instead, Chambon de Montaux privileges the work of Aretaeus in order to offer a new interpretation while maintaining the same reference to sexual desire.

This reference to an animal, or an animal in an animal, also allows Chambon de Montaux to reroute interpretations of the imagination. The imagination, which had been predominant in the mid-eighteenth century to explain hysteric fits as the effect of moral passions, is now associated with the memory of sexual pleasures:

> The authors who have written on the suffocation of the womb have said nothing about the influence of the imagination on this malady. Yet one cannot doubt that all thoughts that recall to the mind the memory of the pleasures of love accelerate the outbursts of hysteric passion in women who are anyway disposed to this affection. Either the uterus is in a state close to plenitude that gives birth to a crisis of hystericism, or it is still far from that state: in both cases, voluptuous thoughts excite an orgasm in the reproductive regions that brings trouble to the nervous system, and the suffocation of the womb takes place.[86]

Montaux thus deploys the imagination to explain not the effects of the mind on the body, originating hysteric pathology in the passions or the sensibility, but those of the body on the mind, attributing a lasting effect to pleasure. Sexual pleasures themselves may therefore be the origin of a later pathology. He recommends the use of pessaries, and then concludes:

> It is thus without reason that the symptoms set out above have been confused under the general denomination of hystericism or hysteric affections, since to put it briefly, the latter incidents depend essentially on the unnatural state of the womb, while the others are the immediate effect of the trouble or the malady of other organs or the exhaustion of the nerves.[87]

While he offers two theorizations of symptoms here, hystericism and exhaustion of the nerves, Chambon de Montaux does not allocate any space to nervous hysteric affections or define them in the *Encyclopédie méthodique*. The reader's attention is directed entirely toward hystericism as a condition in which the woman has reduced her own womb to a state of counternature by indulging in thoughts contrary to moral order. Woman corrupts her physiological nature by surrendering to ideas that she should have rejected, making it practically impossible for her to reassert control over her body. As such, it is not the foreign and disruptive organ of the womb that is at fault, but woman's indulgence in recalling specific memories. The imagination appeals to and agitates the womb.

Reference to the animal metaphor now took on a new momentum, leading to new discrepancies in its understanding. In 1804, Gabriel Jouard even wrote: "The ancients seem to have had a good idea of its incomprehensibility, which, in truth, they rendered in a somewhat burlesque way when they said that it was an animal locked up in another."[88] Jouard's comment indicates the emergence of a new use for Plato's text at the beginning of the nineteenth century. A negotiation took place in response to a new demand for rigor in scientific discourse; the metaphor could now be cited only as laughable imagery. It was no longer antiquity's prescient knowledge that was invoked, but its innocence and its capacity to predict correctly even in its ignorance. The citation functioned within a strategy of proof, as two different approaches, one modern and the other ancient, were deployed to confirm the validity of the diagnosis.

Increasingly, women were portrayed as the accomplices of their desires: their desires had to be satisfied if they were not to be afflicted. In such writings, women became a threat because of their hyperbolic desire. Julien-Joseph Virey turned back to the conceptions of Hippocratic treatises, predicting a cure for hysteria through sexual relationships in *De la Femme sous ses rapports physiologique, moral et littéraire* (On Woman, under Her Physiological, Moral, and Literary Relations), published in 1823 and again two years later. He recalled once more the wild image of the uterus—"an unruly animal, as

an ancient said," that "enters into fury, agitates and shakes the whole body"—
before casting it as "the center out of which the multitude of nervous irradia-
tions depart, in diverse circumstances, above all during the nubile period."[89]
According to this doctor, a whole series of spasms arose with each emotion;
whether out of a desire to please or out of frustration, out of jealousy or
melancholy, all were bound together. The uterus governed woman and her
animality. The role of the organ explained the range of her eccentricity, her
instinctive character, and her vulnerability. In his "Dissertation sur le Libertinage et ses dangers, relativement aux facultés intellectuelles et physiques"
(Dissertation on Libertinage and Its Dangers, in Relation to Intellectual and
Physical Faculties), part of the same volume, Virey further develops the image of the animal:

> Let us consider woman's genre of capricious, singular mind, which lights up in
> flashes, sparks, and fits and starts; sometimes it is carried away and improvises
> impetuously, like a sparkling wine in its effervescence; at times it is doleful,
> silent, sagging, incapable of the least idea, and even plunged into a complete
> void. Such a mind is above all the attribute of these utterly frail, mobile, and
> nervous constitutions, attacked by hysteria or hypochondria, in one sex or the
> other, but it is much more frequent among women. Indeed, this organ that
> is so sensitive, so irritable in them, to the point of seeming to enjoy a special
> life—this fiery and indomitable animal, as a philosopher calls it: the ovaries
> or the uterus—according to the irritations it experiences, according to the
> menstrual periods, jolts of sensual pleasure, and hysteric spasms it under-
> goes, excites not only the economy of extraordinary emotions, but also carries
> to the brain strange, irregular emotions, caprices of enthusiasm or antipathy
> over which a woman has no control.[90]

Hysteric spasms, just as menstruation, are here inscribed into the biological cycle of women. The regularity of their occurrence makes them part of
the natural cycle. Virey's citation is striking in that animality has become the
point of departure for considering woman's interiority. Hysteria is an animal manifestation, and a woman is doomed to bear this other that disfigures
her. A whole mythology was emerging of the unsatisfied woman, insatiable,
frozen in desire, existing only in need. Hysteria became the synonym for a
radical and tragic incompleteness. The devil had disappeared, but desire remained, animality remained, and temptation remained. They moved within.
Not only did the womb define woman's pathological identity as woman, it
also determined her identity in health. Treatises on women made hysteria the
ultimate expression of womanhood. Woman was equated with hysteria in full
force; hysteria was woman actualized.

Whatever the *Timaeus* passage's original function may have been, over time references to it functioned as substantiation, something spoken from outside the respective texts. As we have seen, three phases in the understanding of hysteric pathologies can be identified through different usages of the citation of Plato. A first period, between about 1575 and 1675, was characterized by presentation of the womb as an organ independent enough to rebel and assert control over women, regardless of their will. In the 1730s, at the heart of the Convulsionaries crisis, the citation was instrumentalized to associate the Convulsionaries' practices with lustful desires, even when the pathology was separated from the womb. The pathology indicated desires that ran counter to nature, as the affected body displayed unforeseen endurance and resistance to pain. Physiological exigencies mirrored a departure from the theological, moral, and social order. A third period, which began with Diderot and continued into the nineteenth century, established sympathy, synchronicity, and equivalence between a woman and her womb. Woman's pathology was indistinctly associated with her desires. A woman seemed condemned in advance: animality was inscribed in her body and, in turn, her soul.

The historicization of this animal metaphor of the womb further suggests the limits of Thomas Laqueur's interpretation that a "one-sex" model existed up to the eighteenth century.[91] In fact, a whole tradition of medical texts very early claimed women's difference in the name of her physiology. The lineage of Plato's observation enabled writers to posit a range of visions of women's singularity, transforming the view not only of pathology but of physiology itself. The process moved from Plato's explanation of reproduction to the notion that woman's sexual desire damaged fertility. The introduction of Plato's metaphor had two essential functions: it constituted hysteric pathology as an object of knowledge, and it established a corporeal specificity of women. There was a constant back-and-forth movement between the nature of the body and the pathology of the body. And in physiological understandings of sexual difference, pathology often appeared as woman's natural state.

This chapter has highlighted how the reproduction of metaphors and citations, originally proposed as a way of imagining the body, gradually inscribes them at the heart of medical understandings of physiology. Texts play on the confusion allowed by such references, pursuing illimitable shifts and entering into the world of connotation, of the equivocal, of the supplement. Imaginary configurations become the basis for a categorization. If on the face of it doctors invoke metaphors as a tribute to history and sign of culture, on another level the mention of those metaphors comes to function as a determinant foundation of knowledge.

3

The Writing of a Pathology and
Practices of Dissemination

To speak of illness is to replicate, linguistically, the process of its transmission from one subject to another. Once disease enters into language, its transmissive potential multiplies; the nurse who describes one patient's symptoms may discover other patients, carried away by the imaginative potential of disease, incorporating those symptoms. . . . To respond to a narrative emotionally or physically may produce illness; conversely, to be ill is to produce narrative.

ATHENA VRETTOS[1]

The insistent resort to metaphoricity and citation discussed in the previous chapter casts doubt on a widespread assumption: that one of the hallmarks of the eighteenth century was an increasing prominence of observation and classification in its medical treatises. These two types of writing often stand as the only points of entry for historians of science when attempting to account for the dearth of therapeutic achievements despite the wide scope of medical undertakings.[2] According to these historians, in medicine a turn toward objectivity—objectivity as it was to develop in the nineteenth century[3]— emerged toward the end of the eighteenth century, a period dominated by a need to make observations, delimit symptoms, and classify diagnoses. A number of studies have recently qualified this approach by highlighting quite different rhetorical formats.[4] Some have analyzed the role that correspondence[5] played in the development of the doctor–patient relationship, while others have pinpointed the strategic use of narrative in the prevention of masturbation, for example.[6]

Certainly, far from taking a uniform approach to medical knowledge, a great number of physicians moved away from traditional models (such as typically including a summary of knowledge from the ancients, the assertion of a thesis, and the presentation of illustrative examples) and adopted a wide range of different genres. As this chapter will show, that genre diversity made it possible to insert the patient as an interlocutor in the medical treatise, so that physicians could emphasize the patient's sensibility, the role of moral affections, and their own role in the apprehension of the illness. Doctors showed patients transforming the vapors by interpreting them in light of an image of themselves. These texts suggest that the idea of a fixed pathology is ill suited to understanding the vapors; instead, they give rise to a process of continuous redirection and redefinition of being sick.

In the mid-eighteenth century, at what may be the first important moment in the history of hysteria, treatises on medicine proliferated and were rapidly reprinted.[7] They were mostly published in vernacular languages rather than in Latin, and although translations abounded, physicians did not limit themselves to their own languages, collecting both Latin works and those in other vernaculars.[8] A key characteristic of these writings was an eagerness to address the lettered classes of their time as much as their colleagues.[9] Targeting different readerships for their work—colleagues, students, the lettered elite—required different ways of presenting ideas, which in turn led to different ways of approaching and discussing knowledge: physicians wrote from the standpoint of the experienced practitioner, the professor, or the courtly doctor. The desire to reach a highly literate audience transformed many physicians' goals in writing their treatises: when expounding vaporous affections, they focused on the form of the treatise itself—their writing departed from the traditional format and now borrowed from literary genres in order to enunciate medical knowledge. The necessity of new frameworks arose from physicians' concern to fashion and disseminate a new type of knowledge about a pathology they thought of as being bound to sensibility—namely, the hysteric or vaporous affections. For physicians supplementing their therapeutic role with that of a writer, what mattered increasingly was how they addressed their audience and how they presented their own role in that process. When they spoke to patients, physicians sought to present an affection in terms comprehensible to, and likely to please, all its potential sufferers. The use of language was now gracious, the style of writing literary, and the tone courteous.

This chapter examines a variety of writings by physicians that emerged in this period of the conceptualization of hysteric and vaporous illnesses. The key literary genres adopted were dialogue, commonly used in the sixteenth and seventeenth centuries for treatises on rhetoric and all fashionable topics at court; autobiography, a genre increasingly present in the eighteenth century as the use of daily journals and memoirs spread; and correspondence, either in imitation of the era's fictional writers or as a way of carrying out consultations with patients. Finally, anecdotes had a slightly different role, providing a commentary, like an offstage voice, on the pathology; their mockery often targeted the patient, the physician, and the illness alike. This wide spectrum helped to disseminate interest in hysteric affections among the aristocracy and men and women of letters. The purpose of this chapter is to investigate how these texts created new expectations among their readers about their own bodies, about their physicians, and about medicine.

Dialogue

In the early modern period, the dialogue form had often been used in literary and philosophical writings, but also by some physicians. Such was the case for *Bulwarke of Defence Against All Sicknesse, Soarenesse, and Woundes that Doe Dayly Assaulte Mankinde*,[10] published by William Bullein in 1562, which presented a dialogue between the two allegorical characters "Syckness" and "Health." At the beginning of the eighteenth century the format experienced a new efflorescence, prompted by the scientific dialogues of Bernard le Bovier de Fontenelle.[11] The success of these writings, notably the thirty-three editions of Fontenelle's *Entretiens sur la pluralité des mondes* during his lifetime, made the genre a fashionable one in the realm of science.[12] Such dialogues gave voice to a savant who shared his knowledge with aristocrats, recreating the feeling of a salon conversation, a vital source of knowledge dissemination in the eighteenth century.[13] In Fontenelle's footsteps, the format was also favored by two physicians writing on hysteric vapors. The first of the two was Bernard Mandeville. Born in Rotterdam, Mandeville studied philosophy and medicine in Leiden under Herman Boerhaave before settling in London as a physician. He published a series of fables, including his famous philosophical work *The Fable of the Bees* (1705). *A Treatise of the Hypochondriack and the Hysterick Passions*, published in 1711 with a second edition in 1730, took the form of a dialogue between two aristocrats and a physician.

According to his foreword, it is humility that leads Mandeville to set his conversation among fashionable characters: he renounces any form of intellectual jousting with his peers, and seeks instead to speak directly to those who suffer. Mandeville's work features a physician, Philopirio, and an aristocrat, Misomedon, who are joined in the last chapter by Misomedon's wife, Polyteca. In the dialogue, Misomedon undergoes a number of unsuccessful therapies, alternating descriptions of these with portraits of the respective physicians. The dialogue form allows Mandeville to retrace famous physicians' theses on hysteric and hypochondriac illnesses. By demonstrating his familiarity with medical theories, Mandeville remains faithful to the criteria of accreditation through citation, while the failure of the therapies put forward by other physicians justifies a new therapeutic approach.

Mandeville's dialogue paints a portrait of a male patient, who is later contrasted with a female one. Misomedon is a man of letters, well versed in philosophy, medicine, and ancient literature. His tone is one of grace, nobility of character, and care to avoid speaking inopportunely. The physician, Philopirio, replies in the same tone, appearing in the new character of a man whose study of medicine has not led him to disregard experience[14] or the ability to express himself with distinction. He does not hesitate to criticize his colleagues, outlining the various types of physician and his own sharp distinction from them. In the third chapter, Misomedon's wife, Polyteca, arrives to talk about her daughter's affliction with a hysteric passion, later admitting to the same affliction herself. Her approach differs radically from that of her spouse: it is a practical one, and her words steer clear of physicians' systems to address instead her own experience with apothecaries. "Tormenting and Throbbing Pain"[15] in the head soon forces her to withdraw, a fact interpreted by her husband as resulting from her lack of arguments. The moderation and distance that initially characterized his words are now replaced by an unexpected virulence directed against apothecaries. Philopirio points out the change in his patient's attitude. The interference of personal conflicts and affections is thus superimposed onto the narration of physiological imbalance, giving each character psychological depth while presenting pathology as occupying dual dimensions: one encompassing disorders of the body, the other its origins in the social circle.

Writing in dialogue form allowed the physician to draw out each protagonist's interests. For the patient in this configuration, narration is a means to interest the physician in his case; he can lay claim to the physician's trust in the name of a unique and irrefutable experience. Thus, the dialogue begins with the patient recounting his own biography in great detail, and Mandeville depicts the pathology as arising from the withdrawal of a lonely life and indulgence in excessive reading. Each time the aristocratic patient interrupts himself, the physician encourages him to pursue his narration in the service of understanding the pathology. From this insistence, the reader must deduce the importance of personal narrative.

The genre thus enables Mandeville to emphasize the importance of patients' narrating the physical and moral torments they have lived through—narrative appears as a chance for them to appropriate their own bodies. The patient's sensations, the patient's identification of his or her symptoms, and the deciphering of the patient's body by way of his or her narration are what characterize hysteric and hypochondriac affection, rather than any theoretical system. Establishing the links between the different stages of the patient's suffering is an essential component of the therapy. As such, the dialogue also establishes the importance of conversation as part of the overall treatment method. The publication of a dialogue of this kind created a new dimension for the hysteric and hypochondriacal illnesses: patients were presented as people who should be able to share their point of view and to qualify the physicians' perspective with an account of their own experience, which would be both unique and revelatory of the pathology. Doctors, conversely, were presented as those who must listen attentively to their patients and grasp the specific features of their experience.

Some years later, the themes introduced by Mandeville assumed an even more accomplished form in a dialogue by Pierre Hunauld, physician-regent at the Faculty of Medicine in Angers and a member of the Angers Royal Academy of Letters. On his death in 1721, he left some works that remained in manuscript form until Bruhier d'Ablaincourt published the *Dissertation sur les vapeurs et les pertes de sang* (Dissertation on the Vapors and Loss of Blood) in 1756.[16] This book would earn its author posthumous fame, with a second edition published in 1771. Hunauld picked up on Mandeville's model by indicating in his preface that he would speak to patients and not to physicians, who did not need his help. But he exploited the possibilities of the dialogue form further, by choosing to ignore the theories of his day and instead to focus on the negotiation of viewpoints that must take place between the physician and patient. In the following discussion of this book, my chief interest is in Hunauld's presentation of his work as a dialogue, his description of the pathology, the negotiation between the interlocutors, and the question of what allowed this genre to be used as a means of enunciating medical knowledge.

Hunauld sets up his dialogue just as he would have presented a literary work. He turns to full account the possibilities granted to fiction. The reader has the impression of reading not a medical treatise but a play, with vivid characters and a plot revolving around a hysteric affection. To establish the framework, the dialogue is preceded by two forewords. Both are written in the first person, but they do not refer to the same narrator or speak to the same addressee.

The first foreword legitimates the work, indicating its educative goal: the author explains the choice of the genre as a strategy adopted to convince women of the importance of their health regimen.[17] He insists on the difficulty of his enterprise, since he has had to renounce the use of technical terms in order to reach an uninformed audience. The writer also mentions that addressing a female audience has compelled him to respect proprieties and to remain silent on some points so as not to offend decency, and that he has developed an approach apt to arouse the interest of what he calls *le sexe*, the "the fair sex," for the sake of instruction. This insistence on pedagogical value is clearly intended to solicit the interest of the author's colleagues. Hunauld states that the dialogue form facilitates the presentation of difficulties in understanding the vapors, a more detailed discussion of certain aspects of the illness, and an avoidance of unnecessary abstraction and monotony for readers. He distinguishes his treatise from those following a fixed system, and reproaches his peers for attending to internal coherence in their writing merely as a means of gaining celebrity: whereas the display of theoretical complexities is, he writes, driven by a vain desire for glory, his own choice is to devote himself to the knowledge of use to people suffering from the vapors. He opposes the project of copying nature faithfully, asserting his readiness to give up "verisimilitude" in favor of "veracity." Rejecting all poetic tendencies, he invokes the art of painters concerned with imitating nature.[18] The foreword promises a series of portraits of the pathology, denying any attention to the final composition that could valorize his talent. On the contrary, Hunauld presents this talent as merely the instrument of his patients' autonomy, for with the aid of this book, his female readers will be able to take care of themselves without consulting a doctor. Advice that would feel like a constraint coming from a physician will thus become the expression of a choice.[19] In view of his desire to spread knowledge, which may be rethought and adapted by readers now in a position to judge what is best for them, the dialogue is a suitable genre to mark the singularity of his contribution.

The second foreword is cast in a fictional mold and gives voice to a doctor. The text addresses an aristocratic woman, and here Hunauld's didactic tone yields to a more refined style:

Madame,

I pick up the quill to obey you.

I feel that, if I succeed in satisfying you, you have inspired I know not which ways of expressing myself that I can no longer retrieve. The mind, the agreements communicate themselves; it seems one is filled up by persons who possess them as much as you do, Madame.

I prepare myself to spend a great part of the night in writing; but at this moment, it seems that the quill falls from my hands, and that I had better recant. Would you not forgive me, Madame, if I had the honor of telling you that I will never succeed in preserving in this dialogue all the truth, all the graces that glittered in your discourses?[20]

The tone is not that of a physician, but that of a courtier. The foreword presents itself, in the instantaneity of writing, as having a dual relationship to the time of a work to be written, already failing in comparison with a past conversation. The marchioness was a muse; more than that, she was the mistress of polished language. The narrator concludes: "Why did you not mention it earlier? We would have had our conversations written down: but all this talk does not serve any purpose; you want to be obeyed. Let us then make our efforts even if today it only means my scrawling all over a leaf of paper."[21] The text is given as a substitute for a previous oral exchange. By claiming clumsiness, this second foreword tips the medical treatise into a fictional work, and addresses it to a female audience.

In the first dialogue the same narrator begins, still addressing the marchioness, to indicate the fictional figures that he attributes to the interlocutors. He himself takes the name of Asclepiades to honor his intellectual forebear,[22] and renames the marchioness Sophie, paying her homage by the allusion to her wisdom. Hunauld sets out a series of voices: the narrator of the first foreword, addressing his colleagues; a second fictive narrator speaking to a fictive marchioness; a third one, who is an interlocutor in the dialogue; a fourth, the interlocutor Sophie; and a fifth, "Countess***" (a typical eighteenth-century anonymous epithet in fictions that claimed to describe reality), who barely intervenes. Only the status and the sex of this last character are indicated: woman and aristocrat. Just emerging from a hysteric fit, the countess witnesses rather than participates in the discussion, apart from underlining certain points by praising or judging retorts, in which she anticipates the reactions of the reader. Finally, Asclepiades sometimes speaks in the name of physicians—the seniors or contemporaries of Hunauld—and sometimes against them. And Sophie speaks at times in the name of women, at times in the name of vaporous men and women, at times in the name of one or other of her friends; the dialogue then gives way to indirect discourse.

The exchange between a doctor and a female patient presented here is reciprocal and evenhanded, giving the woman the role of transmitting a specific type of knowledge, that of experience. The patient thus actively participates in the identification of her pathology. In Hunauld's dialogue, the woman claims to have an irreducible knowledge of the pathology through her experience, and expects the physician to endow her sensations with meaning. By including a female voice so prominently in the dialogue, Hunauld casts women as his likely interlocutors. He consistently legitimates their physiological troubles and values the knowledge they have derived from experience. His dialogue establishes an atmosphere of confidence and postulates an exchange of knowledge in which women can respond to doctors' generalizations by affirming their singularity.[23] Sophie's replies to the doctor give female readers the impression that they are listened to, valorized, and able to explain themselves. Surprised at encountering a woman so well versed in contemporary medical theories,[24] Asclepiades commends Sophie for her knowledge and wit, further cultivating the trust of female readers.[25] Even if this is a temporary step—Asclepiades praises Sophie only to reaffirm hierarchical roles later on—Hunauld is ready to risk Asclepiades's position as man and doctor in order to depict him as empathetic and affirm his competence.

Both Hunauld and Mandeville present the vapors as a pretext for self-explanation and as a key to aristocratic life. But the motive for the exchange between their protagonists differs. In Mandeville, the patient is a man who has withdrawn into an erudition accumulated over many years; in Hunauld, the reader encounters a woman, familiar with the medical theories of her time and used to discussing them in salons. If for Mandeville "hypochondriack passions" are an experience of solitude, for Hunauld the vapors are a plural experience, plural in the diversity of people afflicted by them and the conversation between those people. Both writers stress how little control they have over their body. But with the figure of Misomedon, Mandeville presents the hypochondriac, obsessed with his vulnerability, whereas Hunauld's Sophie casts light on the fascination of the vapors. She admits her attraction to the pathology and claims that the only true knowledge of it comes with experience. The reader discovers the lure of the extraordinary character of the fits as the marchioness describes her admiration for the pathology rather than her willingness to reject it. She finds in it a means to reflect on her own body, drawing on the correlation of moral affections and sensations. The vapors are regarded not as a series of fits, but as a way of being, worthy of interest on its own account. If the narrator aims to explain, his explanation proceeds by adding constantly new features liable to increase the mystery. The reader is sent from portrait to portrait, with a claim to summarize the features of the affection.

Each conversation takes place in a different setting, briefly specified at the beginning of the chapter as a play would indicate changes in scenery. The first is set in the countess's bedroom as she lies in bed; another in a park; yet another in a private dining room while a party is going on elsewhere in the house, another indication of the fascination of the topic. The elements of the frame remain sufficiently vague for readers to be able to transpose them to their own familiar circumstances.

What kind of description of the pathology lends itself to being provided in this frame? With his dialogue, Hunauld converts the theoretical character of diagnosis into a shared experience, and knowledge into a discourse that is easy to appropriate. The dialogue takes the shape of an infinite negotiation of points of view, a contrast of experiences, a constant accumulation of Sophie and her friends' cases. Such are Sophie's vapors:

> My vapors are most irregular even though they are less strong; but it is their bizarreness that is the most incredible thing in the world. For example, I told you of my vapors yesterday, one must look at how they are today; there is no longer that despondency, that languor, it is now vivacity, which becomes a new kind of torture. I do not even know what kind of gaiety this is. For it

seems to me to be born from a sad and extremely painful source; it makes me laugh at anything; & often without any reason, & yet the bursts of laughter are such that I remain in rapture with pains of exhaustion in my chest and sides, that I would not know how to convey to you. Soon after I feel agitated with bizarre worries and cannot contain myself, I become agitated, I get up, I want to walk. Those who would hope to restrain me would do me unbearable violence. Nevertheless, should I start to walk or run, I immediately feel that my knees and my trembling legs give way, & that I would fall if I were not supported.

To tell the truth, in these moments I am little present to myself & so little that if, by chance, I become aware of the disorder in which I am, I afflict & irritate myself, which considerably augments the excess of my vapors. I observe all who approach me: I worry if I notice a serious and sad air around me, or smiles that produce my crazed harassments: I then believe myself ready to die or to lose my mind forever.[26]

The narrative compiles a variety of terms to articulate the expression of the self, swaying from the evocation of sensibility to the discussion of pathology. The account takes the form of an endless sequence of languor, despondency, and emptiness. An indulgence in detailed description of the fits contributes to a new vision of the body and a constant attention to oneself. Sophie speaks of herself with absolute passivity, becoming her own spectator. The vapors act on her like a wave, invading and then negating all of her sensations. She is constantly in flux, moving between consciousness and self-abandonment, pain and insensitivity, worry and carelessness, laughter and tears. While analyzing the slightest variations to her humor, Sophie asserts her ignorance of what causes such fits, implying her innocence. Her perception of the malady moves from the occurrence of sudden fits toward an awareness of her self. The point no longer seems to be the identification of a unique cause for the pathology, but rather the identification of distinctions between the patients' expectations and between different ways of describing the afflictions. Sophie's words highlight impressions, reactions, emotions, and sensations that sharpen her perception of each day.

Hunauld develops an art of detail to captivate readers. Each crisis is an episode to recount. There are the vapors that only "harass" healthy people, contrasted with "true sickness in which other illnesses are intermixed."[27] There are forms of the vapors dependent on sexual difference, and others indifferent to it. There are the vague, errant vapors that can arise from either side of the body, and incomplete vapors, composed of gas and tingling sensations. The vapors that contribute to epilepsy and catalepsy are the most violent of all, followed by the hysterical vapors. The reader also learns of the common vapors, the vapors of intrigue, and the reasoned vapors. A great number of possible causes are discussed, as well as reasons why women might willfully indulge in the illness. Asclepiades distinguishes different varieties of vapors without ranking or classifying them, leaving the reader with the impression that the possible forms are infinite. The aim is to weave together a catalog of references, amplifying the modes of comprehension of the body. The movement of a roving mind seems intent on adopting the pathology, accepting it in all its forms, interiorizing all its manifestations the better to ward off future attacks.

If literary works often portrayed hypochondriacs as having a need to share the narrative of their pain with a confidant or doctor, here the vapors seem to exist so that they can be recounted openly in salons. As Sophie explains, "It is like ghosts; everyone knows a story, & you go from one more incredible to the next, up to the infinite."[28] Drawing comparisons between cases, then, reinforces the inclination to linger on afflictions. Elation emerges during these narrations, and distinguishes the vapors from other pathologies by associating them more with mystery than pain; nevertheless, in speaking about vapors one risks becoming their victim. Sophie mentions the possibility of contamination through discourse, as though an interest in the vapors made one vulnerable, and as though the discovery of new kinds of vapors instigated their spread.[29] Willingness to understand the malady means a willingness to accept its dangers.

The dialogue offers a glossary of expressions that diverges according to the specific patient, what has provoked the vapors, and when they arrived. However, the fits are not seen as a direct and transparent expression of the subject, but often as forms of reaction and provocation through which the subject frustrates all expectations.[30] "Have you attentively observed the character of this *vaporous* laughter & crying?" Asclepiades asks. "Yes," Sophie responds, "I admit they have all the appearance of sincere laughter, & true crying. But do not believe that there is any real joy & real pain. There is more machine than mind there. These laughs rather resemble *convulsive* motion."[31]

Hunauld's treatise also introduces new forms and uses for the word "vapor." In addition to describing patients, the adjective "vaporous" is used to characterize a series of emotions, and Hunauld creates an adverb as well:

Without a doubt, . . . when *Sappho* made these beautiful verses, she had, according to you and to me, the most tender of vapors. You know it, nothing could be more *vaporously* expressed:
Yes, Madame, answered Asclepiades, I remember them now:

"I feel a subtle flame from vein to vein
Running through my body as soon as I see you;

> And in the sweet transports in which my soul is lost
> I could not find tongue, or voice.
> A confused cloud spreads before my sight;
> I no longer hear, I fall into soft languor;
> And pale, breathless, disconcerted, overcome
> A shudder seizes me, I tremble, I die."

These are vapors well described. There are the frights, shivers, sweats, despondencies, languor, tears, and cries that customarily bring a scene to its close, Asclepiades concluded.[32]

Having earlier cited the authority of Hippocrates, Asclepiades now recites Sappho. Even for the doctor, the vapors have become an aesthetic rather than a pathological crisis, seen through a prism of sensibility rather than pathology. The body's signs are interpreted not as symptoms, but as privileged gateways to exceptional feelings. In this dialogue, the body in crisis appears to attain new heights of sensitivity and a peak of sensation, and Hunauld's work thus extricates the vapors from social stigma. By presenting them as an aesthetic experience, he attributes to them a personal and inexpressible quality; they became both the experience and the expression of a particularity. At the same time, their association with pathology is counterbalanced by the displacement and obscuring of the suffering body . Excretions, trembling, and suffocation are forgotten in the fascination the vapors arouse—a fascination emanating from the strangeness of the vapors' manifestations, the mystery of their origin, and the unpredictability of their progression.

The series of vaporous portraits invites the audience to find itself reflected in one or other of them. Hunauld thus entices his readers to interrogate their physical sensations, and transforms these into symptoms. The dialogue's process of replies and silences leads the reader to identify with the vaporous, and this contamination is intended. The treatise works to mediate between readers and their bodies, especially as the interactions of the dialogue form allow readers to think and feel that they could have participated in it themselves. As they read themselves into it, Hunauld constructs them as future patients: "It seems to me that all mankind is vaporous," Sophie observes.[33] Putting this retort into Sophie's mouth is a strategic move, for if Asclepiades had said it, readers might have considered themselves too quickly condemned, whereas coming from Sophie it expresses feminine intuition. It is experience that speaks in her assertion, not academic knowledge. Here again, widening the experience of the vapors to mankind legitimates the interest they arouse:

> If these vapors are such that they make you fall in love, & that one could
> mistake the vapors for excessively passionate sentiments, could it not be the

same with the illness as with feelings? Each person expresses them in his own manner. One languishing, another lively, quickly carried away; this one fearful, worrying, unbalanced; another audacious and daring; and another who is finally quiet & has become half dumb.[34]

Symptoms here appear as an expression of the self, and suffering as a way to bring oneself into play. Literary language transfigures the vaporous pathology, as readers are given the possibility of dramatizing their identities through quotations and metaphors. They discover ways to name their discomfort, specify the character of their suffering, and justify their anguish. Their relationship to illness is cast as an individual experience and an instrument for articulation of the self.

In fact, the genre of dialogue enables Hunauld to do more than present a pathology using literary means: it gives him the opportunity to present the negotiation taking place in that pathology's identification. Suspense increases as readers encounter numerous different descriptions of hysteric affections, and because the dialogue form invites them to read between the lines, they can weigh the different causes of the vapors by themselves, discovering that Hunauld offers more than simply a series of descriptions to be read. The two characters are not mere mouthpieces, and the dynamic of their conversation affects readers' understanding of the pathology. Moments of dissatisfaction emerging in the dialogue prompt a reading of alternative ways of explaining the pathology. The singularity of the *Dissertation sur les vapeurs* lies in the author's construction of the relationship between doctor and patient, the doctor and the pathology, and the patient and his or her affection. The dialogue takes the shape of a sharing of ideas, emotions, and perspectives, while sexual difference adds a further source of division, alternately accentuating divisions and enabling their negotiation.

The mutual interest that binds these readers is in unraveling the mystery of an illness that nobody knows how to eradicate or even fully account for. The value of the dialogue differs for each of the two main characters, but a shared fascination with the vapors paves the way for their exchange. Complicity develops between doctor and patient through the mystery of the malady, legitimating their respective roles—the doctor is able to speak with a marchioness, the marchioness to claim a contribution to science—and building a complementary but unequal relationship in which Asclepiades is distinguished by knowledge and accessibility, Sophie by curiosity and susceptibility. For the doctor, the discussion of the pathology offers access to the aristocracy as he ingratiates himself with a marchioness; to this end, he deploys a body of literary references and poetic metaphors, using the particularity of the vapors to exhibit his wit and depth of knowledge. His attentive and patient listening nevertheless appears at times as a rather labored kindness, part of a role he plays to make a favorable impression. On the other hand, Hunauld makes Sophie keen to access medical and scientific knowledge, something usually reserved for men. For her, the vapors offer a chance to be at the center of a conversation by allowing her to claim knowledge gained through experience. However, in order to take up this central place, she must agree to be characterized by the vapors. In the conversation, all of her sensations, no matter how minor, are cast as symptoms.

Sophie's vulnerability emerges in the variations between her statements. Her questions alternate with expressions of susceptibility, remarks too hastily uttered, movements of impatience. Her reactions indicate that the vapors are not a neutral subject. They operate as a sign revealing the fragility of her position. Asclepiades often, despite interruptions, returns to his thesis that not everything can be said without hurting the patient. The reader discovers that this conversation, oscillating between banter and teaching, highlights the status of women as much as the credibility of the pathology and the image of physicians. From the very first dialogue, Sophie's contributions are defensive, or even more: she is ready to attack physicians for their loquaciousness, which she considers a mask to hide their ignorance. Hunauld uses Sophie's voice to criticize his colleagues, and by the same token notes the gap between women patients and doctors (all male at the time[35]), which leads to reciprocal misunderstandings. Asclepiades accepts Sophie's objection with good grace and seeks to distinguish himself from his peers through his sophisticated vocabulary and precise descriptions of cases. When he presents the vapors as a fashionable disease, Sophie insists on their existence throughout the whole population; when he limits their presence to the gentry, she identifies their existence across all ages and social classes. Although eager to assert the particularity of the vapors, she does not want to risk it losing its legitimacy as a pathology by making it too specialized a diagnosis. Naming the wide range of people affected by the illness is thus a way to overturn the conception that the vapors are an idle women's whim; accordingly, she insists on the growing number of men affected. Describing the attention they pay to their clothing and such "little nothings," she likens them to women: "In these matters, they are a hundred times more women than we women ourselves. They like good food & everything must be exquisite, but of *studied* presentation and cleanliness. In a word, if we are reproached for indolence & being self-centered, they go far beyond us; & to tell the truth, the only mark that distinguishes them from our sex is that they have beards, & look like men."[36] Having Sophie speak about vaporous men in this way allows Hunauld to indulge in caricature. Sophie's description is punctuated with irritation, frustration, and intransigence, while Asclepiades reduces femininity to ostentatious marks of excessive fragility. Sophie lets him speak, but it is a tacit or temporary acceptance, necessary to the dialogue. The vapors, whether experienced by men or not, retain a feminine character. Their universalization signals not an equality of the sexes in their risk of being affected by the pathology, but a feminization of manners.

The feminine humility that Sophie affects results now in an impetuous question, now in contradictions opposing the physicians, now in a display of frustration regarding certain interpretations. During her first meeting with Asclepiades, she asserts: "I digest . . . perfectly, & I never lacked appetite, I can assure you."[37] Yet when, at their third meeting, he mentions indigestion as a cause of the vapors, she identifies this as her "most assiduous ache."[38] Upon closer reading, it becomes clear that Asclepiades has enforced this admission through his manner of questioning: "But Madame, you, so regularly the victim of vapors, would you not have felt heartburn in your stomach? Would you not have expelled bitter pituite,[39] which the stomach at times feels so strongly? Do not doubt that they come from the excessively precipitated development of the salts essential to our food."[40] Specifically, Asclepiades distinguishes two types of vapors, some resulting from indigestion and independent of sexual difference, and others depending on sexual difference. Sophie's reaction is immediate: she situates her vapors in the category indifferent to sexual difference. Considering that she has previously denied suffering the symptoms of indigestion, the immediacy of this choice indicates the stigmatization then taking place in the perception of "feminine" vapors. Hunauld has the anonymous countess—in one of only two interventions in the dialogue—laugh at Sophie's quick association of herself with genderless vapors. By uniting with the physicians, the countess draws attention to Sophie's choice and invites readers to suspect its motives. The converse implication is that the only vapors indifferent to sex are the ones afflicting men. Hunauld fashions the role of the learned doctor as being the appropriate suggestion of symptoms: mentioning the existence of possible troubles is a solicitation of their presence. Language becomes performative in this phase of the treatment, when patients are seeking a vocabulary to describe their sensations and definitions are under negotiation. An anticipatory tension develops in the dialogue's silences and interruptions. Displacing Asclepiades's questions, putting off certain questions so as to better ground them later on, the author varies Sophie's responses based on what is likely to trigger her mistrust. In this way, the dialogue gradually lays out the stakes of the vapors. The reader realizes that Hunauld's intent is not so much the definition of the pathology, the description of its symptoms, as a concern with perceptions of it, or, more precisely, with shattering those perceptions. His treatise stages both the attraction exerted by the vapors and the scorn they inspire. The roundabout manner in which patient and doctor approach and affect one another indicates the precariousness of their conversation. Through the hesitations and ellipses of the dialogue, intersubjectivity emerges as a means of delineating controversies around the characterization of the vapors.

Another subject of conflict emerges as well. Asclepiades mentions fashionable vapors and the feminine vapors summoned on demand: "On numerous occasions, women have escaped scandal with the help of vapors," and "with vapors appearing just at the right moment, they have succeeded in many of their designs."[41] Sophie immediately objects, taking it upon herself to defend women: "I hear you, Sir, . . . but believe me, we are of better faith than you men believe & are absolutely not in the least capricious enough to alarm others unnecessarily."[42] Asclepiades leaves the topic of a possible simulation, but not before showing an appreciation for Sophie's words: "I do not even want to believe what I have heard of the great art of vaporizing on purpose."[43] By pretending to yield to Sophie, he regains her support and leads her to concede, speaking now of willful fits: "The art is new, . . . now that I think about it, I can easily believe that it could certainly succeed at times."[44] The manipulation often described as the key to the hysteric's relationship to her doctor is here echoed in the doctor's cunning. When Asclepiades later returns to a discussion of a feminine variety of vapors, he calls them "*reasoned* vapors."[45] He describes how women are "dupes" of their afflictions, cautioning against "feigned illnesses or mediocre illnesses" that are "followed by real illnesses."[46] Again, the doctor retreats only to convince. As the dialogue continues, Sophie herself initiates argument: "Continue, Sir. Why wouldn't you speak just like the others who always want to make the patients themselves guilty of their pains?"[47] And when Asclepiades generalizes the causes of vapors without citing individual cases, she accuses him of adopting doctrines that can easily be disproved. Reversing the criticism regularly directed at hysterics, of being victims of their own imagination, she calls the doctors' theoretical systems "ghosts of the imagination that impress only those with a weak curiosity."[48]

Despite these ruptures, trust does develop between the two characters. Because of her persistent curiosity, or perhaps because of a newfound distance from the illness, Sophie confides the emotional turmoil that vaporous women undergo, advising the physician to remain on his guard when listening to them: "Don't rely on them too much . . . since I am a philosopher today, & it is in good faith that I speak to you, these sorrows, these angers come from the despair of not being able to obtain, or of not even daring to desire.

It is the struggle of reason against the heart. Inside, one feels scenes so tragic that the more reason triumphs, the more the heart suffers."[49] Projections, avowals, concessions, and lies chart the contours of the pathology. The patient withdraws and advances as Asclepiades varies his manner of approaching her. The dialogue remains open; the many attempts at interpretation do not culminate in conclusions.

In Hunauld's text, the dialogue format thus plays a crucial role in the enunciation of knowledge. The genre allows the physician to devote great attention to civilities, which here are the very medium for configuring the pathology. The diagnosis becomes inseparable from a way of being, a relationship to existence, and a status in society. In his portrayal, the pathology fascinates by offering access to understanding the body. It is not the opposite of health, and is not related to life's finitude but to its heightening, a more sharply drawn manner of perceiving and feeling. Life with the vapors is described as living in instability, a life composed of present moments that immediately vanish. One cannot talk about the symptoms alone, implies Hunauld, but only of specific patients, and the physician is not so much the one who heals as the one who knows how to offer the consideration those patients need, how to understand their suffering and adapt to their needs. He devotes himself to being their witness and confidant.

Accordingly, Hunauld's text never defines the vapors, instead describing some of their manifestations and playing out the attraction they generate. Asclepiades moves from one qualifying adjective to another, taking care not to negate the existence of physiological phenomena in the "vapors on purpose" and tracing the possible link between whim and what he calls "true illnesses." The dialogue collects replies as so many diffracted glimpses of the pathology. The body's disorders are described without being classified. This way of characterizing the vapors sheds light on how patients view their illness, how they present themselves as patients, and how they wish or agree to be seen by others. It also shows how physicians may guide patients' articulations of their troubles using suggestion, generalization, and comparisons with other patients. The uncertain trust between patient and doctor is shaped in a vision of the malady that vacillates between fascination and stigmatization. Any temptation to attract attention to oneself through sickness withers before the fear of being irreducibly associated with the illness.

Secondly, by presenting the vapors through a dialogue, Hunauld accords them an intersubjective dimension. In a game of advances and retreats, he offers readers a stream of ideas that they are free to interpret themselves. The architecture of dialogue allows for an exposition of conflicts between doctor and patient, and permits Hunauld to participate in the polemics. But the dialogue also serves him as a mask. Exempted from having to state his own position on the illness, he can simultaneously solicit the complicity of both women and doctors. He does not lay blame or propose hypotheses in his own name. Readers are to make their own judgments by analyzing the stakes that weave through the discursive meanderings of the treatise. The genre frees the author from the need for a conclusion and leaves a series of conflicting perspectives to stand. Rather than incorporating all symptoms into a unilateral gaze or establishing a panorama of pathologies, Hunauld's dialogue positions symptoms in a space between categories: as phenomena that may have no place in theoretical discourses but are instead individual, unique. He thus asserts that it is by listening to personal circumstances in all their particularity, and not by generalizing those circumstances into medical categories, that doctors can comprehend the vapors. Each crisis is, so to speak, the occasion for a new theorization, and treatments are to be adapted to the experience of the individual.

In refraining from comparing the vapors to other pathologies, Hunauld refuses to make them a neutral object of study, and rejects the idea of knowledge as an activity to be pursued by the physician only once alone in his study. Interpretation of the malady takes place at the intersection of a series of dynamics in which the patient's identity and the doctor's authority are at stake. The vapors vary depending on the modes in which individuals wish to define themselves. Importantly, the conversation between Asclepiades and Sophie is held at the bedside of a countess emerging from a fit, and both characters note that they habitually rush to see their patients or friends entering into crisis. Hunauld's dialogue highlights the role of intersubjectivity, as the positioning of the affected person switches between affectation and affliction. The author establishes how visions of symptoms are recast in a web of narrations, where past crises are evoked and subsequently rethought through the prism of new episodes. His treatise demonstrates that illness is fundamentally a process of negotiation between patients and doctor, and that every assertion points the way to a new "becoming"[50] for symptoms as much as to new understandings.

Writers such as Mandeville in England and Hunauld in France proposed a language to pour out feelings. Talking about the pathology became a chance to open up, to look for words capable of accounting for the turmoil of the body and mind. The mise-en-scène located it in a privileged everyday life where physicians give meaning to patients' experience. A catalog of terms aimed to account for the variety of possible states of the body, and drew new lexical forms from the diagnosis of the vapors: to vaporize, vaporously. The exposition of knowledge left aside the criteria of coherence in favor of an effect of literariness, including repetitions, to-and-fro movement in the themes of the questions, anticipations, contradictions, and variations on a theme. The writing was persuasive as much as analytical, and the dialogue allowed the argument to proceed in a zigzag pattern, giving naturalness to the treatise. The doctor's prolixity and constant amplification gave the dialogue an exhaustive scope. Habits, taste, education, and sensibility occupied more space than symptoms in the diagnosis, and pathology became a mode of reading society.

The body was to be deciphered without a preexisting order of knowledge; the aim was to follow the order of physiology, and the temporality of the narrative obeyed that of the fits. Writing was put to work to describe the felt experience of physiological disorders. Mandeville and Hunauld stressed their interest in the patient even more than their knowledge as physicians. They lingered on the slightest details. Rather than allowing an assimilation of different cases, the diagnosis became the start of a personal encounter. The competence of the physician did not lie in the identification of pathology but in the work of recording ever more irreducible manifestations, and it emerged from his narrative and stylistic talent. Curiosity was repeatedly named as the starting point for knowledge, and considered a way to enhance wonder as much as to explain. The paradox of this writing was that it proceeded in denial. It denied itself as writing in order to assert itself as encounter with the patient. It contextualized the knowledge that it established, thereby stressing the importance of singularity. And it aimed to capture instantaneous speech, to address readers directly, to reconstruct the confusion of sensations.

In other words, these writers' borrowing of the genre of dialogue for an epistemological work cannot be reduced to a desire to make their texts agreeable for aristocrats. Rather, it presented the physician as a figure who knew how to move beyond the walls of the faculty and enter the court. In this sense, it might seem to have displaced the pathology out of the purview of medical authority—but in fact it displaced the limits of medical knowledge themselves. It refused to reduce the pathology to the excretions of the body, and made it instead a space proper to the investigation of the mind's turmoil, thus redefining the scope of medical knowledge.

Autobiography

Contemporaneously with Mandeville's and Hunauld's use of the dialogue format, another genre emerged: the physician's autobiography. The rise of this genre mirrored the development of the practice of writing private diaries, memoirs, and confessions in the eighteenth century. Francis Fuller, George

Cheyne, Claude Révillon, and in a briefer fashion Pierre Pomme, Robert Whytt, and the Viscount of Puységur,[51] all introduced a biographical text within their treatises on hypochondriac vapors to describe their own experience of the pathology—a form of account that had previously been extremely rare. Girolamo Cardano's autobiography, published in the sixteenth century, remained a famous example, its autobiographical narrative describing the aches of a puny body in dramatic contrast to his successful career.[52] But a new stake inhabited this new generation of texts: the diffusion of their authors' therapeutic practice in an aristocratic milieu. This will now be illustrated using texts by Francis Fuller and George Cheyne, the most striking examples of such writings.

The integration of these narratives into medical treatises was profoundly discordant with eighteenth-century images of physicians. From the seventeenth century, an etching would portray the famous physician on the facing title page or the page after, to introduce the reader to his complete works. With the development of academies of science in the eighteenth century, paintings of physicians were commissioned for display in the institutions, in recognition of scientific and professional achievements.[53] Portraitists flattered doctors by giving them an imposing physique, smooth skin, a fashionable wig, and velvet clothing—more than the man himself, they represented a status acquired through knowledge. But if paintings presented doctors with a stillness that signified immortality, narrations of affliction disrupted such reverence. The luxurious imagery was overturned by the episodes of fainting, suffocation, and vomiting recounted by Fuller, Cheyne, and Révillon. The myth of the invulnerable body of the doctor—such as those tending to plague victims without being infected—disappeared. Such autobiographies instead presented physicians in pain and long unable to find the proper treatment to cure their own illness. In the treatise, the doctor appeared just like his patients, struggling against affliction and confronted by his own mistakes, ignorance, and excesses. Readers encountered him in moments of infirmity, weakness, and even grave danger. Knowledge did not, in other words, shield him from life's adversities. Narratives of this kind answered the reproach most commonly made by patients: that doctors spoke out of a knowledge acquired in books, in complete ignorance of what pain really was—an ignorance that led to their notorious lack of both intellectual grasp and understanding. At a stroke, they made the doctor into a man who, just as much as his patients, experienced suffering, dubious therapies, and the anguish of uncertainty about the future. The autobiographical narrative relocated the value of the discourse in testimony and the singularity of the experience, its uniqueness.

Francis Fuller's *Medicina Gymnastica; or, A Treatise Concerning the Power of Exercise with Respect to the Animal Œconomy; and the Great Necessity of it in the Cure of Several Distempers*, published in 1705, was a posthumous success, and saw eight reprints as well as a 1750 translation into German.[54] The inclusion of an autobiography in the appendix to his treatise, then an uncommon practice, called for justification.[55] In his vindication, Fuller first insists on his shyness and a fear of prejudice, which, he says, long held him back from publishing the autobiography. However, he soon begins to list abundant reasons for having included his narrative. He mentions the insistence of friends who urged him to make it public, adding that the efficacy of the treatment made its revelation necessary, and that the lack of such an account of its benefits outweighed his hesitation.[56] Additionally, he notes that the unexpected form of treatment, horseback riding, was one more reason forcing him to give an account of himself; without a detailed example, he feared this cure might be mocked and dismissed by readers. The autobiography is, Fuller writes, more suited than any other genre to attesting the efficacy of such a treatment—what he calls the "gymnastic method"—because of the degree of detail it allows. His afflictions gradually emerge as the price he has paid to better perform his professional part. Finally, Fuller rewrites his malady as a mark of God's providence:

> Thus I have given a succinct and true Account, of a Long and Severe Distemper, which it has pleas'd Almighty God to lay upon me; by which it is Plain, that as some Men are distinguish'd by Riches, Honours, and the like; others may be as remarkably in the degrees of their Affection and Anguish, and may be forc'd to pass not only Days, but Years of what we call Life, after such a manner, that if it were not for Higher Consideration, it would be far better not to be.[57]

In such a frame, the autobiographical account is no longer to be interpreted as a sign of egotism. And far from being the wages of slovenliness or insufficient competence, the disorders are presented as a sign of vocation. At the same time, the length of the physician's torment proves his responsibility to teach others how to eradicate it. It positions the physician as an intermediary in God's design, and his whole experience of suffering and relapses is to be read in a theological dimension. Sharing his discoveries despite the mockery they are likely to provoke therefore results from the physician's sense of duty.

In this process, the doctor places himself in abrupt and unexpected proximity to his audience, unveiling the hesitations and missteps that preceded the discovery of effective treatments. Describing his experience with unequaled humility, Fuller completely abandons the image of the savant, but the autobiography format allows him to include many arguments that would not have found their way into a traditional medical treatise. The therapies he prescribes—horse riding and emetics—were not new, and Fuller acknowledges having read Sydenham's[58] advice on the subject; the presence of an autobiographical narrative was all the more important in order to appropriate these therapies. It also exonerated Fuller from the need to refer to the medical tradition and to distinguish his work from that of his colleagues. Refraining from attacks, he simply positions his work outside the sphere of any celebrated research. The simplicity of the therapies he advocates is thus compensated by the strength of his account of their effects. The description of suffering guarantees the truth of his words; his suffering is a unique experience, and as such incontestable, giving him the authority to speak. The autobiographical narrative anchors the quality of the work, in that the author records a practice he has tested on himself. The claim to be affected with vapors as a physician made possible a continuous observation of the illness and of the effects of the therapies undertaken—the experience of suffering is thereby transformed into a case at the physician's disposal. The admission of previous ignorance, in turn, operates to support the authority of his judgment and as a supreme justification for writing the book.

Beyond the autobiographical narrative, this text's singularity resides in the specific bonds of credibility and understanding that it generates by offering patients an image of themselves through its description of pain, suffering, and disarray. Not unlike the dialogue form, this approach allows Fuller to put forward a very different message than that of traditional treatises: for readers, it means that each of their sufferings must be listened to, each of their experiences taken into account. Speaking in the first person displaces the approach to the body in terms of symptoms, and raises the question of suffering. Therapeutics thus operates outside the application of a particular diagnosis; the point is not to name the pathology, but to write the experience of pain as an intermingling of physical sensations and moral afflictions that culminates in the patients' distress. Symptoms are enumerated in order to reconstitute the experience, and not for their value in identifying the diagnosis. Instead of using a formal language, studded with technical terms proper to theorization, Fuller privileges a language that reinforces the authenticity of narrative. He renounces the flair of the savant, choosing to relate a proliferation of details. The rhythm varies according to the emotions generated by the pain; the physician interpolates his own reactions, at times using interjections and exclamations that seem to be taken from life. The temporality of the pathology is imagined in terms of slowness, delays, or relapses, and not solely in terms of steps toward recovery.

By temporarily relinquishing his position of authority, the physician appropriates what being sick is all about. Of course, he nevertheless still distinguishes himself from other patients by his knowledge, which gives him the capacity to describe his suffering precisely. As a result, he is able to move back and forth between an inside gaze, describing the pain, and an outside gaze, describing the symptoms. The physician and the narrator meet in the suffering and the continuity of their concern for the body. All these elements attract readers' sympathy and compassion, transforming the relationship between physician and patient into a potential for reciprocal testimony. Denying movements of shyness or shame, the text induces the speaker to make a narrative of his own body. His body becomes a mirror of the affections of his patients.[59] Readers learn that each move of their body can be read as a symptom, and the narrative asks them to question their body and find a pathological sign in every disagreeable sensation. Equally, the trust that the physician places in his readers by confessing his fits calls for their confidence in return.

When George Cheyne picked up on Fuller's model, inserting an autobiography at the end of his 1733 treatise *The English Malady; or, A Treatise of Nervous Diseases of All Kinds as Spleen, Vapours, Lowness of Spirits, Hypochondriacal and Hysterical Distempers, &c.* he developed its potential further.[60] Named a fellow of the Royal Society in London, Cheyne was one of Europe's most famous physicians in the first half of the eighteenth century (fig. 6).[61] He published a dozen works, most of which were reprinted during his lifetime and many translated into French or Dutch. *The English Malady* had been reprinted five times by 1735. Cheyne is encountering his second phase of fame today, when some of the most important historians of science and of eighteenth-century literature have considered him an exemplary doctor of the time.[62] Steven Shapin argues that Cheyne used iatromechanism as a way to position himself within a new ideological current: "Just as the micromechanism of the natural philosophers was identified as a radical break with Scholasticism, so iatromechanism could provide an intellectual license for radically new medical *practices*. George Cheyne, buttressed by the cultural authority of the new ontological expertise, took on a drastic reconfiguration of dietetics, perhaps the most stable and traditionally entrenched of all domains of medical practice."[63] Much research on Cheyne's life and his works has also focused on his combination of a religious with a medical imaginary, his elaboration of sensibility, and his artful approach to his patients. The brief analysis proposed here is in many ways methodologically similar to that of Wayne Wild, who has examined Cheyne's rhetoric of sensibility mainly in his correspondence. Reading Cheyne's letters in relation to his biography, Wild powerfully shows how this rhetoric allows and conditions Cheyne's

FIGURE 6. Portrait of George Cheyne. Image courtesy of the Truman G. Blocker, Jr., History of Medicine Collections, Moody Medical Library, University of Texas Medical Branch at Galveston, Texas.

self-fashioning and his relationship with his middle-class and aristocratic patients:

He literally and symbolically embodied the transitional changes occurring in medicine after the heyday of iatromechanical medicine, appropriating to himself the themes of sensibility, of individual experience, and of deep spiritual fervor as part of medical cure. His prescription of a moderate lifestyle tapped into socio-economic changes of the period in which the middle class saw themselves as achieving the refined sensibility of the aristocratic class but without succumbing to the moral laxity caused by excess.[64]

In the following, I investigate not the correspondence but Cheyne's treatise *The English Malady* to consider how he disseminated a new image of hysteric illness that was strictly associated with the upper class, leading him to conceive of a new role for the doctor and reframe what medicine is about. By focusing specifically on the way Cheyne intervenes as a narrator throughout his treatise, I wish to show how Cheyne inscribes nervous and hysteric illness as an effect of life, a reflection of diet and habits, and a result of modern society's tastes. Pathology is both a constant threat and something that can be warded off by constant attention to temperance.

Cheyne's influential treatise lists a variety of names for convulsive, melancholic, and hysteric illnesses. The author refuses to link these disorders to humors or to associate them with the risks of the melancholic "temperament." Nor does he see fantastical whim, folly, or deprived animal urges in the illnesses; for Cheyne, hysteric and hypochondriac maladies have a decidedly different origin. In the preface, he reappropriates the traditional jest that the English are hypochondriacs, declaring nervous maladies to be a characteristic feature of the English nation and in this way turning them into a prestigious illness.[65] He does not contradict the old diatribes against English weather and the dampness of the island, but adds an argument that soon outflanks the older narrative. In a portrait flattering to his readers, Cheyne presents England as an unparalleled maritime empire, carrying products from all over the world back to its ports:

> Since our Wealth has increas'd, and our Navigation has been extended, we have ransack'd all the Parts of the *Globe* to bring together its whole Stock of Materials for *Riot, Luxury* and to provoke *Excess*. The tables of the Rich and Great (and indeed of all Ranks who can afford it) are furnish'd with Provisions of Delicacy, Number, and Plenty, sufficient to provoke, and even gorge, the most large and voluptuous Appetite. . . . Not only the Materials of *Luxury*, are such as I've describ'd, but the Manner of Dressing or Cooking them, is carried on to an exalted Height.[66]

Victims of nervous illness have thus been overcome by the wide diversity of goods and refinements available to them. With the pronoun "we," far from condemning his fellow citizens Cheyne deliberately signals that he shares their risks. He associates himself with the practices of his contemporaries and becomes one with them in avowing an excess of consumerism and possessions.[67] Hysteric illness turns into a pathology that reflects modern England's way of life and its flourishing economy, as citizens are constantly urged to consume. The pathology is a burden borne by those whose wealth has freed them from physical activity, and actually comes to serve as a sign of luxury. His explanations of the pathology's origins remain very cautious; he seems at times to praise the possessions newly available to English citizens and thus to justify desire for such goods[68] even though it is an excess of such indulgence that has caused the physiological imbalance and an oversensitivity of the nerves. Cheyne encourages his patients to care for themselves, seeing illness as nothing more than a warning of the body's exasperation signaling that pleasure has led to irritation. Here, listening to and caring for oneself are not indications of egotism, but a duty incumbent on those susceptible to spoiling their bodies through excess.

Cheyne establishes his will to understand the temptations that confront his patients: inaugurating empathy with a patient's taste is a key to curing the afflicted. He thus demonstratively refuses to speak in the name of moral superiority, and presents himself as the first to fail in discipline, as a man prey to the weaknesses of his time. Gradually, readers are to be convinced that only a physician who appreciates such a way of life can truly grasp the temptations of his patients and take up the commitment to cure them.

The presence of the autobiography justifies Cheyne's speaking in terms of "us." Establishing a community of experience is a means to claim similarity beyond the difference of status, so that the function of this text seems to be the promotion of the physician as much as the curing of the pathology. Mockery of hysteric illness is replaced by a respect for social status; criticism is erased by flattery. The pathology becomes a sign of wealth; it now seems a title reserved for the privileged. It operates as a representation in a society replete with signs, and here the physician takes up the role of attesting its existence.[69]

But wealth and class are not the only possible origins of hysteric affection. Cheyne writes that nervous maladies are also intellectual in nature. He associates them with minds that are particularly stimulated: "It is a common Observation, (and, I think, has great Probability on its Side) that *Fools, weak* or *stupid* Persons, *heavy* and *dull Souls*, are seldom much troubled with Vapours or Lowness of Spirits."[70] Intellectual activity here exhausts the nerves and fibers as much as do excesses of luxury. In an astute reversal by Cheyne, the illness has become a symptom not of vice but of intelligence, sensitivity, and status. The malady is therefore flattering, testifying to fine qualities in its sufferers. Cheyne's invocations of "common sense" allow his readers to imagine themselves capable of evaluating conceptions of their malady. And by subtly interposing "I think" into his analysis, and coming back to the first person, Cheyne discreetly elevates himself to the essential role of weighing others' ideas and acts.

Cheyne's crisscrossing of pronouns reinforces the identification of the speaker with the addressee. The author constitutes his work in a relationship to truth, in the name of the narration's sincerity, as constantly recalled by the pronoun "I." The enunciation posits the unique location of the speech, the "I" implies the singularity of the experience. As Émile Benveniste shows, such deictics introduce the other who faces us,[71] and in this case the other is an aristocrat or a man or a woman of letters. The first-person pronoun claims to open debate and establish the importance of intersubjectivity, but what it in fact does is to imply that a consensus already exists.

However, Cheyne also uses the third person, and does so to talk about the distress provoked by the experience of the pathology. From the outset, his treatise lingers on the mounting difficulties faced by its audience:

> The Spirit of a Man can bear his infirmities, but a wounded Spirit who can bear? saith a Prophet. As this is a great Truth in the *Intellectual World*, so it may allude to the *Human Machin*, to insinuate, that a person of sound Health, of strong Spirits, and Firm Fibres, may be able to combat, struggle with, and nobly bear and even brave the Misfortunes, Pains, and Miseries of this mortal Life, when the same Person, broken, and dispirited by Weakness of *Nerves, Vapours, Melancholy*, or *Age*, shall become dejected opress'd, peevish, and sunk even below the Weakness of a Greensickness Maid, or a Child. . . . To expect *Fortitude, Patience, Tranquillity*, and *Resignation*, from the most *Heroick* of the Children of the Men, under such circumstances, from their natural Force or Faculties alone, is equally absurd as to expect to fly without Wings, or walk without Legs; the strength of the *Nerves, Fibres*, or *Animal Spirits* (as they are call'd) being the necessary Instruments of the former, as these Members are of the latter.[72]

According to Cheyne, illness invades body and mind. When it overcomes the mind, one is unable to think of anything but illness. Patients are seized by feelings of submission and defeat; all desire to be healed leaves them. The doctor's choice of the metaphors of the child and the maid emphasizes the patient's loss of control and responsibility. The practitioner's role is thus to provide moral support, uplifting patients and reorienting them despite their confusion. Cheyne was not the first to describe the torments undergone by hysterics and hypochondriacs, whose despair had already been portrayed by Sydenham:

> But their misfortune does not only proceed from a great indisposition of body, for the *mind* is still more disordered; it being the nature of this disease to be attended with incurable *despair*, so that they cannot bear with patience to be told that there is hope of their recovery, easily imagining that they are li-

able to all miseries that can befall mankind; and presaging the worst evils to themselves.[73]

A comparison of this passage with Cheyne's words illuminates the transformation at work in the latter, demonstrating how profoundly an observation can be altered by a shift in agency. From "they" to "a person," the tone changes completely. Sydenham describes the lack of patience of the sick from an objectifying distance. He segregates himself from them, casting the doctor as far removed from bodily troubles and his patients as pathological cases. The dichotomy he establishes is reinforced by repeated use of the third-person pronouns. Cheyne, in contrast, affects interest in the experiences of his patients and depicts their situation as universal, thereby suggesting the possibility that anyone may become a patient, even when he least expects it. No one can pass moral judgment on such patients, since the same dangers threaten all.[74]

The narration of Cheyne's own illness seems to arise as a natural conclusion to a treatise in which the author has so frequently associated himself with his readers, and its location at the end of his treatise is a strategic one. As Shapin notes, if Cheyne's patients "were repeatedly given to understand that he had suffered what they were suffering," this implied two things: "that his experience was not merely theoretical, and that the efficacy of his method could be vouched for by his own now-healthy, but once seriously deranged, body."[75] Despite being entitled "Case of the Author" and following the case of another physician, it takes the shape of an autobiography—it depicts Cheyne's childhood, education, and later life, not simply his experience of illness. To justify the inclusion of this autobiographical narrative, Cheyne cites a controversy circulating about him that accuses him of "enthusiasm."[76] He says that his recommendation of moderate therapies has been presented by his adversaries, trying to frighten away his patients, as the effect of religious practices. In the circumstances, a detailed account of his own life is necessary for the victory of truth.[77] He explains his concern for restraint and his desire to promulgate a diet based on milk as being the result of his own painful experience of the cost of gastronomic excess.[78] Using an argument already put forward by Fuller, Cheyne presents his experience of illness as a sacrifice, something that sets him apart and endows him with the obligation to help humanity. From the third page of the treatise, he insists on the satisfaction he hopes to gain from relieving his contemporaries. And who better to help them than he, a man who has tried all the available therapies to fight symptoms with multiple causes: lack of discipline, unreasonable habits, and fragility of the nerves.[79] Here again, the number and types of symptoms are infinite, and their resistance has long thwarted the physician's attempts to heal. Their description does not serve the identification of a diagnosis, but is a testimony to the suffering experienced.

Like Fuller, Cheyne asks for the reader's commiseration with his history, a commiseration that can then be extended to any other patient who experiences the pathology. The references to Cheyne's friends and their flight when he began to suffer are frequent enough to increase the readers' compassion. He also arouses their interest by avowing his own weakness, reporting how he let himself be gradually recaptured by his old habits after having recovered. Refusing himself anything but green tea on waking up in the morning, he would then indulge in excesses at dinner.[80] An initial infatuation with his own therapy yields to a period of abandon, and Cheyne puts on even more weight. Cheyne's relapse creates a tension in the text by postponing the return to health. Echoing Fuller's focus on deceleration, it acts as a delay inside the text, while the physician's body is exposed to new risks. Cheyne stirs the compassion of readers likely to sympathize with someone subjected to harsh trials. Doubts and weaknesses are only stages that the physician himself went through. The success of the therapy after the lapse fortifies its validity more than ever, and readers can imagine themselves partaking in the therapy without fear of failing to meet the doctor's expectations. Here again, it is by making himself assailable that the physician can become a model.

Cheyne's self-portrayal as a repentant sinner takes up a strategic role in the narrative. When describing his profound melancholy after a religious reawakening, Cheyne appears to mirror many of his patients. The correspondence he maintains with one of them, the Countess of Huntingdon, for example, indicates that "religious melancholy" is a frequent diagnosis among his patients.[81] Spiritual questions are the frame within which concern about the body is formulated. Cheyne's text is unique in soliciting sympathy from his public by means of extensive self-criticism. He confesses his "*Carelessness* and *Self-sufficiency, Voluptuousness* and Love of *Sensuality*, which might have impaired my *Spiritual* Nature,"[82] and talks about the melancholy that followed his relapse into bad habits. The tactical manifestation of a concern to repent appears in his comparison of illness to punishment: "I went about like a *Malefactor* condemn'd, or one who expects every Moment to be crushed by a *ponderous* Instrument of Death, hanging over his Head";[83] foreshadowing hints of a dire future increase the reader's suspense and fear in contemplating the therapy's possible outcome. Cheyne's fallibility illustrates mankind's fallibility, and his confession of weakness makes him more accessible to his readers, while powerfully suggesting the need for them to be on their guard against themselves. He thus indicates that rather than a treatment, it is a discipline that must be followed, with the daily attention this implies. Cheyne's approach redirects the focus of medical knowledge: it is not in building systems, nor in anatomy, nor in the definition of categories, but in discipline. And what counts is not the identification of a specific diagnosis or the exact description of each organ, but the identification of a web of relationships between the patient's way of life and his physical experiences, allowing a dialogue between the individual and his physician.

In this way, the autobiographical narrative does not present only the physician or his persona, but a conception of what medicine is all about. It exemplifies a way of dealing with the illness, the expectations one should have, and the individual's duties toward his or her health and soul. Illness is not something exterior to the patient but is presented as bearing a direct relation to his food and way of life. Hysteric illness is not, thus, a state outside of normality, but is the continuation of the direct correlation between body and soul—a correlation experienced every day with breathing, appetite, or digestion: "Only *Temperance* and *Abstinence, Air, Exercise, Diet,* and proper *Evacuations* can preserve *Life, Health,* and *Gayety,* or cure *Chronical Diseases.*"[84] In the conclusion to his treatise, which comes just after the autobiography, Cheyne talks about a "metaphysics of a regimen." He advocates simplicity in taking care of the body, which should be equated not to ignorance but to the avoidance of excesses, both in everyday life and in fighting disease. This places the doctor as a tutor, whose role is to prevent illnesses even more than heal them, an approach that will be accentuated at the end of the eighteenth century, as is discussed further in chapter 6. It positions medicine in analogy with religion: medicine provides the metaphysical knowledge necessary to the care of the body, religion the metaphysical knowledge necessary to the salvation of the soul. Complementing and reinforcing each other, both are transmitted by means of narratives, using models that are meant to be reflected in the patient's trajectory, teaching the patient what should not be done and what should be imitated, in the tradition of *De Imitatione Christi.* Such a conception of medical knowledge, focused on each singular body and individual, found its most effective enunciation in narration. It gave the physician and the patient a story by which they could abide and to which both could contribute. Medicine here is not the tool to identify a diagnosis but the constantly renewed exercise of seizing a life as a whole within a teleological frame and negotiating every aspect of it.

Can one measure the reception of such a work? The numerous editions indicate the substantial size of its readership, and the practice of its reading can also be studied thanks to marginal notes in a 1734 edition (fig. 7), part of

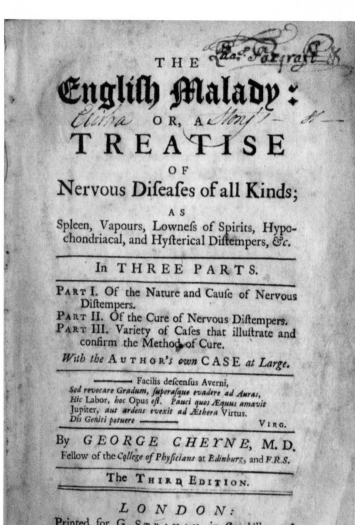

THE

English Malady:

OR, A

TREATISE

OF

Nervous Difeafes of all Kinds;

AS

Spleen, Vapours, Lownefs of Spirits, Hypochondriacal, and Hyfterical Diftempers, &c.

In THREE PARTS.

PART I. Of the Nature and Caufe of Nervous
Diftempers.
PART II. Of the Cure of Nervous Diftempers.
PART III. Variety of Cafes that illuftrate and
confirm the Method of Cure.

With the AUTHOR's *own* CASE *at Large.*

———— Facilis defcenfus Averni,
Sed revocare Gradum, fuperafque evadere ad Auras,
Hic Labor, *hoc* Opus eft. *Pauci quos Æquus amavit*
Jupiter, *aut ardens evexit ad Æthera* Virtus.
Dis Geniti potuere ————
VIRG.

By GEORGE CHEYNE, M.D.
Fellow of the *College of Phyficians* at *Edinburg,* and *F.R.S.*

The THIRD EDITION.

LONDON:

Printed for G. STRAHAN, in *Cornhill*; and
J. LEAKE, at *Bath.* M.DCC.XXXIV.

FIGURE 7. Title page, *The English Malady*, by George Cheyne, 1734. Image courtesy of the Truman G.
Blocker, Jr., History of Medicine Collections, Moody Medical Library, University of Texas Medical
Branch at Galveston, Texas.

the Blocker Collection,[85] which make it possible to examine the appropriation of a work by some of its readers.[86] Two readers marked the book, distinguishable by their different ink and handwriting.

That these markings date from the eighteenth century is suggested by the style of handwriting, as well as the fact that Cheyne's work was completely disregarded in the nineteenth century, when a different relationship to knowledge was developing. On the first page, a long handwritten citation inscribes the reading in a bibliophile context. The extensive passage copied from Cornaro's *Sure Way of Attaining a Long and Healthful Life* corresponds perfectly with Cheyne's vision (fig. 8). Thus, if Cheyne rarely invokes other writers and physicians, the reader her/himself has inserted the work into a supporting tradition.

On approximately one page out of three, passages are underlined, at times the same passages with the two different inks. The lists of symptoms are annotated with explanations, advice, remedies, and types of therapy, as are Cheyne's good resolutions. The process of appropriating and reappropriating the book can be retraced as one reader and the other selects in turn important passages, annotates different parts, occasionally selects again what has already been underlined, varies the lines, or doubles the underlining of particular words. These markings were clearly not designed only to facilitate a first reading. One of the readers has annotated the text in order to use it more easily at a later date, using a series of *nota bene* marks. The appropriation of the text includes the table of contents, with the addition of two subdivisions, one of them specifying the page range of the author's autobiography (fig. 9). Inside the "Case of the Author" itself, a series of cross-referenced page numbers is given.[87] One of the readers jots a calculation at the bottom of a page, comparing Cheyne's weight with that of a bullock.

However, at the end of the work one of the readers attacks the book (fig. 10), and this with a simple piece of information: "It is reported that the author had return'd to a plentiful diet & free life not withstanding all."[88] By adding this note, the same reader who has patiently underlined so many passages and added a *nota bene* with such frequency suddenly casts doubt on Cheyne's authority. This model reader, ready to use Cheyne's case as a manual complete with cross-references and markings, appears now to demystify the attraction of a confession, by referring to what came after the publication of Cheyne's book and hence reducing the therapy to a passing phase of Cheyne's life. The annotations, lodged inside the book, undo the published words. The limits of autobiographical narrative come into sharp focus: if the narrator fails to fulfill his promise, the narration becomes a failure or a lie. By annotating the book and informing readers to come, the reader puts the book into

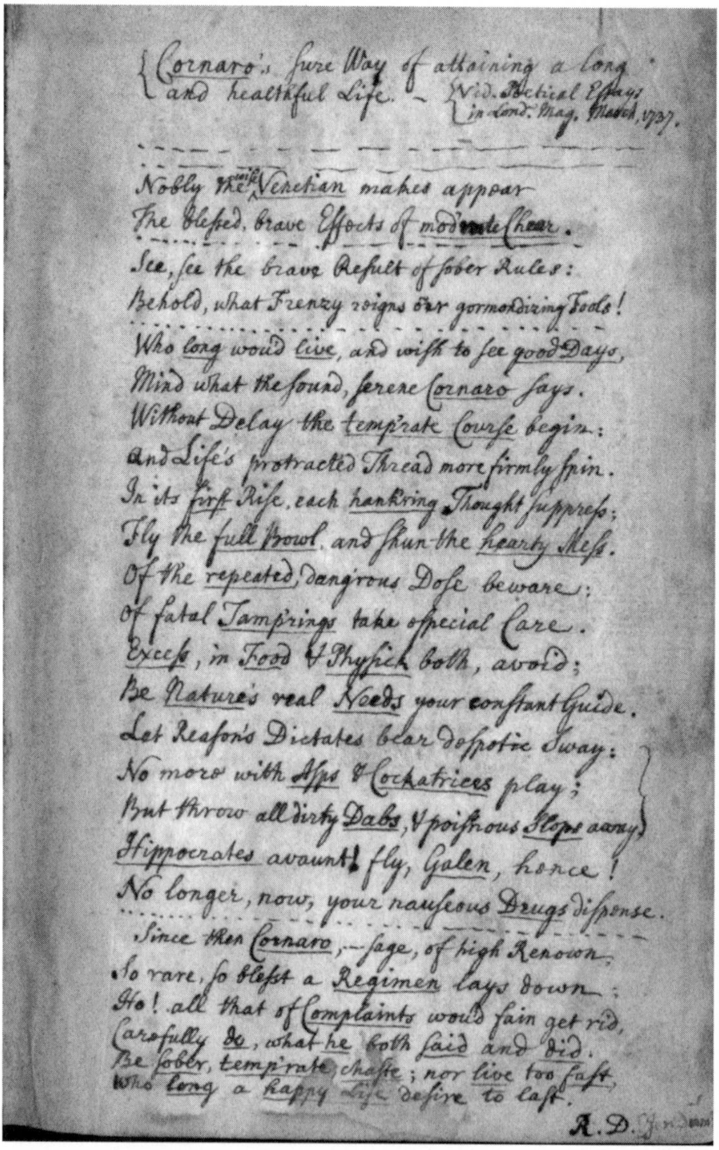

Cornaro's sure Way of attaining a long
and healthful Life. — Vid. Medical Essays
in Lond. Mag. March, 1737.

Nobly the Venetian makes appear
The blessed, brave Effects of mod'rate Chear.
See, see the brave Result of sober Rules:
Behold, what Frenzy reigns o'er gormodizing Fools!
Who long would live, and wish to see good Days,
Mind what the sound, serene Cornaro says.
Without Delay the temp'rate Course begin:
And Life's protracted Thread more firmly spin.
In its first Rise, each hankring Thought suppress;
Fly the full Bowl, and shun the hearty Mess.
Of the repeated, dangrous Dose beware:
Of fatal Tamprings take special Care.
Except, in Food & Physick both, avoid;
Be Nature's real Needs your constant Guide.
Let Reason's Dictates bear despotic Sway:
No more with Asps & Cockatrices play;
But throw all dirty Dabs, & poisnous Slops away)
Hippocrates avaunt! fly, Galen, hence!
No longer, now, your nauseous Drugs dispense.
Since then Cornaro, — sage, of high Renown,
So rare, so blest a Regimen lays down:
Ho! all that of Complaints would fain get rid,
Carefully do, what he both said and did.
Be sober, temp'rate, chaste; nor live too fast,
Who long a happy Life desire to last.

A.D.

FIGURE 8. Citation of Cornaro, *The English Malady*, by George Cheyne. Image courtesy of the Truman G. Blocker, Jr., History of Medicine Collections, Moody Medical Library, University of Texas Medical Branch at Galveston, Texas.

xxxii CONTENTS.

CHAP. IV.

The Author's soft Regimen, p. 361.

Dr. Taylor of Croydon, p. 253, 335.

FIGURE 9. Table of contents, *The English Malady*, by George Cheyne. Image courtesy of the Truman G. Blocker, Jr., History of Medicine Collections, Moody Medical Library, University of Texas Medical Branch at Galveston, Texas.

who has not paffed thro' endlefs *Multiplicity*
in Study, Obfervation, and Experiment in
thefe *Sciences* ; fuch a *Simplicity* is the
greateft Contradiction to *Lazinefs, Foreign
Studies, Negligence, Incuriofity,* and *Ignorance*
in the Profeffion ; but fuch a *Simplicity* (produced by rejecting *Need-not's*) when (if ever)
attained, is worth a *Million* of thefe little *falfe*
and *foreign Arts* fometimes us'd to rife in it ;
for it is, in Truth and Reality, an *Eminence*
of *Light* and *Tranquillity.*

*Difpicere unde queas alios, paffimque videre
Errare, atque viam palantes quærere vitæ.*

LUCRET.

F I N I S.

a vulnerable place where its message will be probed. Its validity wavers, its truth is hollowed out. A demand to gauge and qualify is made, distinguishing the narration of an experience from a statement with the force of truth. In that distinction, the reader has the last word.

But by asserting the failure of the physician's discipline, that reader only contributed to reinforcing what the text sought to state: the similarity between the physician and his patients. The use of autobiographical writing succeeded in constructing a double identity in the collusion of two identities traditionally counterposed, the physician and the patient. Here, the physician is not presented hunched over his books or closeted in a room, dissecting bodies; he lies in wait for physiological variations and the suggestion of appropriate responses. His capacity to listen to his own body positions him as an ideal patient, analyzing his malaises with double finesse. More than transmitting knowledge, the autobiographical narrative authenticates him.

Autobiographical narrative and dialogue were thus not mere changes of style, making medical knowledge enjoyable to the elite: they produced a different kind of expectation of the physician and of medicine itself. The novelty of these texts lay first in legitimating the expression of pain and anguish and portraying the dangers posed by a badly chosen therapy. Their author belonged to a brand of physician directly contrasted with those then named charlatans;[89] he abstained from claiming the unconditional success of any newly invented therapy, to promise instead a well-being gained through attentive listening to one's own body. The identification of common interests was also a means to postulate the narrator's sincerity and his probity, as the autobiography operated as a guarantee of the efficacy of the author's advice. The constant sliding between identities—savant, object of knowledge, subject of experience, writer—inaugurated a new form of medical knowledge, associated with a new image of medicine as knowledge gained through trial.

Fictional Correspondence

Correspondence was one of the most fashionable styles of the eighteenth century. The genre allowed writers such as Madame de Graffigny, Laclos, Montesquieu, and Richardson to give force to feelings, and space to a wide range of impressions. Speaking in the name of personal experience grounded their narration in a search for authenticity. Many physicians, too, used the letter format to defend a model of therapy. The epistolary form had already been used in the seventeenth century, but at that time it was adapted to the genre of the medical dissertation itself. In fact, one of the reference texts on hysteric illness exemplifies that process: Sydenham's work was initially a

letter addressed to a colleague, William Cole, but took the format of a medical dissertation. It was written in Latin, the language was that of a savant,
the tone theoretical. Sydenham did not emphasize cases encountered in his
personal experience, and only the first and the last sentence indicated to the
reader that this was the reproduction of a text conceived in a private exchange between two physicians. When the epistle was included in a volume
of complete works, it was divided into numbered paragraphs, and the original format disappeared. At the height of the eighteenth century, in contrast,
the epistolary genre spread as a fully rhetorical medium, used especially
to convince mothers to inoculate their children. Jean Jacques Menuret de
Chambaud also used it for his 1784 work on local temperaments, claiming
the treatise had arisen in response to a letter from a colleague in Valence
who had asked him for an account of local temperament in Paris. The genre
developed further with the polemics on magnetism that flourished between
the 1780s and early 1800s.

   In those years, the use of correspondence offered the opportunity to write
texts in the voice of the patient or that of the physician. An example of the
former is Jean-Marie Gamet's *Théorie nouvelle sur les maladies cancéreuses,
nerveuses et autres* (A New Theory on Cancerous and Nervous Illnesses and
Others). Gamet's work is mainly composed of letters from patients, and this
to a specific end: it was through personal testimony that he hoped to propagate his remedy. He presents letters from patients of different sexes and social
classes, asserting that they were sent after he expressed his desire to publish a
work. The letters evince respect and gratitude toward the physician. The letter contributed by M. de Rossignol, a secretary in the police force and the victim of "profound melancholia and black vapors,"[90] starts with these words: "I
make it my duty, Sir, to publish the testimony I owe to truth in favor of your
remedy."[91] The Abbé of Chalut, whose vaporous and melancholic incidents
subsided after Gamet's treatments, gives a long list of his previous troubles
and the therapies he has tried. He draws attention to his care in paying tribute
to Gamet: "I am charmed, Sir, that this general topic gives me the opportunity to publicize the beneficial effects I owe you, & and my inexpressible
gratitude."[92] The letter paints an enthusiastic picture of the relief that future
patients will find.

   The use of these letters allows Gamet to indicate the wide scope of the
physician's roles, acting at all levels of society and curing fathers, mothers,
children, and priests. By relieving his patients' suffering, the physician gives
them back to their family, enables them to work, makes them fully active
members of society. He supplements the letters with a list of eleven physicians and nine surgeons "of reputation," stating that "physicians commend-

able for their knowledge and probity have presented a petition to the King based on their own observations" of the therapy's effects.[93]

It is thus by means of a web of testimonies that Gamet seeks to demonstrate his treatment's efficacy. His book stresses his concern to establish his reputation through the channel of his patients and colleagues, who are all given the authority to judge his competence. Statements patiently collected, or shrewdly suggested, are assembled to gain the trust of the reader. The publication aims to be an account of an existing practice, ratifying what has already been proved; the value of theorization is entirely subordinated to personal testimony. It highlights Gamet's care in listening to his patients and following the effects of the cure long after the period of convalescence. The point here was to speak not about the pathology, but about a cure, by collecting narrations of experience in as wide as possible a spectrum of contexts.

Of the writers giving a voice to physicians, Claude Révillon made the most eloquent use of the epistolary genre. A member of the Academy of Science of Dijon and a correspondent of the Société Royale de Médecine in Paris and Mâcon, Révillon combined in one work the formats of correspondence and autobiographical narrative. *Recherches sur les causes des affections hypochondriaques, appelées communément vapeurs* (Study of the Causes of the Hypochondriac Affections, Commonly Called Vapors), published in 1779, is composed of fourteen letters addressed to an imaginary patient, to whom the physician relates his own experience of the malady and his research into appropriate treatments. Révillon claims to address his treatise to his "previous companions in unhappiness,"[94] too rarely taken seriously by physicians. Numerous book reviews in medical journals attest to this work's success on its first publication and in its second edition seven years later. *L'Esprit des journaux françois et étrangers*[95] gave an account of the book; the *Journal des savants*[96] praised the author's effort to start from his own experience; the *Mercure de France*[97] noted the timeliness of this treatise when such illnesses had become more common than ever before. The work was equally well received in England, with a flattering review from Tobias Smollett in *The Critical Review*.[98]

Here, again, theoretical discourse is replaced by first-person discourse, grounding the authority in the circumstances of the enunciation: a status—that of physician—and the need for advice by a patient whom readers must imagine. The authorizing source of truth is no longer the ancients, but experience. From the start, Révillon attacks works of medicine that fail to provide proof of what they theorize. Despite references to physicians who have written on the topic, the physician does not position himself as a savant but as a man, who draws his ability to understand from his own past life. By adopt-

ing the letter format, he authorizes himself to include numerous biographical details without further justification. This strategy allows the physician to
speak directly to his readers and take up a position even when describing
symptoms. Interspersing his own remarks in the description, Révillon can
avoid a professorial tone and create a vivid narrative. Révillon's voice draws its
legitimacy from this double "I," one based in experience, the other identifying pathologies in the name of competence. Révillon's treatise further gains
in impact by presenting each of his assertions as a response to a letter not
reproduced in the work. By including only the doctor's letters, the treatise
invites readers to interpolate their own reflections, imagining the replies and
inserting themselves into the position of the vaporous addressee.[99] Like Pierre
Hunauld, Révillon begins by establishing a contract with this interlocutor:
"You know I have committed to cease writing to you as soon as my letters no
longer interest you; and it is in the trust of your promise that I start to keep
mine."[100] References to the reader proliferate as the physician starts each letter with "Monsieur," making abundant use of apostrophe. The reader of the
treatise is to become the letters' addressee through the very act of reading.

The physician frequently offers portraits in which the patient can recognize himself, insisting that it is his experience that gives him the capacity to
portray vaporous fits: "You have probably just recognized yourself in this first
portrait: you will see, by the second one, if my own experience has not put
me in a position to be a faithful painter."[101] The pathology's symptoms are
sufficiently flexible to encompass all patients, and far from claiming to be a
unique case, the letter writer indicates that all his patients have enabled him
to verify his own hypotheses.[102] He admits having included cases drawn from
accredited observers, justifying this by saying that he could not have experienced all symptoms himself and aims to present an exhaustive description of
the pathology. Yet all these sources appear without reference, rewritten in the
first person, and the reader has no choice but to assimilate all the anecdotes
to Révillon's life—"I," as Benveniste wrote, implying a testimony.[103] By presenting himself as a speaker, the physician makes his reader an interlocutor.

However, the epistolary narrator here does not position himself only as
a patient, but also as a man of letters. He inscribes his illness in the tradition of the melancholic, situating it outside his own responsibility or error.
Rather, Révillon seeks the cause of the vaporous pathology in a diminution
of transpiration. Right from the start, in the preliminary discourse, he insists
that the vaporous patient should not feel guilty about his state. The physician indexes the body as responsible for the wandering of the soul, arguing
that men of letters—who can, he flatteringly notes, certainly not be accused
of pusillanimity—often fall victim to fits.[104] He describes his own feeling of

vulnerability, even disarray, conceiving of the bodily state's variations as a constant source of worry and urging increased control through discipline in food and sleep. The sense of complicity generated by this means of address seems to stimulate a return of confidence.

Révillon's description of suffering reinforces that complicity by siding with his patients in a range of different ways. First, localizing the origin of vaporous fits in physiology allows him to combat a common perception that such accesses are simulation or mere indulgence. Here, the physician transforms his competence in physiology into a moral authority, taking it upon himself to dispel guilt. He also stands up for female patients, accusing education, social customs, and men of originating a way of life apt to increase their vapors. Attacking colleagues was another way for the doctor to engender complicity with the reader. Révillon pursues a tactic used by Hunauld, extending his accommodating attitude to the point of deliberately taking the side of the sick against the physician: "I was for several years the dupe of means recommended to me; I understood that the inconstancy given as a characteristic sign of this illness came less from the patients than from the inadequacy of the reliefs offered by art."[105] The origin of fickleness, a determining feature of the so-called vaporous, is thus located in the incompetence of physicians. The experience of suffering claimed by the letter writer allows him to exonerate patients, by the same token demonstrating how much this has changed his own perspective.

In this treatise, the transmission of knowledge is presented as a process, and temporality plays a very particular role in its writing. The exposition does not follow a theoretical progression, but is organized toward the reading and toward creating the space necessary for the acceptance of new ideas. Neither does Révillon follow an autobiographical chronology, instead articulating his experience in terms of the difficulties undergone by the imaginary patient. The narrator uses the correspondence format to make the temporality of the exchange between two people his framework, guiding the addressee with statements such as "Before going further, let us determine with precision the ideas that this word presents."[106] Far from a systematic order, the letters evolve according to the point of view asserted by the epistolary writer, introducing delays as a source of suspense.[107] Each letter is conceived as part of a series, and stages the thinking that took place in the act of writing. Révillon announces that he will explain the cause of particular fits in the next letter, indicating that this postponement is intended to give his addressee time to reflect. The narrator also refers to what his patient reportedly said, the questions he reportedly asked, thus regularly anticipating the reactions of his reader. By referring to the act of reading itself, the writer introduces a series of deictics that potentially transform his statements into a series of performatives. The temporality of the letter format suggests that the doctor and his patient have been brought together, that trust has been established in the course of this exchange.

The "I" appears as one who reflects and weighs his judgments, posing himself in a continuous relationship with a "you," his interlocutor—whom he supports, whose reactions he anticipates. He organizes his reasoning into steps, indicates his choices and duties, and comments on the words he employs.[108] He calls his interlocutor to bear witness[109] and speaks to him as if their shared suffering created complicity and mutual understanding.[110] In a conversational tone, the physician recommends various readings and shares his impressions of these. He even admits the possibility of being wrong, offering to accept contradiction if he is provided with evidence of his mistake— though it would be peculiarly difficult to prove him wrong, since everything he says is grounded in idiosyncrasy. The physician accentuates his discursive relationship with a whole series of attitudes around the knowledge that is being enunciated: commiseration, admiration, surprise, worry, even irony and distrust.

The second edition, in 1786, still includes only one side of the correspondence, but it adds two new letters, inserted at the center pages. The style of all of the letters is revised, and, most strikingly, new theorizations are developed, adapting the treatise to the latest medical models: Révillon now attributes the lack of transpiration to a lessening of electricity. Days of malaise are, he writes, the days of a decline in electric fluid,[111] resulting in a state of languor, vertigo, or hallucinations.[112] Such variations indicated to readers the artificial character of the editing and, even more to the point, the fact that the author did not judge it necessary to hide this artificiality: to disclose the work's artificiality is, it seems, not to undermine its validity. The physician's ability to construct a narrative of this kind signifies his capacity to understand the patient—whether the narrative itself is true or false. As with Hunauld's treatise, the work operates not as a truthful testimony, but succeeds because of its verisimilitude.

The complex enunciation generates the plausibility of the words and their appropriateness. Révillon finds multiple ways to position his identity to that end, speaking not only as a physician but as a man. The work is built on an oscillation between the suffering "I," the patient "I," the physician "I," and the writing "I"—a composite "I," then, that describes the pathology according to his experience and gathered knowledge. From the preliminary discourse to the treatise onward, the narrator associates himself with distinct categories: vaporous people, physicians, inhabitants of a province with a particular climate. In the preface itself, there is a constant sliding of pronouns:

> For a vaporous person, it is comforting to be able to tell oneself: I owe to the illness the fears, the worries that incessantly surround me. Born with a weak constitution, my imagination is disturbed when physical causes, to which I am subject like all other beings, differently modify my organs, my tastes, my sensations, my sometimes strange passions. All these things are the product of my organization, & the accident of my state. The vaporous of my Province will know that the north and north-east [wind] purify the air, that they clear up the sky; . . . they fortify the fibers & animate our circulation, & . . . finally, in this disposition of the atmosphere, the inconstancies, the inconsistencies of which we are accused disappear, we become physically stronger and morally more calm.[113]

In this passage, the epistolary writer slides from a "vaporous person" to an "I," introduced in the form of direct discourse, that embodies the vaporous; then to an "us" that encompasses the author and the vaporous of his province. In the course of the book, the shifting subject produces a displacement of the categories and typologies used to position knowledge. The distinction between doctor and patient gives way to that between the inhabitants of his province and the others, then to that between the vaporous and the others. Just after making a first-person biographical statement, the narrator insists on the exclusivity of the experience of the vaporous:

> I can say it has happened to me a thousand times to start reading or writing with a north wind; I would continue my occupations with pleasure & success as long as it held; but if it turned to south or west, in an instant I would recognize the difficulty of following the ideas of the Authors & of connecting them with mine. This sensibility will only ever be understood by the vaporous.[114]

This time, the narrator implicitly addresses the vaporous, claiming to be understood only by them. Yet a few pages later, in a move similar to Cheyne's, Révillon refers to luxury and idleness as a cause of vapors. Here the "us" that he uses assimilates men and women living in the eighteenth century, in a way likely to attract his readers. The first letter then starts, opening with a sentence as suggestive as it is short: "I have been as unhappy as you are";[115] the physician appropriates the position of the patient in the name of the suffering he has experienced. However, the letters also include numerous generalizations on "the vaporous," and these occur at specific moments. When less flattering statements are made about the vaporous, the narrator withdraws from the intersubjective relationship and talks about patients in the third person plural: "The whole of nature is, in the eyes of the vaporous, covered with a funereal crepe. . . . These poor souls, unceasingly busy with the preservation of their being, and often finding themselves hoping for its destruction, are tormented by a cruel despair."[116] The demonstrative "these" further accelerates the estrangement. Just a little later, the writer chooses the third person singular to talk about the most trying moments of the pathology and cases where the vapors have been provoked by a reprehensible way of life: "Let us consider the vaporous as he is, that is, a weaker being in his primary organization, or due to his constitution debilitated by sleepless nights, by pleasures, at last by great passions; & we will have the cause of an unequal, difficult, languishing circulation, which is enough to deprive him of the necessary quantity of fluid & to give him a tiresome existence."[117] This time, the epistolary writer and the addressee are on the side of the observers. The "us" gathers the physician together with his interlocutor and anyone else interested in the vaporous. Switching pronouns allows Révillon to appropriate some symptoms while keeping the most disagreeable elements of the pathology at a distance.

At the end of the second edition, Révillon also added twelve tables of meteorological observations, each table around thirty by fifty centimeters once unfolded. The tables record for each day, three times a day, the temperature, barometer, hydrometer, the level of electricity in the air, the wind direction, the state of the sky, and the state of a vaporous person. The state of the vaporous is quantified as "suffering," "weak," "just all right," and "better."[118] In Révillon's view, meteorological parameters affect the electricity of the air, which in turn affects transpiration,[119] and he presents himself as ready to test this hypothesis by provoking an access of malaise.[120] Hence a further addition, his "Journal of the State of the Body," as well as four references to the state of the physician himself, his excretions, and his sweat. The second edition thus adds to the autobiographical language a tabulation of numbers that certifies experience by giving it the impartial character of something measured. The neutrality of a table with measurements complements the biographical presentation of suffering in the letters, offering two forms of writing and, as such, two approaches to the pathology.

With Révillon, a moment of change in scientific knowledge had arrived: enunciating knowledge came to mean taking a stand and carrying it into one's own body, in a way opened up by Santorio Santorio.[121] Besides the fascination with scientific instruments that typified the second half of the eighteenth century,[122] Révillon's experimentations translated into a new attention to the body as such. The physician took his body as a first object of study and experience, advising careful choice of aliments as the means for the sick to counter meteorological problems and maintain their health. Much further advice is given in the letters. Révillon states, for example, that the best moment to exercise the body is in the morning, when transpiration is heavy, and that the patient should not be troubled by abundant sweat.[123] He also recommends breakfasting on bread and wine, since this lends strength without weighing on the stomach, and advocates waking at six in the morning and retiring at ten at night. The experience of pathology becomes an opportunity to experiment. Provoking his own fits, this narrator has inscribed the figure of the physician as a man able to interact with his own physiology, and his relationship with his own body as one of constant transformation resulting from his own intervention. The writing of correspondence in the first person thus made it possible to open up a new dimension: that of experiment, which was to supersede observation as a scientific method and genre of writing and which would dominate in the nineteenth century.

The Epistolary Consultation

In writing his treatise in the form of a correspondence, Révillon was mimicking an established practice: the epistolary consultation. Laurence Brockliss has shown that the practice of consulting by letter was already established in late medieval Europe, though before 1500 it had fully taken root only in Italy.[124] By the second half of the eighteenth century it had become a familiar and increasing practice in Europe, as physicians' and patients' letters held in archives indicate. Théodore Tronchin, Samuel-Auguste Tissot, Robert Whytt, Pierre Pomme, and William Cullen, all famous physicians focusing on the cure of nervous and hysteric illnesses, carried out epistolary consultations with patients they had never met.[125] Tronchin is a particularly interesting case in this regard. His only publication, written when he already had a large clientele, never enjoyed any success and was later denounced as plagiarism. Nevertheless, his celebrity is attested by the appearance of his name in the correspondence of Voltaire, Jean-Jacques Rousseau, Denis Diderot, René-Antoine Ferchault de Réaumur, Georges-Louis Leclerc de Buffon, Louise Florence Pétronille la Live, known as Madame d'Épinay, and Friedrich Melchior Grimm.[126] He was also mentioned in satires, in which Parisian women ran to welcome him whenever he arrived from Switzerland. It was neither works published nor therapies invented that caught the attention of his contemporaries, but rather his capacity to gain his patients' trust in consultations, which, for a large part, were held in letter form. Yet the high public profile of epistolary consultation is not the only reason these texts are interesting today. At a time when an increasingly specific language was emerging through the publication of medical treatises and nosologies, the level of detail provided in correspondence constantly grew, creating an arena for a radically different language. Letters were often characterized by a claim to share unlimited information with the physician in order to better understand the body. Expressions of feelings and sensations were interspersed with vocabulary from the medical treatises. In the letters, the narration of suffering—its precise combination of the said and the unsaid—played a determining role.

Studying this type of consultation, the historian is faced even more than in the case of published work with the problem of uneven access to source material. As Brockliss emphasizes,[127] such analysis cannot draw firm conclusions about a practice, but only comment on remaining fragments. Neither is it possible to analyze particular exchanges between doctor and patient in full, as only partial correspondences have been found—either the doctor's letters, in the case of Tronchin, or the patients', in the case of Pomme and Tissot. It is tempting to see in this partial archive a selection motivated by the different physicians' different uses of the letter form: Tronchin, a medical practitioner with little interest in publication, copied his own letters and prescriptions in order to evaluate the later steps of the treatment. Tissot and Pomme, who both became famous through their publications about curing nervous illnesses, most probably kept these letters as case studies to be used when writing further books.

The patients' attempt to set down on paper their suffering or turmoil was an initiative strongly encouraged by the physicians themselves. Many insisted on the necessity of patients' retracing their education, their history, their habits in order to pin down the pathology and its origin. Since the physicians tended to believe that hysteric and nervous affections could have a variety of causes, it was through the patient's history that the relevant ones could be determined and the most suitable therapy prescribed. In that process, it might seem that everything had to be said, that everything played a role in the pathology. Tronchin, for example, often begged his patients to confide their moral states, their sorrows, and their worries. Here is one of his letters to Madame de Belzunce:

> I can see, Madame, that I must come to terms with your modesty, I don't want to fight against it, that would mean always starting afresh. Each time you do me the grace of writing to me, you repeat that you don't know if I still remember your existence. This reproach, Madame, would overwhelm me, if I did not imagine that I am not the only one to whom you make it. All appearances are that it is quite a habit of yours, it stems from your character, I no longer take fright. I do not, Madame, easily come to terms with your silence in relation to your health. For you to understand this, you would have to imagine the interest I take in it, but that is what, also by character, you will never

understand. I am mortified about this, Madame, because you may suffer from it. . . . I read your letter with all possible attention. One appears to perceive more resistance, but that the nervous sensibility is still very great, and that the winds, which are the ordinary effect, are as indisposed as ever. I hope, perhaps because I wish it with all my soul, that if you once unfortunately had sorrows, you now have them no longer. Of all the causes of this sensibility, Madame, having such sorrows would be the most difficult to vanquish and the one I would fear the most. You have never given me any clarification with respect to this; yet it would not be of indifference for you to tell me their nature, the knowledge of their cause, I would in fact need no detail, I would simply need to know if you had sorrows. The interest, Madame, that I place in your cure, but of which you have no conception, is what leads me to press you much on this matter. The law of silence you have prescribed to yourself, since you have not answered, cannot prevent you from acknowledging what I had the honor to tell you, what I repeat to you now, that sorrow was and will always be what is to be feared the most. It seems to me, Madame, that a beautiful soul such as yours—your modesty will perhaps forgive this thought—itself alone has the right not to sorrow. This advantage is reserved to those who think as you, who act the same way. But I fear that the excess of your modesty may hide this advantage from you, it is nevertheless yours, Madame, death itself cannot take it away from you. I conclude this by saying that physical remedies would be insufficient if you did not join with them the moral ones. Your virtue will give them to you. As for the other remedies, Madame, which belong to my department, I take the liberty to remind you of the regimen we agreed on, I have nothing to change in it, I have made only a few changes to the internal remedies. Morning, lunch, and dinner, for three months, half an hour before the meal, you will take, Madame, three of these new pills, and afterwards two spoonfuls of the new draft in your mouth. I hope that no pretext, for you may have reasons, will prevent you from giving me your news. It seems to me, Madame, that you owe this kindness to the lively interest, the singular esteem, and to the deep respect that I have vowed to you.[128]

The physician's necessary discretion obliges him to measure his words carefully when stressing the role of moral afflictions. His tactful approach honors the request for interest implied in patients' letters, carving out an appropriate tone that combines the position of knowledge with that of the confidant. Reading Tronchin's prescriptions sheds light on the importance of the phrasing of the cure itself when writing to an aristocrat, as the doctor skillfully intermingles therapeutic instructions with advice about lifestyle, making sure to phrase them as cautious suggestions and not commands. He advises following the natural, not the Parisian, way of life, eating simply, sleeping at night, and waking after seven hours. He also prescribes a more active life and to have the body rubbed with flannel. He insists on the role of sorrows, asking to be told if there are any, though assuring his reader that he does not need details. If medicine is prescribed, it is usually only a potion for three months, at times with pills to be taken for one to three months.[129] The point here is not to level accusations of opportunism against a physician who replaces bleedings with a ban on chaise longues and a few spoonfuls of potion. What Tronchin shows in these prescriptions is that the first concern of the physician must be not to do violence to the body. He is propagating daily attention to the body, aligning himself with a long tradition of writings about the daily regimen at a time before the terms of hygiene and discipline dominated medical discourse.[130]

The patients writing to Tissot, Tronchin, and Pomme were mostly aristocrats, sometimes bourgeois or members of the clergy: men and women living mainly in smaller towns in France, Switzerland, or Italy with no access to renowned physicians. Such epistolary consultations were not easy, and their style varied considerably from patient to patient. At times, patients asked their own physician to write to renowned colleagues in search of advice. Other patients would take the liberty of seeking advice from a famous physician directly. A plethora of details would often follow, and was welcomed as a sign of the proper humility of a patient hesitant to select for the physician what was a sign and what was not. A woman named Mallebois de Sourches, originally of Mans, wrote to Pomme to tell him about the stiffness of her jaw. She specified: "My Menses arrived with a lateness of eight days; They had difficulty deciding themselves on the first day; presently they come strongly." The patient immediately adds: "I quite fear that you find this detail less than Intelligible; not being accustomed to giving an account of my ailments by letter, if nevertheless you do understand something of it, I beg you Sir to convey to me whether I should change anything in my Diet, or if I should keep on taking the same quantity of Veal Stock as usual."[131] Correspondence explaining bodily embarrassment or pain was unusual, and patients typically found it difficult to put their disorders into writing. Also, their attempts might be rebuffed. In the fiftieth consultation collected in Paul-Joseph Barthez's compendium, the doctor tells the patient plainly that he is incapable of writing precisely enough about his ailment and avoiding redundancies. Since writing was likely to fatigue him, he should return to relying on the authority of his own physician for any further correspondence.[132]

A family member, or someone in the patient's immediate circle, would sometimes describe the case on the patient's behalf. A woman writes for Madame de Chanterelle's tax collector, worried about the surgeon's prescriptions, which alternate bleedings and purges. In another example, Dehaussay, a cavalry warrant officer, writes regularly on behalf of a female relative, toward whom he begs Pomme to be understanding without omitting any advice necessary for the cure. The topic of the exchange is initially a prescription of baths, which the patient refuses absolutely, and reduced consumption of wine. In the following letter, this step has been successfully completed, and the letter writer explains how he cheats the patient by adding water to her wine. He goes on to mention his hesitation in following Pomme's prescription of pouring cold water on the patient's head: "When we take it upon ourselves to contradict her, which considerably grieves us, the agitation it gives her can only be a greater source of evil than the good she will obtain from the cure we would like her to undertake."[133] These letters were a space of negotiation, where a drastic therapy might be softened or attention to particular symptoms expedited.

Some of the letters were also attempts to grasp the truth of the body beyond the prejudices that an informed gaze might have. Monsieur Bon's letter to Tissot, sent from Perpignan, a small town in the south of France, to Lausanne, is one such case. It concerns a young woman:

> Her parents gave me the responsibility of addressing to you in the greatest detail the report of her illness. They preferred my pen to that of the people of the art, who cannot prevent themselves from letting their opinion break through and presenting things from their own viewpoint. I could not, Sir, fall into this trap. The report I have the honor to address you is the result of what I saw, of what the parents of the afflicted—who never leave her—have observed, of what the patient felt. For fear of omitting interesting details, I have fallen into prolixity, but I thought that you would forgive me in view of the cause.[134]

Whereas the specialists would want to be recognized, and might underestimate certain symptoms because of their desire to assert one diagnosis and to gain the assent of the famous physician, ignorance becomes the voice of truth. The purity of the writer's intentions, devoid of ambition and prejudices, presents itself as the mouthpiece of innocence. At times, a correspondent outlines the case a second or even third time, with a letter just as long as the preceding one, reminding the physician that he has not yet replied. Others ask the physician to keep their letter, anticipating that the treatment will be made in a number of steps.

Some patients participate actively in negotiating the therapeutic intervention. They make specific suggestions to their physicians, not forgetting to justify their view, and are eager to undergo new therapies—although the letters sent to Pomme and to Tissot, often by aristocrats, all display the greatest humility toward the physician. A female patient from Besançon, for example, wrote Pomme to ask him to complement the prescriptions of her own attending physician. She goes as far as to suggest: "In order to put an end to my miseries, it might be advantageous to apply to the posterior for a few months half a dozen leeches eight days after my menstruation, and again ten days after these first ones." It is striking that the correspondents take part in the writing of their case, right up to the enunciation of the diagnosis, or reinterpret their case in light of the physicians' publications.

Under the patient's pen, the idea takes shape of a coincidence between meeting the physician and being healed, reading and speaking, initiating the therapeutic act and heralding the end of the illness. Expectations of a meeting with the physician were high, but they also clustered around the act of writing itself. The description of the symptoms, and their association with lived experience through a narrative, indicate that far from folding themselves into passivity or ignorance, many patients actively want to know their bodies. Their narration of suffering, of its progression, of the slowness of the cure, of the renewal of the illness, anchors a truth of the individual. Describing sensations in an epistolary form appears as a way for them to narrate the uniqueness of their experience. Throughout the relationship with the physician, patients' writings about their bodies become a writing of the self, and an occasion to encounter their own bodies. Letters are the space in which patients can grasp their affection on their own behalf, adopt a pathology, and by naming their affliction, begin to heal it. The task of writing about sensations and bodily disorders is a means of understanding the singularity of their own journey and the mistakes in education, habits, or morals that have been made during its course.

Letters addressed to Samuel-Auguste Tissot often took the shape of bundles of paper closely covered with cramped writing. The ample correspondence in which male victims of the vapors confessed to masturbatory habits suggests how eighteenth-century readers saw their own troubles in light of published observations.[135] They uncovered the origin of their own physiological troubles, and believed themselves obligated to admit their guilt. Such confessions indicate how the hope of attracting the doctor's attention and obtaining a cure for the body displaces the anguish and frustration tied to the malady. In all these letters, the vocabulary and gaze of the doctor are deployed by patients in the discovery of their own bodies.[136] They consider their biography through the prism of the pathology. Describing their physiological troubles and proposing possible causes, they willfully incriminate themselves and confess past habits in a tone of repentance. They testify to their own awareness of the needs and limits of the body, affecting a new sense of responsibility. It seems that these correspondents seek to convince the doctor of their goodwill, pledging to be disciplined patients devoted to recovery. They demonstrate readiness to accept the doctor's suggestions, even anticipating his questions and cautions. Their writings convey their determination to play an active role in their therapy, rejecting the passivity to which their malady seems to condemn them. Shame and a guilty conscience give way to the demands of the narrative, as if confession were necessary to recovery. Only rarely do these letters try to diminish the patients' direct responsibility by evoking the ignorance of youth or the origin of pollutions in dream.

Toward the end of the eighteenth century, while physicians were constructing the pathology of the body as a sign of harmful habits, worried readers recognized themselves and watched for symptoms. Were their troubles and sensations also symptoms? What about their own bodies? Were they vaporous themselves? Here, medical discourse's claim to truth caught readers in a trap. It ensnared them in doubts about their sensations, their education, their habits, turning these into meaningful signs. In their letters, while examining their pains, patients remembered the excesses to which they had succumbed. Assorted indulgences, gastronomic excess, libertine practices, and masturbatory habits became the sources of their troubles. The doctor urged readers to listen to their bodies, but in the attention he gave to the body, he was also instructing them to interrogate their souls.

Further reading of correspondence and consultation reports, however, establishes that the space of speech itself had boundaries, and that physicians and patients edited what could be said. It becomes clear that both sides delimit what can be told, on the basis of reciprocal expectations. The therapy implies a selecting of words, both the diagnosis itself and the words to explain that diagnosis. Often a withdrawal, a detour takes shape in the descriptions of symptoms and suffering. What, then, belonged to the *to-be-said*, what to the *to-be-kept-silent*—what were the constraints and limits that structured medical exchange?

When patients limited their confidences, the reason was not always modesty. For the patient, to be silent was a way to express a laudable lack of complacency in one's narrative of oneself; in addition, anyone wanting the attention of a renowned physician would treat him considerately to demonstrate respect. A letter sent by an aristocratic woman to Pierre Pomme manifests this form of reserve. The patient vows "a quite tender and quite true friendship . . . until the grave,"[137] and, talking about past afflictions, indicates: "I did not write to you at the time because I was too poorly not to be in your eyes a whole new school of Jeremiah." Here, the pact of trust between physician and patient implies not saying everything, and knowing how to say what must be said, and beginning to speak only once one can distinguish what should be said from what should not. The letter is written in awareness of the need to preserve the patience and interest of the physician. The patient sees her role in effortfully expurgating her discourse and excluding all excess from it. Moderation in the narrative of the self functions as proof of discernment, so that the description of symptoms will, it is hoped, be received more attentively. The relationship between physician and patient proceeds in a restraint seen as proper to the therapeutic act, a condition of trust and credibility. And, through the measured constraint she imposes on herself, the patient learns to look at her state with an exterior eye. She becomes accustomed to being a patient, to evaluating and accepting the limits of the therapeutic act. However, the physician too sometimes attempts to limit the patient's narrative. Tronchin, for example, did not always present himself as a confidant. Here is another letter he sent to Madame de Belzunce:

Your weak imagination, Madame, frightens me even more than your ailments, it will make you the victim of those who want to abuse it. You have already fallen victim to it so many times, you will again, I am sorry for you with all my heart and this may be all I can do for you. Most of the ailments you have taken the trouble to recount to me, and that, out of compassion, I read from beginning to end, are pseudo-ailments, which consequently do not resemble anything. The number of these actual or possible ailments is almost without end because of the weakness of the body and soul of the sick, because of the multiplicity of bad remedies and of the ignorance and mischief of those who prescribe them. The factory of bad remedies is nowadays the greatest of all European factories. How you are to be pitied, Mademoiselle, for not having kept to Mr. Carisot, you would have spared yourself a quantity of real pains & and those worries of the soul that are even worse than such pains. Is this poor life worth so much pain? Indeed, at the beginning there was something scrofulous, that's very likely, a poorly handled fever that, being undecided, occasioned new episodes, that's very common. Did that justify taking advantage of the weakness of your soul, and of your pusillanimity in conscripting the contribution of your body, this, Mademoiselle, I cannot think of without horror. The art which should have been the most salutary became the most grievous. The series of cruel torments to which it led you affected your poor soul, and upset the spleen. This is the common effect of sorrow. If, abjuring your past mistakes, you can transcend your uncertainties and worries, the upset will dissipate. Make the effort to forgot this succession of miseries which perturb you, lead a gently active life, distract yourself, and for three weeks adhere to the following recipe.[138] It is now the moment of the year most proper to it. At seven and at nine o'clock in the morning you will take one third of the potion, which will therefore have to be renewed every day, fifteen minutes after taking it you will drink a glass of warm milk, and at the end of three weeks you will tell me your news. I hope, Mademoiselle, that your news will be wise, that the imagination will not speak.

I will send you back your sad papers.[139]

The letter was followed by a prescription. The particular interest of this letter lies in the fact that in many others, Tronchin suggested possible afflictions and begged his patients to tell him of them—whereas here Tronchin's role is more that of a gallant doctor who hopes to induce trust by refusing bleedings. The elegance of his style is not just empty verbiage that could be contrasted with François Boissier de Sauvages's knowledge or Pierre Chirac's authority.[140] On the contrary, he brings to light the price of patients' and doctors' indulgence, the risks they have undergone by taking their role too seriously. The message indicates patients' habit, already established in the eighteenth century, of recounting at length their body's journey in the land of suffering.

Tronchin voices the dangers involved in speaking about the pathology: such speech may exacerbate its violence. In this view, expressing anguish does not disarm but favors pathology, as a reaction catalyzed by words becomes malaise. Interpreting the slightest physiological responses as symptoms, continually reading in one's body the genesis of pathologies, is more than a mere indulgence of one's griefs. Recounting one's pains to a physician is here regarded as a way of making oneself sick. The individual invents her/himself as a patient in the detailed description of sensations, and gathers them into the teleology of a body desirous of a history. To make the body into a case is, in Tronchin's analysis, a temptation faced by both patient and physician. If the body risks losing its own health by becoming the theater of therapeutic efforts and semantic explorations, its health can be found only in restraint of treatments and of words. Refraining from pouring out one's feelings becomes the condition for healing to take place.

With his pledge to "send you back your sad papers," Tronchin further reinforces his point. He does not want to read them again, and by sending them back, he will undo their power and reject their message. Tronchin's words push against the reach of other words, and their role is to undo other roles. He attempts to create a space of trust, a space of dialogue that serves the therapy but does not dominate it. Words must kept at bay, not inspire patients. Tronchin writes to Madame Thellusson, for example:

I surely believe I would be moved by your indisposition, Madame, if it were of a nature to cause me anxiety about its course, but you will recover and everything will have been said. Its seat is in the nerves that are distributed around the whole area. They are of a peculiar mobility, and their irritability is proportional to their extreme mobility. But as these nerves are precisely the

ones that have the most intimate connection with the immediate organ of feeling, they act upon the feelings, and the feelings act upon them, in a peculiar
fashion. The slightest pain or the smallest worry produces in them a disagreeable impression. . . . The momentary tension which at times results from this
makes you susceptible to the smallest noise; it is joined by a feeling of anguish
which alarms and saddens your soul. I am sure, Madame, you have often experienced this. At such moments it is as if the soul feared for its own existence.

The weakening they cause to the nerves, as I was telling you, is what you
call the vapors, for this word—to which the one who utters it and the one
who hears it rarely attaches a fixed idea—expresses a pain one feels but cannot explain. It is a mysterious word which expresses quite clearly that no one
understands anything about it. What one would want to express is a Hydra
with a hundred heads. It is an incomprehensible monster. Do you not recognize it, Madame? I wager you do. I would be quite glad if what I mean to say
to you does not appear mere verbiage. Here is its conclusion. [The treatment
is then explained.]¹⁴¹

This passage is followed by the treatment prescribed. The physician develops the diagnosis indicated by the patient—the vapors—and deprives it
of its aura. By describing physiological movements, he evaluates and neutral-
izes the role of mystery. Indeed, although he indicates that nothing can be
understood, this failure to understand is itself fashioned as an explanation,
followed by the means of vanquishing the affection. Tronchin does not accuse his patient of indulgence; he draws together all her moral and physical pain before dissipating its power by enunciating a treatment. The doctor
gives the patient the tools to think her relationship with her body, the words
to seize control of her afflictions. The reformulation proceeds to eliminate
the turmoil.

The force of Tronchin's assertion "you will recover and everything will
have been said" resides in its ambiguity: it invokes simultaneously the need
to say everything and the need to close the discussion, to end the speech once
everything has been said. This underlines the importance for the physician
and the patient to steer clear of words and phrases, misleading formulas, or
inopportune expressions that could turn against the speaker or his interlocutor. The speech will conclude along with the suffering; it is speech as *pharma-
kon*, both poison and remedy, to be shunned as soon as suffering disappears,
in order not to reactivate that suffering.¹⁴²

By calling the treatment a "conclusion," Tronchin playfully makes his language performative. The conclusion of his letter will conclude the affliction;
the physician notifies his patient that her illness has an end, and that this end
is set out at the end of the letter. Here, the language of treatment establishes

a new language to talk about the body, replacing a series of uncertainties and vague terms with a series of terms that indicate the temporality of recovery. After the narrative and the evocative phase, language now becomes specific and irrevocable, marking the end of the disorder. The physician phrases recovery as a legitimate expectation, and even more so as something that already began, irreversibly, in the act of writing the prescription. Tronchin's letters aim to be more than a message—they are to act. Tronchin endows the words with the power to cure by offering the patient a new perception of her body and disorders.

In the letters from Pomme's patient and in Tronchin's correspondence, the enunciation of the pathology must be circumscribed in order to limit its progress. Tronchin undoes the uncertainties associated with the term "vapors," stabilizes the shifting ground governed by imagination, and declares obsolete the addition of other words about other symptoms. He aims to have the last word. A parallel emerges between a suffering body, readable within a pathology of symptoms that become its expression, and a patient articulating his state with a throng of terms. Exercising the power of restraint through language signals the arrival of a new power against suffering. To weigh the terms is not only to describe the pathology, but also to cure it.

The restraint in these letters' choice of words was motivated by therapeutic, epistemological, and political stakes. It was a need for circumspection that developed in the 1760s, as attention to speech and silence became the core of a new therapeutic power. In the letters, reserve crucially fashions the role of the patient and physician, just as in other genres the narration or avowal of experiences plays the central role. The correspondents' use of language diverts the previous perception of the disorders as immediate suffering, and substituting words aims to produce a new relationship with the suffering body. Tronchin's moderation and discretion are means to construct the conditions of the health of the body as he tries to assert the act of curing through performative utterances. Circumspection is not only an exercise in diplomacy; it is an act of power that rejects the disorder of the suffering body. The writer's restraint in formulation acquires an epistemological and healing function. And while they exchange words, or listen to each other through a shroud of circumlocution, doctors and patients play with the power of the affection, unfolding and refolding the experience of the body.

Anecdotes

The eighteenth century delighted in moral anecdotes, a genre that, if it makes no claim to truth, nevertheless effectively disseminates particular perceptions of society. In the name of entertaining the reader, anecdotes transport opinions and stigmatizations, and enable prejudices to be confirmed on the pretext that these are only tales without value or import. They offer the historian an insight into the interplay between scientific imaginings and popular and aristocratic culture, for, playing with picturesque situations and colorful characters, such anecdotes turn small incidents into the truth of a particular period. Specifically, they were what some eighteenth-century doctors used to speak with irony about the vapors, basing their humorously framed critique on the excessive and unexpected character of the fits. In the second half of the eighteenth century, following a growth in the publication of medical treatises on the vapors in the 1750s and 1760s, physicians began to deploy anecdotes in an attempt to articulate a different approach to the pathology. Physician François Amédée Doppet, for example, published a work entitled *Le Médecin de l'amour* (The Doctor of Love) in which he tells twenty-five anecdotes relating to the dangers of love, citing the genre's efficacy in transmitting moral precepts.[143] Deceived husbands, disappointed lovers, cheated masters, and youngsters fooled by the illusion of love are so many portraits presented to awaken the reader's apprehension and curiosity through a series of clichés. The vapors of an aristocrat's wife upon the death of her parrot are presented to the reader in the form of a scene from a play: the lady's maid arms herself with a flask of sal volatile, and the male servant withdraws in embarrassment.[144] Whether they stage a female patient and her doctor, a sick woman and her lover, or a lady and her entourage, anecdotes crystallize the tensions and power relationships at play in vaporous affections, caricaturing sexual and social difference.

While philosophers and moralists mocked their contemporaries' eagerness for medicine, some doctors, such as Julien Offray de la Mettrie,[145] Louis Henri Bourdelin, and Pierre Sue, focused instead on denouncing the practices of their colleagues, accusing them of borrowing the tactics of charlatans. In 1768, La Mettrie published *Ouvrage de Pénélope; ou, Machiavel en médecine* (Penelope's Labor; or, Machiavelli in Medicine) anonymously. The sarcastic narrator addresses himself to his peers with some advice on promoting their careers:

> May gravity unfold its sad cloud on your Physiognomy, & eclipse all its wit. Nothing marks so well the interest taken in the malady as suddenly changing one's expression & appearing stupid and stupefied. . . . Indeed what is more serious than illnesses, especially in the fair sex? It is true that if anything is tedious in Character, it is listening to a story that never finishes about vapors and imaginary pains. But do not imitate the laconic *Chirac*, who, to cut things

short, denied sick people their proper feelings. On the contrary, listen to them and appear to be greatly interested in them; prescribe remedies and vary them incessantly, as also your words, which must be cheerful and entertaining in tone. In return for which, you will have the department of the Vapors, formerly held by the hard-to-replace *Sylva*. Follow *Finot*; he had a repertoire of remedies to amuse the Sick for an entire year. Read the eyes of the Vaporous woman; perhaps they will inspire you more than Hippocrates.[146]

La Mettrie proceeds with his mockery in a series of portraits reminiscent of La Bruyère's character sketches. He effects a reversal: sciences usually associated with medicine, such as anatomy, chemistry, and physics, are presented as useless, and other fields—literature, philosophy, rhetoric, music—are henceforth to be the basis of the physician's success. Rather than attacking his colleagues' ignorance, La Mettrie is concerned to portray their unscrupulous motivations. A picture emerges of accelerating corruption, born from the imitation of colleagues either in a desire for fame or in the sheer need to imitate them in order to avoid becoming their victims. Later La Mettrie borrows a particularly acerbic passage from Alain-René Lesage's *Gil Blas* about physicians, presenting it as an anecdote and thus giving it an accent of truth.[147]

Art iatrique (Medical Art) appeared anonymously a few years after La Mettrie's satire, and was attributed to Louis Henri Bourdelin. In it, poetry is the format chosen to paint a mocking portrait of physicians, and in particular those devoting themselves to the vapors:

> We squabble over him, he could not suffice
> For the appointments everyone wants him to attend
> And sometimes called upon at court
> He casts a hundred beauties into despair . . .
> —Forgive me, Madame, I am abashed,
> One hundred times forgive me! You were waiting for me?
> I have just seen two dukes, a countess,
> A marshal, a certain duchess
> Whose discourse long and dull
> This morning made me perish before her eyes.
> I am exhausted. But you are charming,
> Sweet sick one, kind and interesting,
> And being near you gives me recompense.
> For all the tedium that overburdens me.
> Let us then remove the causes of your vapors.
> Let us feel your pulse. The rosy-fingered dawn
> Would envy the whiteness of this beautiful arm.
> And your mouth? Oh heavens! What freshness![148]

In the seventeenth century, men of letters deplored physicians' use of Latin, seen as the pretext to assert a sibylline knowledge that could not be contested. A century later, it is their eloquence that prompts sarcasm. This time, physicians' gallantry, transfiguring the fits of aristocrats into signs of sensibility, is identified as a way of currying favor. Humor provides a mirror or anamorphosis to reflect in an entertaining and critical form the figures of doctors and patients, aristocrats and men of letters.

Pierre Sue, master of surgery and librarian at the School of Health in Paris, was another detractor of the increased interest in the vapors. He regarded it as a fad and fits of the vapors as a form of emulation through which sufferers sought to prove their special sensibility. In his *Anecdotes historiques, littéraires et critiques sur la médecine, la chirurgie, & la pharmacie* (Historical, Literary, and Critical Anecdotes on Medicine, Surgery, & Pharmacy), published in 1785 and reprinted four years later, he enumerated a dozen cases of the vapors. Here are two of them:

> The vapors have only been known among us since the beginning of the last century. It is written in the Eighth Volume of the Historical Dictionary . . . that Abbé Ruccellai, Florentine Gentleman, brought the fashion to France. He was indeed of an unequaled refinement of morals; a mere nothing could injure him, sun, clear weather, heat, cold, and the least disturbance of the air would impair his constitution: he would only drink water, but a water that one had to fetch from far away & to select, so to speak, drop by drop. At his table, gilded basins were served loaded with essential oils and perfumes &c, and others containing gloves and fans for the guests. The Marechal d'Ancre was his main protector at court; Le Vasseur says, in his history of Louis XIII, that when the King believed he had an attack of the vapors, all the courtiers, even down to the bourgeois, believed themselves to be suffering an attack as well.[149]

> The Count de Bussy, entering the Petites-Maisons one day,[150] found in the yard a man who seemed to be less mad than the others: he asked him what was the madness of most of the people there. Well, Sir, it is quite a small thing: they pass us off as mad, because we are poor; were we people of quality, it would be said that we have the vapors, & they would let us run around the streets.[151]

This latter anecdote was mentioned in the *Dictionary* of Trévoux, a reference that Sue appropriates and expands.[152] In his vignettes, patients perform a complete staging of their bodies, reproducing a series of signs learned by observing their neighbors. Sue makes the vapors the result of ulterior motives: whereas melancholia is the price of excessive solitary meditation, the vapors are provoked by inactivity and the desire for social acknowledgment alone. He challenges the legitimacy of a pathology that is so fashionable and ostentatiously displayed, and denounces it as an affectation of refinement directed at laying claim to a particular social status. In fact, only those already in possession of that social status can affect to be stricken with the vapors without risking the label of madness. Far from being unlimited fits, accessible to all, the vapors function inside the rules of propriety. They appear here as a circulation of indulgence between doctor and patient and between patient and society. Fits are perceived as strategies, in which patient and doctor alternated in playing the interlocutor, the advocate, the defense, and the beneficiary.

Between Cheyne's writings and Sue's, the interpretation of symptoms has changed diametrically. Sue substitutes idleness, pusillanimity, and whim for the affluence, sensibility, and intelligence that Cheyne identified in the physiological disorder. By presenting the vapors as a recent historical phenomenon, Sue also differs from Raulin, who had considered them an ailment affecting humanity "from the birth of medicine."[153] Yet although the precise content and context of the expectations have changed, the act of setting limitations persists. In Sue's anecdotes, the vapors are a reserved code, to be handled as a form of etiquette. This idea emerges with even more force in the following anecdote:

> Sir, a vaporous man with a healthy constitution said to me one day, You have vapors just as I do; but you don't want to admit it, because you don't understand anything about it. Well! Know, Sir, that Hippocrates did not understand anything more than you, & yet did not abandon his belief in them. My Doctor has assured me that he said in his Works that there was something divine, Δειον τι, in this illness. Such were his words. He added, I will not forget them.
>
> I was careful not to disabuse my man: I would rather let him believe that I was besotted with the same illness as him: it is such a consolation for the poor wretches to find fellow creatures! And how would you prove to a blind person that it is daylight at midday? Don't we know that most people with the vapors, similar to the ones Molière portrayed in the Imaginary Invalid, become angry when one refuses to credit their illness? Don't we see some who, when told they seem to be well, become as furious as if they had been told they were scoundrels? It is of these that Montaigne said: *They let themselves be bled, and purged & medicinized for pains they feel only in their talk.*[154]

The anecdote achieves a complete reversal of the image of the pathology. The dialogue refers to the *Sacred Malady*, a text from the Hippocratic corpus written to demystify the convulsions associated with oracles by proving their physiological origin. In contrast, the vaporous gentleman enlists it as a further testimony to the disorder's mysterious and fascinating origin. Sue makes of the vapors a mystification whose first victim is the patient, and which the doctor aims only to perpetuate.

The irony permeating Sue's anecdotes indicates that the authority of the diagnosis is by now well established. Hysterical affections no longer need the support of physicians to be accepted; they are now claimed by the patients themselves. The diagnosis of the vapors, long seen as a way to combat the illusions of religious fanaticism, is now itself accused of engendering a new illusion, this time at the mercy of fashion. From the experience of possession to the experience of the vapors, the role of the imagination remains intact: it is only the appropriate occasion that has changed.

The writing of anecdotes is only the converse of the approach taken by preceding medical works: by presenting hysteric illness as a simulation or a mere indulgence, the anecdotes reverse the meaning given to it in medical works, addressing the same public but turning the same references to opposite effect. The anecdotes thus confirm the success of those predecessors. A pathology has been born, along with a rhetoric to describe it and a type of physician suited to curing it. All three figures are now sufficiently widespread to be recognized in the sparse brushstrokes of the anecdote.

*

By focusing on enunciation, this chapter has attempted to show how a conception of the body may be articulated in such a way as to imply a pathology, a perception of the readers, and a conception of medical activity. Different kinds of language can be distinguished, inflected differently for different objectives. For physicians, selecting the appropriate form of writing about the body helps them to propose new approaches. Genre and style become an opportunity to assert a fresh understanding of the body, one in which diagnosis is reached no longer primarily by inspecting excretions, but by examining lived experience and sensibility.

In these texts, the use of rhetorical turns of phrase is as important as diagnosis in the construction and diffusion of medical knowledge. In particular, crucial to these works are the ingeniously constructed slots that allow physicians to address their peers, their patients, and people from the court whom they may not know personally. This conception of patients as addressees was a singularity of the eighteenth century. It enabled the physician to present himself in the position of listener. That role would disappear in the following century, when physicians referred to their patients as case studies—but in the eighteenth century, doctors aimed to attract aristocratic patients by presenting the vapors and hysteric affections as something reserved to the nobility, strongly hinting that they could be successfully healed only by physicians who shared, at least partially, their patients' way of life. Each physician vied for selection on the basis of his peculiar understanding of his patients, an understanding that—more than the catalog of remedies he offered—became the key criterion of his ability to cure. The trust that a physician was able to inspire was at the heart of this competence: it was on trust that the future reciprocal availability of doctor and patient depended. He was the one who could heal because he was the one who understood, and he understood because of his shared sensibility.

In these treatises, the rhythm of knowledge diffusion does not follow methodological steps but the unfolding of trust, the developing conviction of a shared perspective, the adoption of a common language to furnish comprehension and comfort. The physician weaves a web of shared philosophical, literary, and biographical references capable of functioning beyond the difference of status. He emphasizes objectives common to his patients and himself: health, comfort of life, piety, peace of mind in freedom from the fear of relapse. Referring to Sappho, Virgil, Descartes, or Rousseau, these texts locate the ills in evocative settings: palaces, city walls, baths, the royal court. A sense of familiarity is staged.

These writings established a space to conceptualize the pathology, a space from which the vapors made sense to patients and doctors. Far from setting the protagonists in opposition, they demonstrated how each side participates in affirming the same understanding of the body. The conflicts, frustrations, and deceptions that divided them are minor compared to the models of comprehension that unite them. In a mirroring movement, physicians and patients showed each other an image of the pathology to which each could relate, for as Jan Goldstein writes: "It takes two to make a diagnosis—a patient with a set of symptoms, and a physician who gives them the label he deems appropriate—and hence it takes a multiplication of *both* parties to make an epidemiological trend."[155] The mutual acknowledgment of doctor and patient facilitated by this new writing changed the basis for approaching the body and pathology. It transformed the epistemological relationship with the body and with medicine and its actors.

Code, Truth, or Ruse? The Vapors in the Republic of Letters

A writing project is first only a way to participate in an enigma, rather than resolving it. . . . And this always puts into play more than a desire: a new or unforeseen anatomy that signification can gather together.

JEAN LOUIS SCHEFER[1]

Focusing on the same time period as the previous chapter, the "Enlightenment," this chapter proposes another trajectory. While chapter 3 asked how physicians assimilated literary genres in order to mold their therapeutic objectives and to attract an aristocratic audience for the pathology, the question addressed in this chapter is how the Republic of Letters itself adopted and challenged the theme of vapors. Some of the writers discussed here played with the ways of vaporous patients and their entourage; others hinted at new ideas of the body prompted by the vapors. In their works, vapors and hysteric affections were called on to name inner turmoil, distress, embarrassment, and physiological disorders; the terms allowed them to evoke modernity, the human species, women, or even literary creation. Using these notions, men and women of letters were participating in a move to fashion sensibility. They contemplated the vapors without systematically regarding them as an affection to be cured, and certainly not as the exclusive province of the physician—in fact, even some physicians, such as Julien Offray de la Mettrie and Antoine le Camus, chose to present themselves more as men of letters than as doctors, and did not view the vapors only as a pathology. They discussed them as an affliction of sensibility that offered insights into human nature.

All these writers worked with a whole range of literary genres outside the therapeutic field. As Dena Goodman points out, "The Enlightenment Republic of Letters was true to its name, for there was no hierarchy of genres, no queen of the arts, in a republic whose citizens engaged in all the variety of literary practices, stretching the limits of the literary itself."[2] It was through a very wide range of writing practices, then, that they investigated the vapors as a topic suited to people of fashion and as promising privileged access to the workings of the imagination.

The salons, crucial in the development of the Enlightenment, provided a space for the construction and circulation of such ideas.[3] The theme of vapors enriched discussions on aristocratic self-fashioning and sensibility, feminine energies, enthusiasm, and imagination. It provided opportunities to depict, suggest, or think through extreme experiences of the subject. Looking at writings by Paumerelle, Buffon, La Mettrie, Diderot, and Chassaignon, this chapter will address the construction of the fascination with the vapors, their unequaled dissemination and influence compared to other medical categories, and the distinctive frameworks used to think about them before the advent of the French nation.

Well-Timed Fits

In the sixteenth century, when physicians were using the term only to explain the origin of afflictions, it was men and women of letters who first coined "vapors" as the name of a pathology. Rabelais's *Tiers-Livre*, d'Aubigné's *Tragiques*, and Honoré d'Urfé's *Astrée* all adopt this term, along with "melancholia" and "hypochondria," to designate an affection of the mind. Boileau's *Épîtres*, many of Molière's plays, and works by Tristan l'Hermite, Corneille, Voiture, Scudery, Guez de Balzac, Pascal, Bossuet, and Retz continued to employ it in the century that followed, and the fact that they never explain the term suggests it was already in wide use. It is equally often to be found in the correspondence of Père Mersenne, Bussy Rabutin, and Madame de Sévigné.[4]

Seventeenth- and eighteenth-century French, English, and Italian writers of satires, tales, moral anecdotes, and comic theater drew on the notion of the vapors when detailing the morals of the day and the latest fashion at court. The term is used in a more or less ambiguous or ironic manner[5] in texts that evoke an aesthetic of existence typifying the century. Even before the spread of medical treatises addressing the aristocracy and presenting the pathology as an illness of sensibility, the "doctor of vapors" had already become a recognizable comic role in the theater; neither doctors nor patients nor the pathology was spared ridicule.

All these texts present the courtly world with irony, aiming primarily to entertain the reader but at times also claiming a moral dimension in accusations of affectation and manipulation. Others in the same line mock the self-important physician, promulgating an image of the vapors that is already on the verge of becoming a cliché.[6] They associate the vapors with women much more often than contemporary physicians did, and make the pathology an attribute of capricious, egoistical, and needy characters, playing with the imagining of the vapors as a laughable feminine temptation. If we believe the playwrights, such disorders deceive only those who want to be fooled—those who are too conceited to listen to others and be disabused. The vapors soon also become a dramatic device.[7] In *Le Barbier de Séville*, for example, a play marking the beginning of Beaumarchais's success in 1725, the vapors work to progress the plot. Rosine's tutor Bartholo tries to gain possession of a letter he believes she has received from a lover. Rosine refuses to comply, and discreetly swaps the incriminating letter with one from her cousin. To encourage Bartholo to take this letter, without seeming to have given in to his command, a swoon is the perfect solution.[8] Bartholo, sure of his success, seizes and reads the letter while Rosine affects to faint. Pretended vapors become an instrument to reverse the roles: they are a trap that snaps shut on the tutor at the very moment of his supposed triumph.

Literary works like these display the era's interest in the body and in sensibility, and ratify the role of pathology in the staging of the self. The vapors become increasingly popular elements of tales and novels, in which they characterize women and court society. They are also associated with propriety, seduction, and intrigues, as, for example, in Marie-Françoise Abeille Kéralio's tale *Visites* (The Visits),[9] the Duc de la Morlière's *Angola*,[10] Claude Joseph Dorat's *Malheurs de l'inconstance* (The Misfortunes of Inconstancy),[11] and Diderot's *Bijoux indiscrets* (The Indiscreet Jewels).[12]

In this tradition, Louis-Antoine Caraccioli presents the vapors as an art, and dedicates one of his works, *Le Livre à la mode* (The Fashionable Book), to vaporous people. His affected characters, defined by their whims, sculpt their own way of being and their reactions with a lightness of touch that is predictable in broad outline and yet constantly changing. The better to address a society of whims, the artifact itself becomes a whim, as Caraccioli's book changes color more rapidly than the seasons: a first edition is published in green, a second, shortly afterward, in pink.[13] For a society in love with appearances, the book must please by its presentation. More precisely, it must imitate women's rosy hue, and illustrate the delicate flush animating their faces during fits. The second edition of his book, dedicated to "Gentlemen and Ladies with the Vapors,"[14] begins: "Gentleman and ladies, the green color having lasted only for eight days, just like all fashions, I offer you the most beautiful of vermilions, just the color that glows on your magnificently & furiously illuminated faces & just as it must be according to the opinion of the most gifted Tronchins, to embellish & brighten up your vapors."[15] The tone is lighthearted and playful, not claiming authority but seeking to entertain. In the preface, the narrator notes: "Nowadays continuous doses of hypochondria are necessary, especially for our twenty-year-old Wise ones, &

warehouses of vapors for our seventeen-year-old Prudes. We are sick without knowing where we hurt; we suffer without being aware that we suffer; but we say it anyway; & the face adjusted to the discourse, we die every fifteen minutes while eating, yet still living."[16] The book praises women's capacity "to arrange an order of illnesses, as they would arrange ribbons."[17] Pathology becomes a cult of the body:

> In the morning she rings the bell for the vapors; since she has around her bed bells for all her needs and all illnesses. Domestics flock in, & immediately our Lady, voluptuously sick, restores her soul with a delicious broth; amber broth, broth more excellent than ambrosia itself. . . . The Physician enters, feels her pulse, which he always find a little quickened, checks the regularity of the heartbeat, banters a little, & tells some anecdote from the previous day. At last the time to get up arrives, that is to say midday: one lets oneself be led in the arms of two maids, who transport the idol to a magnificent *resting chair*. Here one yawns four to five times, closes one's eyes again as if wanting to doze; for the custom is that one always needs an hour to rest, having slept ten. One wakes up all over again, asks for a mirror, soon cries out that one is a perfect fright of ugliness; one changes one's adornment.[18]

Far from dramatically suspending everyday habits, in Caraccioli's account the vapors are firmly embedded in a whole aesthetic of life. Existence has become an art of passing time. The repetition of gestures, of sighs, offers a glimpse into an infinite modulation where the slightest detail is built up as a singularity. That leaves space neither for unforeseen events nor for empty time—reversing the common idea of "untimely" fits. The vapors are instead an art of distributing one's time. This form of expenditure, of loss, never fractures existence: it does not cast existence into question but inscribes in it a new linearity. In fact, Caraccioli praises the skill "of leaving illness its intervals. There is only one day of the week, & a few quarters of an hour every morning, that are reserved for it, & that's as much as one needs for fashionable indispositions. I much prefer this routine; because one knows what to expect, & does not have the pain of hearing a Lady shriek, in the middle of her game, that she is dying, & that everything is lost, as happens to so many Ladies who do not know how to apportion their time."[19] The rhythm of the fits obeys the order of a rite rather than that of a crisis. The continuous exaggeration of the signs transforms all emotion into pathology, functioning as a regimen of values. Hyperbolic emotion becomes the convention for all expression. The vapors signal a peak of luxury in the possibility of total abandonment of oneself; that carefully controlled loss functions as an emblem of aristocratic life.[20]

The Practice of Vapors

Fits inspired an enormous number of literary works that mocked them as a fashion and interrogated the truth of their symptoms. Particularly ingenious in investigating the vapors' power of manipulation is a fictional correspondence published in 1771 by C. J. de Bethmont de Paumerelle. Mapping the ruses made possible by what one might call vaporous practices, *La Philosophie des vapeurs; ou, Lettres raisonnées d'une jolie femme, sur l'usage des symptômes vaporeux* (The Philosophy of the Vapors; or, Reasoned Letters of a Pretty Woman, about the Use of Vaporous Symptoms) cannot be classified within either libertine tales, treatises on women, or moral works.[21] The author depicts the vapors with irony and presents them as an art whose mastery has ensured numerous successes. The republication of this work ten years later indicates that it did not go unnoticed, even if it has since fallen into oblivion. In the second edition, Paumerelle adds a direct reference to magnetism. He recommends the book as a remedy apt to comfort all those stricken with distress after the prohibition of Mesmer's therapeutic method: "Before despondency becomes incurable, we must counter these developments with the powerful voice of Philosophy; this is the only way to lift victims out of despair."[22]

La Philosophie des vapeurs consists of a series of letters from a marchioness to her newly married friend, a countess, educating her in the use of the vapors. The marquise presents these affections as indispensable tools for any aristocrat worthy of her status. While her writing begins under the guise of light banter, readers soon encounter a complete reversal of common ideas on the vapors. The text moves away from the usual terms of fragility, nervousness, and folly to present in their place a repertoire of strategies. The emphasis on the vapors' uses switches the perspective: whereas doctors undertake to explain the vapors' causes, Paumerelle's treatise invites readers to solve the enigma of the malady as a woman reveals what the symptoms can bring about.

With an affirmation worthy of Magritte's painting,[23] the angle from which the vapors are approached changes; the narrator avoids their referent, the symptoms, to focus solely on their representational force: "You understand vapors to be a real malady, but this only describes the thing: concentrate on the *word* and all obstacles will be cleared; you will be intimately convinced that this frightening expression can be pleasing to the ear."[24] By approaching the vapors as signs, the marquise extricates them from the physiological interpretations that had circulated during the previous two centuries. In considering the word "vapors" itself, she accentuates their use as an instrument.

The vapors take their place within a system of signs. The treatise affirms the benefit of the vapors in courtly society while obliterating the malady's pathological reality, and it does so in the form of a kind of epistolary training manual. The art of the vapors requires an apprenticeship in their signs, for a single symptom can have a variety of significations. Skilled practitioners acquire a full spectrum of effects to be expressed in a particular order: yawning, languor, weakness, dizziness, swooning, and convulsions. The vapors are handled as a language, to be understood as a hypothetical or conditional mode. More than the fit itself, what counts is the moment of its occurrence and its eventual progression. It is the context the vapors generate that takes on the crucial role: creating an event, they simultaneously provoke its interpretation. The simulacrum reigns, and the very question of a truth of feeling seems irrelevant. In a milieu where propriety is paramount, the power of sensibility lies in affected sighs and feigned restraint. The vapors are a vehicle with which to pursue one's interests while observing society's conventions. They permit periodic departures from propriety while an appearance of conformity is maintained. The body therefore becomes a material that the vaporous sculpt to their own ends. They develop the vapors as a bodily discipline that can be used to please others, solicit attention, indicate preference, or command respect.

Adopted as a sign of social class, the vapors reproduce and reinforce social discrimination. Just as, according to Rosen, in the case of madness "rules of social conduct mingled with axioms of morality and notions of social prudery to provide a standard by which all unseemly depths of emotions were concealed and extravagances of behavior avoided,"[25] in *La Philosophie des vapeurs* the vapors allow a subtle display of measured extravagances while simultaneously demonstrating a perfect mastery of the rules of bienséance. Courtly society claims the prestige of the vapors for itself, mocking members of the bourgeoisie who aspire to imitation.[26] As for the young lady she addresses, the marquise advises that the correct application must be learned from a master of the vapors, as from a music teacher or dance instructor. A teacher must be able to parse the infinite distinctions between types of crises, and specify the uses and interpretations they favor.[27] While the marquise insists that many lessons are necessary to achieve proper performance, she also claims that practitioners draw on innate traits:

> How happy one is, my dear, to have fibers as irritable as yours and mine! A little nothing hurts us: we do not have to be molded into irritability, like all those other women whose toughness nothing can soften. . . . You have been destined by nature to appear among the most celebrated of the vaporous women. It would be a crime not to make the most of your extreme refinement.[28]

According to the marquise, the truly vaporous are simply developing what their nature has destined for them. This "nature" is the physiological privilege of a particular social class: the aristocracy. An education in the vapors teaches one how to use possible reactions of the body and take advantage of those that can provide evidence of sensibility. By casting the vapors as an outgrowth of the indolence of the aristocracy,[29] Paumerelle is reiterating a connection that runs through the medical works of his contemporaries, who present luxury, sensibility, and indolence as causes of the vapors. But in his presentation, the truthfulness and plausibility of feelings lose relevance: it is knowing how to fashion their representation that counts.

Yet the vapors are an art of suspense as much as one of exhibition. Women must judiciously select the occasions and settings likely to produce the effects they desire. They are advised to develop a predilection for enclosed spaces where idleness reigns. Outside these protected settings, such as a dinner party or a night at the theater, the vapors might not function. Vaporous women appropriate the forums of theaters and pleasure gardens, diverting spectators' interest for the duration of the fits. Such scenes offer a spectacle within a spectacle, and accordingly demand a heightened level of art. Exhibiting restraint in general merely signals an absence of sensibility, but the truly skilled create suspense while executing movements at the edges of acceptable decorum. The marquise identifies Versailles as another training academy for such expressions of sensibility. "The vaporous epidemic of the court"[30] makes the palace a stage for frequent performances of the vapors, for which the king's antechamber, the space marked by power and exception, is most suited: an arena of dissimulation where the vapors may achieve their full scope.

What the marquise calls the "sanctuary of the vapors,"[31] the boudoir, is privileged as the space of greatest intimacy; its privacy lends an air of exceptionality.[32] Merely being invited in signals recognition and favor. As a metaphor for interiority, this setting conditions the emotions of those who enter: the boudoir furnishes the decor required for scenes of abandon, while proprieties are bent to the urgency of desire. For the marquise, it is the ultimate space of artifice, where power relationships are played out in their full force. Paumerelle's contemporaries shared this vision of the boudoir as an ideal setting for the instrumentalization of desire. Pierre-Ambroise-François Choderlos de Laclos exploits the ambiguity of the space in *Les Liaisons dangereuses* (Dangerous Liaisons),[33] a work published shortly after Paumerelle's treatise. Its account of how the Marquise de Merteuil traps Prévan in her boudoir demonstrates the room's effectiveness in games of strategy.[34] In *La Philosophie des vapeurs* the boudoir is a site of apprenticeship, where patient practice leads to self-control and discipline of the body.

In the marquise's letters, China, a land that inspired particular curiosity in eighteenth-century Europe, offers a counterpoint to Paris in its system of signs that create a system of recognition proper to the upper class. In China, the marquise writes, women undergo the pain demanded by custom, whereas in Paris refinement is just an act: one affects pain, and it is the quality of the exhibition that matters. Paris awaits a production to play on its stage. At first glance, the risk to the vaporous appears minimal, since their sacrifice is only affected, but it soon becomes clear how much is at stake in the manifestation of the vapors. In the course of the correspondence an insistence on complete and continuous control over one's expressions emerges. "You would not be so tactless as to be truly vaporous without rhyme or reason?" the marquise asks her friend; "It would be absurd of me to think you would. I have told you so many times that one must indulge in the vapors only as in a game!"[35] The decision on each performance is a gamble, and the threat of disgrace urges reflection before the next attack. The vaporous stake their entire reputation on a single gesture, and they have to consider carefully the audacity their plan requires.

The precise way in which the vaporous bring their bodies into play is crucially important, and here Paumerelle's conception is far from that proposed by Caraccioli in his books. Trickery can only succeed through an absence of visible affectation; the ruse serves the vaporous well only as long as their expressions are perceived as innocent and uncontrived. To embody emotions without arousing suspicion, performers cannot remain unmoved but must indeed experience worry and weakness during their crises. The vaporous thus have to provoke their episodes rather than mime them, accepting an element of abandon and transport. A further element of the performance is an apparent concern to conceal their malaise. As a result, their performances involve dual demands: vulnerability should be expressed in the body as though against the sufferer's will, overwhelming her in such a way that onlookers suspect deeper troubles. As the pains arrive, she should appear unable to fend them off despite a brave resolve to conceal the symptoms.

The art of vapors is founded on the utility of the perception of sensibility that developed during the eighteenth century. In Paumerelle's treatise, the vaporous harness the coexistence of two contradictory schemas—while decorum mandates restraint, sensibility demands signs of the soul's disquiet. Playing on this dynamic, the vaporous benefit from propriety in two ways: it presents the vapors as constituting a rupture with established order, thus assuring surprise, and it shields the vaporous from inquiry into the causes of their crises.

While the marquise does not cast the vapors as exclusive to women, she

does present them as an eminently feminine form. She writes in her first letter: "You are a woman—that is to say, you are curious about the marvelous,
avid for singularity, crazy about the bizarre. What happy dispositions!"[36] Paumerelle establishes women's talent for vapors with a series of such references
throughout the text. When the vapors are enlisted as a weapon in negotiations, it is to outmaneuver power that is exerted directly through strength
and is often masculine.

Paumerelle juxtaposes the risks in public representations of oneself with
risks taken in times of war, once the most important marks of prestige among
aristocratic men. Though the nobility has been disarmed by Louis XIV, it can
now use the appearance of being disarmed to pursue fresh maneuvers. Nobles no longer risk their lives in tournaments or wars, but they can risk them
figuratively in emotions.[37] Prestige no longer lies in the bearing of arms as
emblems of bravery; it is created in the loss of defenses, in attacks of weakness
unique to the men and women of the court. The vapors are the new emblems.
Nobles must, of course, control their emotions, reconquering sensations to
conform to rules of decorum, but when sensations make themselves seen
they function as avowals of an interior combat. The combat of aristocrats is
largely a struggle between signs, signs concealed or shown with detachment,
or signs displayed as if one did not want them to be seen. In the privileged
circle of the court, vapors emerge as an ersatz rerouting of the warrior spirit.
The risk of death is replaced by the risk of losing one's reputation; honorable
wounds received on the battlefield are replaced by the wounds of a sensitive
heart that is trapped in Versailles. As the lifestyle of aristocrats and conceptions of their role are transformed, so is pathology. As the political becomes
spectacle, so does pathology. And if the wounds of battle were instrumentalized as signs of bravery, the vapors now serve as signs of sensibility and
refinement. In Paumerelle's work, the aristocrat's body is vested with signs to
the point of pathology. The body must be controlled down to the last sigh,
forever working toward mastery of the signals it sends. And if an aristocrat
fails in this control, she risks sickness and servitude to her body.

Despite the suggested display of emotions, the letters relate episodes in a
reasoned, measured, and calculated voice. Contrasts customarily drawn between men and women and corresponding separations between action and
passivity, reason and emotion,[38] give way to an opposition between the vaporous and the nonvaporous. This new division classes the manipulators and
the manipulated to the advantage of the vaporous. Rather than appearing
as pusillanimous, passive characters, the vaporous now belong to the ranks
of the active, employing operations of reason to shape the manifestations of
their sensibility. Women gain in this new partition. They redirect to their

own advantage the schema society has imposed, and appear to emerge triumphant on all counts: sensitive and reasoned, experienced and calculating. The art of vapors is threefold—it is the art of confounding the audience that judges vapors and proposes their causes; the art of using vapors to create vapors; and lastly, the obscuring of those vapors that must be hidden. Vapors are perfected in advance, the better to seize opportunities that may arise unexpectedly. Woman here is a master of discernment and tactics. She gambles with stakes known to her alone. It is her reputation as belonging to the weaker sex that permits her to become such a grand strategist: she knows how to maintain conventions while simultaneously playing them in her own favor. Through her understanding of the conditions of this rigid structure, everything becomes a game in which she deals the cards. Living as a woman is a motivation to be vaporous without becoming trapped in the role. Being a woman becomes a practice of being a woman.[39] This entails the incarnation of a role more than the appropriation of an identity. As much as the actual manifestation of the vapors, the performance of being a woman becomes necessary to the execution of one's project.

The ruses at play in La Philosophie des vapeurs place this epistolary text in a tradition of writings inaugurated by Machiavelli and continued in Baltasar Gracián's Oráculo manual y arte de prudencia (1647) with some adaptations to court society. But while Gracián demonstrates how to succeed despite society's risks and constraints, Paumerelle shows how one can play on society's paradoxes and restrictions. While Gracián recommends limiting dangers, Paumerelle proposes a gamble. The vaporous make a game of lifestyle, weighing the risk of their performances.

Reflecting a division common in seventeenth-century thought, Gracián insists on acting within reason rather than being carried away by passion, counseling readers to work on the self and keep their intentions shielded from view.[40] Paumerelle's eighteenth-century work refuses to negate passions and instead mobilizes their power. In his treatise, passion can displace the propriety and restraint within which aristocrats are normally expected to remain; moments of passion are then instrumentalized by reason, preventing its expressions from becoming potentially harmful. Thus, the vaporous advance not by making passion subservient to reason, but by playing passion and reason off against each other. They exploit passion for its expressive force while maintaining a role for reason in the selection of the occasions and objectives for crises. Paumerelle's "manual" demonstrates how a proper use of passion can obtain indulgences that would have been denied under customary codes of conduct.

The overwhelming majority of Paumerelle's audience would have been

salonnières,[41] aristocrats, and/or the vaporous. The author seems to alternate between an appreciation of strategy and a ridiculing of pretense, and
his layering of irony ensures that readers can navigate the text while feeling protected from direct attack. Thus, doctors are a frequent target for the
marquise. The interpretation of the vapors in terms of "toughening,"[42] for
example, is a thinly veiled reference to the work of the famed doctor Pierre
Pomme, who posited a theory of "toughening of the nerves" in his *Traité
des affections vaporeuses des deux sexes, ou maladies nerveuses, vulgairement
appelées maux de nerfs* (Treatise on Vaporous Affections of the Two Sexes;
or, Nervous Illnesses Vulgarly Called Nervous Disorders) (1760).[43] Many of
Paumerelle's readers would have recognized that the terms had been lifted
from Pomme's treatise, a work addressed to aristocrats as well as doctors and
circulated so widely that it was published in six editions.

But Paumerelle goes beyond the caricature of medical diagnoses or therapies. In her ironic barbs, the marquise suggests that doctors unwittingly aid
the vaporous in their enterprise and become ornaments and instruments of
vaporous strategy. Blinded by the ruse, physicians go so far as to condemn the
entire nation to this infirmity, and the vaporous seize on their pronouncements to grant legitimacy to their crises. Medical theories suggesting that the
vapors originate in the emotions and imagination also furnish a favorable
reading of symptoms. The marquise even creates a satirically tinged neolo-
gism, *faiblomanie*, which she puts into doctors' mouths. The word epitomizes
their willingness to view all conditions in terms of the vapors, even when the
diagnosis is unwarranted by the symptoms. The tone of gravitas in doctors'
pronouncements is humorously twisted to cast symptoms in an advantageous light. Paleness and swooning are explained by the attribution of sensibility, and understood in terms of sympathy between the body and soul,
again legitimating the behavior of the vaporous.

The treatise's view on the relationship between the vapors and the imagination also satirizes the projections of contemporary medical literature.[44] As
indicated in chapter 2, the medical profession often considered the malady
to be caused by an excessive imagination and frustrated desire, but for Paumerelle, the exercise of the imagination does not translate into pathology.
Rather than submitting to incurable exaltation, the vaporous befuddle others
with their shrewdness. They channel imagination's creative force into crafting
their plots. Here again, while much eighteenth-century literature depicts victims of vaporous illness as prisoners of their thoughts, Paumerelle presents
them as in control of their reflections and actions. Women, so often viewed as
overwhelmed by the objects of their caprice, here appear unconcerned with

the material benefits obtained through their symptoms and more interested in the chances of success in the social game. Successful vapors are not only a means to an end, but an end in themselves: success lies in a virtuoso performance at least as much as in winning a man's attention or gifts of jewelry.

In measuring and manipulating the limits of their bodies, the vaporous obtain a depth of knowledge that medicine cannot claim, and look upon doctors with disdain. Far from requiring advice from scientific treatises addressed to aristocrats, such as those by Bernard Mandeville, Pierre Hunauld, and George Cheyne, women emerge in Paumerelle's work as being in a position to instruct doctors. However, this knowledge is to be fostered in a specific format: letters aimed at specific readers, who read them in a specific context. In the marquise's letters it seems that in order to explain the vapors properly, to explore the philosophy of the vaporous, it is paramount that they are discussed between women, in privacy and confidence. Constructing his work as correspondence, Paumerelle thus assumes not only a feminine voice, but also the feminine genre par excellence. A letter differs from a treatise in being addressed to one individual, conveying the object of a moment and even caprice. Correspondence also creates an interlocutor without describing her fully, without introducing her to the reader. This invisible woman emerges in the letters, which act as a mirror of her personality for the readers. She appears and flees through advice, remarks, criticism, memories evoked, and the second addressee, the reader, is taken in a movement of identification and rejection, as if donning a mask to better understand the scope of the message and the interlocutor it addresses.

While the marquise foils her peers using the vapors, the treatise's epistolary construction suggests that she herself has been thwarted by the publication of confidences that were intended for a sole addressee, in the belief that they were protected from outside view.[45] The work's format as personal correspondence lends an air of secrecy and scandal. It positions the reader as a kind of voyeur, and orients the text to be read within a frame of disparagement. At the same time, the architecture of the text excuses Paumerelle from speaking in his own name. Readers are left to reach their own conclusions and to determine whether their reading is an exercise in mockery or fascination. Given the scarcity of biographical information on Paumerelle and the dissimilarity of his other surviving works, his own stance is impossible to ascertain. The author's elusiveness contributes to the impact of *La Philosophie des vapeurs*. His treatise offers revelations while shielding him from scrutiny. It is as though Paumerelle were heeding Gracián's advice, advancing his agenda while remaining shrouded in obscurity.

The Force of the Imagination

Because the disorders associated with the vapors were so varied and so suggestive, they could be considered outside of any medical agenda as a characteristic of human nature. Beyond the theater and the literary letter, therefore, the glossary of vapors also inspired texts belonging to the genre of the essay, from works as important as Buffon's *Natural History* to anonymous short works now long forgotten. Despite the diversity of their aims and of their positions in the later canonization of the Enlightenment, such essays join in seeking to understand human nature through physiological disorders—of which the vapors became the fashionable choice for such purposes. In works of this kind, vaporous fits are not a matter of therapeutic concern but exemplify the intense and reciprocal relationship between body and soul. Among them, the 1749 *Discours sur la nature des animaux* (Discourse on the Nature of Animals), published in the fourth volume of George-Louis de Buffon's *Histoire naturelle* (Natural History), holds a singular position. The naturalist and polymath took a very early interest in the vapors, and this text makes them the object of an analysis of the human species.

In his pioneering *The Life Sciences in Eighteenth-Century French Thought*, Jacques Roger noted that in Buffon's work "details of an animal's form or habits are closely accompanied by reflections on general problems of science, and often of philosophy, politics, and morality besides. From these dispersed morsels, it would not be hard to reassemble Buffon the *philosophe*, giving that word all the meanings it had in his day."[46] Roger points out that the fourth volume plays a special role in the architecture of Buffon's natural history, being devoted to the superiority of man over other species. But regarding the soul as what constitutes the inalienable character of man for Buffon, Roger leaves aside a dimension to which Buffon gives just as much importance in his specification of the human—the vapors:

> The interior man is double, he is composed of two principles different in their nature, and contrary in their action. The soul, that principle of all knowledge, is perpetually opposed by another purely material principle. . . . The existence of these two principles is easily discovered [if one turns to oneself]. In life there are moments, nay, hours and days, in which we may not only determine of the certainty of their existence, but also of the contrariety of their action. I allude to those periods of languor, indolence, or disgust, in which we are incapable of any determination, when we wish one thing and do another; I mean that state or distemper called *vapours*; a state to which idle persons are so peculiarly subject. If in this situation we observe ourselves, we shall appear as divided into two distinct beings, of which the first, or the rational faculty,

blames every thing done by the second, but has not strength sufficient effectually to subdue it; the second, on the contrary, being formed of all the illusions of sense and imagination, constrains, and often overwhelms the first, and makes us either act contrary to our judgment, or remain inactive, though disposed to action by our will.[47]

What distinguishes mankind in this vision is not only the faculty of reason, as proposed by Aristotle, but a state of vapors that Buffon describes without defining it, characterized by a movement with the current of the tide. The passage is set alongside deliberations on matters including the organic senses, health, and dreams, in a section entitled "Homo Duplex." According to Buffon, the vapors convert wills into vague desires and jeopardize each decision, casting people into wavering and uncertainty. He locates their cause in the illusion of the senses and of imagination itself, as a force strong enough to shift priorities and divert projects—a deleterious force that attacks not life but activity. Buffon finds proof of this in the frequency of vapors among the idle. The paradoxical character of the fits gives rise to a reversal, an inversion of the subject, as man becomes an object in the exercise of his imagination: "makes us act," "forces us." The "us" makes this a common experience that binds the author to his readers. The vapors are the frequent manifestation of a force that mankind cannot govern and yet that uniquely characterizes it. Even before the publication of Hunauld's dialogue-form treatise, Buffon depicts vapors as a form of sensibility. He stresses that the vapors are crucially marked by loss of control, incapacity to decide, idleness, and abandonment. Here such passivity is not being in love, neither is it ecstasy or creativity: it is an ordinary state. The vapors are read without any flattering accents, and yet are accepted with no moral judgment as being part of mankind's nature.

Throughout his text, Buffon offers an interpretation of the vapors antithetical to that given by many writers of plays and tales. Whereas for them the vapors were a sign to be used in a logic of simulation, for Buffon the vapors are an experience stimulating reflection: they are "easily discovered" by "turning to oneself." His text makes the vapors a state that one may try to fight, but the existence of which is impossible to deny. Even if its symptoms were to be cured, that would not eradicate it as a principle of the human species. Buffon here replaces a therapeutic approach with a philosophical one, a pathological characterization of the vapors with an ontological one. Interestingly, this proved inspiring for doctors, and Noël Retz quotes Buffon at length in his *Des Maladies de la peau et de celles de l'esprit* (Of the Illnesses of the Skin and of the Mind), published in 1785, 1786, and 1790. When Retz set out his objective to cure the vapors, among other illnesses, he brought the specificity and importance of a vaporous pathology into medical debate. His citation of Buffon also indicates the wide circulation of this text through different fields of knowledge in the eighteenth century.

If Buffon's discussion of the vapors is an opportunity to portray the power of mankind's interiority, Denis Diderot assembles examples to illustrate the force and plenitude attained by a hysterical woman, making "hystericism" a female trait. What for Buffon was a source of drift and languor, for Diderot is a new source of intensity. In his 1772 essay "Sur les Femmes" (On Women), mentioned in chapter 2 for its use of the animal image, Diderot's examples include the Convulsionaries, whose bodies seem to fight for visibility and exist only in tearing themselves apart, and Saint Teresa of Avila, whose state of abandon is presented as one of sacrifice and delight. Well before Charcot, Diderot offers a reading of religious history in terms of the vapors: "Her head still speaks the language of the senses, though the senses themselves are mute. Nothing is more closely related to hysteria than ecstasy, visions, prophecy, revelations, and fiery poetry."[48] Hystericism appears as woman's absolute presence with what happens to her, and this presence is violence, forcing her to abandon herself to events and to live and express without restraint. The sensibility with which a woman lives, thinks, and acts finds a worthy role in prophesying, a role where imagining becomes reality, the verb becomes body, simulation becomes truth, and the extraordinary becomes present. Woman is carried away by the very force of her physiology, torn away from herself in a constant motion of transformation. Hystericism here is what gives woman an unlimited sensibility in both ecstasy and fury: it is the key to her expressivity. "No man ever sat at Delphi on the Sacred Tripod. The Pythian rôle is suited to women only. Only a woman's brain can be sufficiently exalted seriously to sense the approach of a god, to wave, to tear her hair, to foam at the mouth, to cry 'I feel him, I feel him, here he is, the God.'"[49] For Diderot, this disorder connects woman to an imagining that she embodies, displaces, mystifies, and transcends. Hystericism realizes contradictory features in woman, uniting ferocity and sanctity and making the least of her expressions extreme and paradoxical. But the power of Diderot's text lies in his refusal to consider women outside of the civilization they belong to. Intuition and instinct appear at once stimulated and policed by society, as a resource elaborated in response to an unequal power relationship.

Paradoxically, hystericism is what makes women both inaccessible and easy to master. Alluding to Silva, Diderot writes: "But one word is sufficient to destroy this fiery imagination, this spirit one would have thought indomitable. A doctor said to the women of Bordeaux, who were tormented by terrible vapors, that they were threatened with the falling sickness. In an instant they were cured."[50] The philosopher takes the famous example of hysterics silenced by the presence of soldiers threatening them, and notes that frightening the imagination of woman will suffice for her upset sensibility to fold into itself and cease all hysteric manifestations.[51] What is commonly seen as an effect of woman's "temperament" is identified by Diderot as the result of the consideration — or lack of consideration — shown to her. She is caught in a web of interpretations of her gestures, each of which may be singled out as a sign to define her. A woman puts her reputation into play; she risks mockery and contempt at any time. Her alleged vulnerability is thus represented as an effect of the social order, the disruption of her sensibility as a reaction to the power relationships that constrain her. Diderot presents woman as being caught between contradictory movements not because of instability or indecisiveness, but because of the contradictory demands that society imposes.

Yet Diderot does not make woman a passive victim of a blind society. He sketches the force of attraction she exerts on those around her. He describes her as someone who is never identical to herself and remains difficult to predict, finding in her predisposition to hystericism the force to surpass herself, body and soul. In the pathology he sees a clue to the tensions that articulate experiences in each phase of her life, a life of desire in youth and a life of devotion when older.[52] "In her delirium she goes back into her past and plunges forward into the future, both states being all the while present to her":[53] the womb is the source of an ebb and flow and originates woman's relationship to time in her body. She shatters all continuity, undergoing tensions, attractions, expectations, and impatience. A woman in her hysteric momentum combines the intensity of her emotion and the radicality of her stand. Her aptitude for transformation attracts her in a constant movement of self-redefinition. Her approach to the world is through intensity and desire. She conceals herself as much as reveals herself in her interactions. She seems to seek, again and again, to decouple her body from the code, to create from her body a new code. This body waits for definition; it depends on the gaze of others, on their consideration and their verdict. It cannot act without manipulating large numbers of signs. The body exposes itself to the gaze, showing itself too much or too little. It gathers gazes that soon enclose it. In Diderot's text, being a woman, far from the submission it usually appears to involve, emerges as an extreme.

Diderot thus presents a vision in which the vapors provoke an excess. The body struggles, and invents new postures for itself. Hystericism seems to be, above all, the power to be amazed by oneself, to catch oneself off guard, to push the image of oneself to the limit of all agreed representation. Each example indicates a fascination with the body itself, with its resistance, and with its vulnerability, all instruments of a possible freedom.[54] A woman is in a constant movement of becoming; she is a creative being on whom hystericism bestows genius. Diderot concludes his text with these words: "When women have genius, I think their brand is more original than our own."[55] Unlike the "us" in Buffon's text, gathering together all those who do not work, the "us" in Diderot's text, published twenty-three years later, gathers men. A similar transformation will take place in medical discourse in the subsequent years.[56] From Buffon to Diderot, the vapors, whether they are a synonym of lack or of surpassing oneself, reconfigure the priorities of the individual.

Genius was a privileged theme in the second half of the eighteenth century, and women and men of letters frequently associated it with convulsive disorders. For physicians immersed in philosophy, talking about vaporous fits or hystericism also offered access to the effects of sensibility. They presented it as a means to stimulate creativity through the body, mystifying the reader with the power of vapors. In these essays, hysteric affections became a way of awakening the body's faculties, stretching its limits, and reconceptualizing the relationship between body and soul. In an era marked by the desire to confirm the relation between the physiological and the moral, the intensity, regularity, and instantaneity of their reciprocal effects could be exemplified by the mention of fits. The circulation of the term "vapors" in essays of this kind served both to illustrate the ascendancy of the soul and to cast that ascendancy into doubt: for such writers, with their materialist views, pathology demonstrates the role of the body in the soul's affliction. Without attempting here to analyze such doctrines in detail, some specific usages of the vapors may shed light on the strategies used by women and men of letters to question the hierarchy of reason over the senses and the emotions. Julien Offray de la Mettrie, for example, writes in his 1748 work, *L'homme Machine* (Man a Machine):[57]

> In disease the soul is sometimes hidden, showing no sign of life; sometimes it is so inflamed by fury that it seems to be doubled; sometimes, imbecility vanishes and the convalescence of an idiot produces a wise man. Sometimes, again, the greatest genius becomes imbecile and loses the sense of self.
>
> . . . What was needed to change the bravery of Caius Julius, Seneca, or Petronius into cowardice or faintheartedness? Merely an obstruction in the spleen, in the liver, an impediment in the portal vein. Why? Because the imagination is obstructed along with the viscera, and this gives rise to all the singular phenomena of hysteric and hypochondric affection.[58]

In this statement the body and the soul are absolutely vulnerable, touching the individual in his identity. Faculties are nothing other than the equilibrium or the temporary disequilibrium between organs, and the subject depends on his physiology in genius as in foolishness. What constitutes his greatness also constitutes his vulnerability. In these few lines, La Mettrie describes a two-step process: from the organs to a faculty, from a faculty to a pathology. The pathology redefines the order of faculties, increasing or reducing their potential in a new order of which the individual is the passive recipient. The faculty of imagination is associated with the body's health, while the obstruction of the vessels or organs alters its function. The stakes of this new vision of the body are huge: the health of the body becomes the criterion of the intellectual capacities. In this reasoning, too, it is medicine that can provide the competence necessary for the development of man. La Mettrie presents the vapors as a disorder that has an epistemological value per se, a disorder that tells the observer about human nature as a whole and not only about illness. Far from opposing health to illness, such writing is based on the assumption that there are many possible states of the body, each displaying a different set of faculties.

For La Mettrie, the physician holds the authority to talk about human

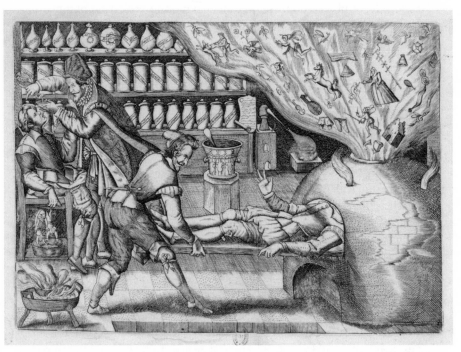

FIGURE 11. Mattheus Greuter, "Le médecin guarissant Phantasie purgeant aussi par drogues la Folie," 1620 (Collection Michel Hennin, Estampes relatives à l'Histoire de France, vol. 21). Image courtesy of the Bibliothèque Nationale de France.

nature over and above the theologian and the philosopher because he can ground theory and systems in facts. If a century earlier vaporous pathology had been the explanation used to extricate the possessed woman and the witch from the hands of the exorcist, it now inspired new discourses, providing arguments to men of letters with a political and philosophical agenda. Mentioning vaporous fits was now a device for radical thinkers to articulate doubt and criticism. The acknowledgment of physiological disorders went much further than the mere diagnosis of a few nuns in a convent; it prompted questions regarding the ascendancy of the soul and the mind's authority over the body. While such writers could not openly contest moral and religious truths such as the existence, immortality, and spirituality of the soul, they could talk about hysteric fits as observers, grounding their statements in experience (leaning on the "staff" of experience, as La Mettrie puts it[59]), and thus redirect attention to the physical. The body, its power, and the impact of physiological disorders on emotions and thoughts undermined all attempts to hierarchize the faculties. Antoine le Camus, physician at the Faculty of Medicine in Paris, further developed this approach in a 1753 essay, ambitiously titled *Médecine de l'esprit* (Medicine of the Mind), that proposes to move from observation to experimentation. The project, as outlined in the preface, is even more ambitious than the title: the object of medicine is not merely to cure the body or remedy the mind's failings, but to perfect the mind. According to Le Camus, the point is to "plot a method with which one could eradicate the flaws that appear to be in the soul, in the same way that physicians heal a fluxion of the chest, a dysentery, a malignant fever, & all the other illnesses that attack or that only seem to attack the body."[60] The 1779 reissue of this work attests to the success enjoyed by such conceptions. Even though the physician professes to be far removed from materialism, and adds a paragraph on the soul's immortality, physiology is set as the origin of the soul's movements, leaving no space for metaphysics. Le Camus presents himself as the herald of a new science. He proposes to extend medicine's role to all healthy people, configuring the use of remedies toward improvement as opposed to healing. According to him, this kind of treatment could facilitate access to both science and the arts—by stimulating the imagination's creative force.

Le Camus makes the hysteric and hypochondriac vapors a clue to the mind's potentialities. He appropriates the idea of a vicious cycle between emotions and pathology, and presents the relation between body and mind in its duality: love and fear disrupt the balance of the body,[61] and the vulnerability of the body accentuates the manner in which the soul is affected. It is only a single step from noting the relationship between body and mind to articulating the imagination's power and the ways in which it can be stimulated. Endorsing the theses of La Mettrie, who did not hesitate to predict the transformation of people considered "stupid,"[62] Le Camus argues that changing the organs' economy changes the subject's dispositions and that, once newly educated, he will appear as a new individual. The physician could thus combat inequality of the talents and mind by combating physiological inequalities.[63] But Le Camus is chiefly interested in developing the mind already gifted with a good constitution; even for such minds, medicine must intervene before education. To justify his argument, he references ancient theories on enthusiasm and inspiration such as the experience of plenitude with a divine quality:

> Enthusiasm is nothing else than the moment in which all the soul's resources are set in play, in which one's knowledge of the subject is even greater than the subject itself, in which the conception of the thing, being lively, bright, & pure, necessarily carries the demonstration away with it, in which finally the subject, considered in all its elevation, in all its scope, in all its beauty, strikes one so evidently that, reason being silent, one surrenders to the transport which stirs, one goes through intervals & reflects on the others with the same force as the beams of light which have struck one.[64]

Since the seventeenth century, the term "enthusiasm" had increasingly acquired a pejorative connotation, resulting from the political stakes within which it was mobilized. During the time of the French Prophets in England[65] and the Convulsionaries crisis in France, for example, enthusiasm was seen as a form of wandering of the mind provoked by noxious vapors. To break with this vision, Le Camus draws on the tradition of Plato's *Phaedrus* and *Ion*.[66] *Ion* portrays poetic creation as a moment of furor in which the poet is deprived of reason to be fully directed by the gods, who are the masters of his inspiration. The poetic work is conceived in the extrication from the self, at the moment when the artist no longer belongs to himself. He is then fully experiencing enthusiasm. Similarly, Le Camus constructs inspiration as the moment when reason's powers are suspended, a moment of "rapture, a delirium, a furor"[67] provoked by the enchanting object. He also refers to the association of genius and melancholy in Aristotle, and takes the example of Torquato Tasso, in whom, like some other contemporary physicians publishing on the vapors, he sees the moment of intense creation and that of aberration coincide. By combining these dimensions from the starting point of a physiological analysis, the physician grants himself the authority to present a philosophical hypothesis. Le Camus does not connect the state of enthusiasm with that of femininity of the body, at the level of the organs or the fibers;

instead he situates its origin in the fibers' vibratility, and sees in the brain's motions a movement of action and reaction, ebb and flow, which agitates the fibers. In his work, the "imagination" itself is never really explained: it seems to function as a faculty, but its relation to memory, perception, and sensation remains vague. Rather, it seems to provide a name for what cannot otherwise be explained. Urgency, vulnerability, reaction, violence, agitation, appear as the qualities conducive to the jolts of the mind that are necessary for creation. Le Camus endows the manipulation of vapors with a hyperbolic force that enormously heightens the power of imagination and catalyzes will, action, and sensibility in a single move.

Writing in the late 1770s, Jean-Marie Chassaignon turned such discussions of genius and imagination to good account, binding together aesthetics and mystification. His *Cataractes de l'imagination, déluge de la scribomanie, vomissement littéraire, hémorrhagie littéraire par Epiménide l'Inspiré* (Cataracts of the Imagination, Deluge of Scribomania, Literary Vomiting, Literary Hemorrhaging by Epimenides the Inspired),[68] published in four volumes in 1779, looks like a manifesto. The preface, a hundred pages long, lays the foundation for what could be called an aesthetic of convulsion. Far more than his predecessors, Chassaignon invests power in the medical, political, and philosophical heritage:

> If despite all my confessions & protests, one still insists on proving to me that my work is extravagant, and that it should never have seen the light of day, I will cry out: cold minds, apathetic geometers, throwing a look of rage upon my relentless persecutors, that have never felt your being tremble, your faculties wince, ferment, that have never endured the shattering of an impetuous soul, caving under the weight of its ideas, tormented by an excessive energy, & by a need for explosion . . . ah! If you knew the painful convulsions of a confined enthusiasm, you would have indulgence, you would pity me, & you would applaud the excesses of my imagination.[69]

Writing makes itself interruption. It is the constant refusal to be weighed down. Writing is breath, pulsation, discharge, deflagration, shock. Chassaignon does everything to convey its physicality. The reader imagines the quill that deletes, tears the paper apart, and splinters. The rejection of authority is played out afresh with each stroke of the pen. The effort disappears in jubilation, only to make itself felt again with renewed violence. Chassaignon likes the tumult: "I never write for more than an hour at a time; often I stop after fifteen minutes, a tensing of the nerves, a dizziness in vision, a palpitation of the heart, and a bubbling of the brain prevent me from keeping my grip on the pen, looking at the paper, or even combining an idea."[70] There is no anteriority of thought in relation to writing. The author is a suffering body, of which crispation, dazzling, palpitation, take possession in turn. The body finds itself in a struggle to give voice to the unknown and unpredictable. In the lineage of Le Camus, the experience of inspiration is a physical experience. The muse is the shaking of the fibers.

This may recall Jean-Jacques Rousseau, for whom the convulsive body is also the mark of the work's inspiration. More than once in *Les Confessions* (The Confessions), and even more precisely in the eighth walk of *Les Rêveries du promeneur solitaire* (Reveries of a Solitary Walker), Rousseau traces the "suffocating palpitations"[71] that take possession of his body when his ideas invade him: "My eyes flash, my face is aflame, my limbs tremble."[72] The reader witnesses the emergence of a phenomenology of writing.[73] Here, the description of a hysterical pathology is a way to delineate the experience of thought and of writing.[74] Risking vulnerability seems to be the only way to create. With these few lines, Rousseau postulates that creativity is letting through a force that weakens the body and reveals its own power. Thoughts shatter the body, and the body rushes out thoughts in a reciprocal movement. In the effort to think, the body becomes tense to the point of tearing itself apart.

Sharing this notion of imagination, Chassaignon projects medical and to a certain extent political imaginings in his approach to inspiration. He instrumentalizes medical theories, turning the overheating of the imagination into a source of the exaltation of ideas and displacing the authority of inspiration from a divine source into his own suffering body. If the pathology was used in this period to castigate the speech of inspired people, with Chassaignon it becomes the origin of a power. The work is his testimony to an experience of rending. Claiming a physiological origin for literary creation allows him to make the experience of being torn a source of knowledge. The physiological reaction is an inscription of truth and the source of a conviction that no one can weaken. To account for this experience, writing sacrifices structure and unity, and is henceforth devoted to becoming a surface for impression, a vehicle of feeling. Writing becomes event.

Chassaignon's recourse to a medical vocabulary has three main roles. Firstly, he averts readers' suspicions about his mental health by establishing his mastery of the medical discourses of the time. He also calls on irony to distinguish himself from the image of the hypochondriac or melancholic writer. Speaking in the voice of a friend and addressing himself, he writes: "One would believe it possible to distinguish, across the darkened vapors of your brain, & the sadness of your desiccated appearance, the black symptoms of a dangerous fanaticism. They would sequester you between four walls for the state's peace & your fellow citizens' security. Make up your mind! You have common sense; a poetics of hypochondria does not dishonor, & and alarms only slightly."[75] Not satisfied with this vision of hypochondria, Chassaignon risks even more: excess, quickly forgotten when ascribed to genius, is otherwise seen as a sign of madness. For society is mobilized by a need to explain disorders, and in the absence of genius, a justification in terms of poetic creation is replaced by the accusation of political fanaticism. Hypochondria is, then, the lesser of two evils.

Secondly, the reference to a medical imaginary safeguards writing from idealization. The pathological is called on to avert the illusion of a metaphysical force—the classical vision of enthusiasm—and ground the practice of writing in physiology instead. With his work, the writer aims neither for the beautiful nor for an ideal, but for a truth underwritten by physiological disorders. His suffering is the work's raison d'être, the seal of its authenticity. No comfort, no response, no essence may be attained: the only plenitude is the experience of being torn apart.

Finally, medical theories about the role of imagination are imported to rethink creation. Chassaignon conceives of creation as a motion that radically distinguishes him from those who approach writing as an exercise of style. He characterizes writing as a flow that runs across him and that he does not try to control: "Our grammarians, our purists, & our rhetors may, their *Quintilien*, their *Batteux*, & their *Vailli* in hand, put my deviations & my incorrectnesses in the dock; with just one flash of my imagination I will strike down this pusillanimous herd of slaves born to align words, symmetrize sentences, & clip the genius's wings."[76] He adds in a note: "I do not write for cold brains, but for fervent heads."[77] Writing is a motion of opposition to an institution, a language, a literature. Chassaignon nevertheless, in the course of his speech, lays claim to a genealogy of authors, ones chosen for their energy: Plutarch, Shakespeare, Montaigne, Milton, Bayle, Rousseau, and Young. He makes his model all those who addressed the experience of writing per se: "A man of letters whose soul is burning & whose brain is fired up must, in the fieriness of his delirium, be unable to put a certain order into his conceptions and a certain harmony into his speech." The author carries his work in his body; the pangs of creation are the pangs of expelling what has developed inside him to the point of becoming foreign to him. He resists and undergoes these trials as so many challenges that could destroy his work, and tears the work away only at the moment when it would have shattered into pieces.

Medical theories on imagination here figure creation as giving birth. Refusing to idealize that process, Chassaignon presents his work as a "superfetation"[78] and offers to keep it "as a doctor keeps in a jar of spirits these

prodigious moles[79] to which some women give birth."[80] Presenting his work
as a mole, a superfetation, a monster, Chassaignon understands imagination
not only as what must shape the work, but as what must bend it out of shape.
Imagination becomes the capacity to tear the work apart with a continuing
élan and to make it the revelation of a lived experience. The work is neither
a work of dreams nor a collection of whims nor an arrangement planned in
relation to conventions or taste; it is the uncontrolled inscription of a disruption. It is not the narration of the affections of the soul, but the monster that
arises from the emotions and the torments undergone. One recalls Rétif de la
Brétonne's use of a metaphor of the time to describe writing as the review of
the soul's affection: "The writer anatomizes himself to unveil the springs of
human heart."[81] For Chassaignon, such anatomizing constitutes a threat for
writing and foils its execution. The writer does not wield a scalpel to dissect
sensibility, but engenders the monster that testifies to the scope of his feelings. Shortly after this, Chassaignon moves from the image of giving birth
to that of vomiting, thus destroying any idyllic vision of the act of creation.

Chassaignon departs from a polite and polished aesthetic. He rips apart
the rhetoric that the reader expects, privileging overshort and overlong sentences, a virulent vocabulary highly varied in its registers and sonorities, foot-
notes that occupy many pages in a row, a preface that stands as a work in itself. He puts into play a writing in which metaphors reign, rhythm hammers
without ever finding peace. Language seems to strive only toward expressivity, not toward communication. It becomes instability, rage, laceration.
Speech becomes convulsive.

Body, writing, representation — Chassaignon ties them together and tears
them apart, then asking: How can one play with this rent? Indeed, how can
one make the rent exist? How can one exist in it? His writing moves from undoing his body's code, to making a new code, to creating a new language with
his body in the constant redefinition of his act. His body wrestles with imagination, with imagination's image. Chassaignon invents himself in poses. He
describes a body made of stupefaction, which stupefies itself while stupefying
the reader, and refuses to vanish in the text. Rather than inscribing its pain,
the body inscribes itself in anticipation of the writing, and becomes the author. The body seems to struggle to attain visibility, giving visible shape to the
inaudibility of its emotions, to its expectations, to its pains. The body loses
itself, the body is distorted. The body tightens, the body breaks up, shatters
expectations, and solicits them anew to live now and listen. Is it a ludicrous
struggle or desperate struggle? Chassaignon writes and rewrites his body's
frenzy. He offers the reader a rumpled body, a body in suspense; the tragedy
of a body surprised between flinching and manipulation.

Writing is indissociable from a writing body, from a body collecting ideas, overhauling them in effort and pain. It is difficult to read this text without thinking of the literature that will develop in the nineteenth and the twentieth centuries: it seems to mark the emergence of a writing later given shape by Friedrich Nietzsche, Georges Bataille, Antonin Artaud, a writing that constructs a relationship with a torn and undone[82] body. To cite Bataille citing Maurice Blanchot at the heart of *L'Expérience intérieure* (Inner Experience), "Experience itself is authority (but . . . authority expiates itself)."[83]

A work published anonymously in 1782, indicating only the initials M. J. F. followed by *medecin anglais* (English physician), devotes itself to the effect of imagination in love. If I interpolate it after Chassaignon, this is because its author tries to exploit the suggestive powers of the imagination and remains far removed from medical approaches. While most doctors adopt a moralist approach and advise avoiding anything that might reinforce the power of the imagination, such as novels or solitude, M. J. F. makes a radically different set of suggestions. To be sure, he insists on the dangers of imagination, locating in it the source of passion and warning especially of its capacity to evoke memories at the wrong moment. However, M. J. F. concludes that attempting to deny the influence of the imagination is a lost cause—such denial will only make its suggestions return in greater force. He advises, instead, nurturing it so as to better direct it:

> [The imagination] is what provides the poison, & it is where one must find the antidote. Supposing that the image of objects which have enough activity to move the brain's fibers, & excite the passions, produces the effect of the objects themselves, one can change, correct, or slow down this movement by representing an object that excites a different passion. If we examine the objects that one knows, we will be convinced that the person who excites one passion erases, obscures, & diminishes the impression of the one who excites a different passion. The reason is that he excites a different movement in the brain's fibers which slows down the first one, if he does not destroy it completely; the brain will print a contrary movement in the heart.[84]

If the imagination can move individuals until they lose themselves, there is no reason to attempt to tame something so slippery; the solution is the ruse. If imagination corrupts reason, reason must in turn outwit imagination. In the case of love for someone inaccessible, or for someone unworthy of such feelings, reason must cause the imagination to redirect its force. Reason must take a Machiavellian turn, offering the imagination opportunities to catch fire differently. These subterfuges take the form of cross-references: the physician advises inscribing at the core of favorite images other memories, sources of fright or anger, to deflect their force. It is up to the victims of imagination to stimulate and channel its force by giving it new pretexts. In this way, a new mastery of the imagination becomes possible, perverting the emotions it arouses by associating them with most dreaded ones. The game is the apposition of one narrative to another, in the confidence that the strongest of the two will seduce the imagination. As images are superimposed, the power of the first will fade while what disgusts or horrifies will remain. The aim is no longer to regulate imagination's force, but to regulate its field of action. The energy of imagination will be redirected by reason.

With this approach, the author steps out of the opposition that structured moral thought in the seventeenth and part of the eighteenth century: the notion that the imagination was a faculty whose development had to be repressed. For M. J. F., imagination is, instead, a tool, a force whose motivations can be manipulated. Its powers are acknowledged if anything more than before; the point is to put one's faith in the imagination and the possibility of rerouting its force.

Le Camus and this anonymous doctor thus both act as strategists, exploiting a faculty against itself. Le Camus claims to increase its powers, M. J. F. to ward off the risks associated with it. But neither seeks to build a therapeutics or an ethics. Their writing fully adopts the spirit of the courtier, offering him the occasion to indulge in pleasing thoughts and setting up a game of mystification. They thus take advantage of the contemporary climate and the prestige of imagination in salon conversations, presenting themselves as men of letters more than physicians. Ceding power to the imagination allows it to become a mere pretext for pursuing their advancement within the society of aristocrats. The imagination is sufficiently malleable to be lent to contradictory discourses, and these writers use it strategically rather than conceptualizing it. For them, vaporous troubles are not only the evidence of its movement but a clue to how its effects may be accelerated and intensified.

Since they could not attack the soul, these men of letters attacked reason; reason became a physiological activity, and for some of them, such as La Mettrie, Chassaignon, and even Giacomo Casanova, the brain a kind of womb. In contrast to other publications soon to come,[85] they did not claim that women had a specific way of thinking because of their womb, but were merely interested in associating two organs usually thought of as strictly separate.[86]

*

In the Republic of Letters, bodily disorders were a subject of fascination and an opportunity to displace values, installing a nonhierarchical relationship between body and soul. In this sense, doctors followed their elders, such as Willis, in questioning the spiritual origin of fits. However, they were now no longer saving convulsive people from frightening demons, but querying the very order of society by casting doubt on the church's authority. Interrogating and extending the limits of human physiological and moral faculties, they laid claim to a role that was simultaneously political, epistemological, and ontological.

The selection of excerpts I have presented indicates not only the popularity of the vapors at the time, but the elasticity of the conceptions that developed before the medical interpretation of vapors became dominant. In these texts, the vapors are much more than a pathology. Rather than demanding a cure, the fits arouse the curiosity of the interpretive eye, which can watch them reveal wondrous faculties and take advantage of them to overturn understandings of human nature. The wide dissemination of the theme of vapors gave the category an equally wide range of connotations, and the overlaps of meaning reframed it as a crisis that could display human nature: a crisis, thus, with an epistemological value of its own.

Men and women of letters claimed to see the vapors in an almost dilettante way, as illustrating the fashions of the time or the excesses of modernity. Yet while seeming to talk about a delectable object of fancy, they were in fact setting the scene for a radical and far-reaching philosophical shift in views of human nature. Developing a taste for fantasizing about humanity's fragility, imagination, and genius, such writings highlighted the body's potentialities and subverted the order of the body and soul. As such, vapors were no longer the price of civilization but its prize.

By bringing together such varied writings on physiological phenomena associated with hysteria, this chapter has shown that in the eighteenth century the implementation of a new category of hysteria was the outcome of very different positionings toward the interpretation of bodily disorders. These differences added to the complexity of the discourse of the vapors; at the same time, the wide circulation of literary fascination or mockery could only fortify the need to provide "scientific" explanations—explanations that it both inspired and displaced. The concern with such signs in eighteenth-century works contributed to the production of a pathology that could be equated to no other and that no one could claim to master in full.

Relating Fits and Creating Enigmas:
The Role of Narrative

This powerful assurance of the narrative as regards its power of truth, this immediate authorization of history to take the discourse of the real, has provoked suspicion: the suspicion that narrative is also a trap, and a trap all the more effective for not appearing to be one.

LOUIS MARIN[1]

The second half of the eighteenth century saw a rise in the use of the narrative form to address hysteric affection, a shift that embraced both the literary and the medical fields, with the genres of novel and observation, respectively. Novelists described hysteric fits in order to give depth to their characters and mark their stories' turning points. The crises displayed the intricacies of body and soul in a move that contributed to the shaping of eighteenth-century sensibility. Medical writers used the narrative format as a way of endowing hysteric fits with coherence; in their observations, they interpreted the bodily disorder in terms of its place in an individual biography, gathering information about the patient's education, way of life, and habits. The narrative structure helped to vindicate the physician's choice of diagnosis and explain the pathology. In striking contrast to the texts discussed in chapter 4, these literary and medical narratives present vaporous or hysteric pathology in terms not of rupture but of a particular progression of events, from conditions favoring the disorder up to an often spectacular climax of the symptoms and finally a resolution. The narrative framework contributed to mitigating the sense of alienation or paradox associated with the disorder, and inscribed it within the linearity of a series of stages. This chapter looks at three novels, then turns to medical observations, some of them consultations published in medical treatises and others never published but left forgotten or almost forgotten in the archives.

Bodies Awaiting Exegesis

The novel genre presented a rich opportunity to depict vapors as expressions of sensibility.[2] From the middle of the eighteenth century, many made the vapors key to their protagonists' subjectivity and to their narrative turning points. From book to book, disorders oscillate between simulation, revelations of a secret against a character's will, and manipulation. As Evelyne Ender remarks, "To study the texts and textuality of hysteria is to be brought back to an inner state, where the spectacular performance of hysteria represents a call for attention and an invitation to read, to decipher the workings of the mind."[3]

In fact, the term "hysteria" was not used in French and English eighteenth-century novels, but the associated terms commonly describing hysterical symptoms did appear in force: "tears," "paleness," "convulsions," "swoons," "fainting," and "catalepsy." The writers who chose these terms most frequently are English:[4] Pope,[5] Heywood,[6] Samuel Richardson, Smollett,[7] Sterne,[8] Fanny Burney,[9] and, among the early proponents of Gothic literature, especially Ann Radcliffe.[10] In their novels, the vapors are regularly described, with a wide range of different stylistic expressions.[11] Fits mark critical moments and often herald turning points in the plot. Only works labeled as libertine, and erotic literature, are largely devoid of hysteric pathology:[12] there, nymphomania appears regularly, but the vapors almost never, and when they do it is briefly and with irony, characterizing secondary characters. Fougeret de Monbron's heroine Margot is an example: "A mylord, or more exactly a 'mylourd'[13] came to me to pay his sterling homage and his vapors."[14] Monbron describes the prostitute as being far removed from this kind of pathology: "I don't wallow in the vapors: purified and distilled feelings of love are fare unsuited to my constitution; I need stronger victuals."[15]

The study of the period's literature has long been articulated around the relationship of the pathological body with sensibility.[16] Over twenty years ago, John Mullan set out its possibilities:

> In the novels of the mid-eighteenth century, it is the body which acts out the powers of sentiment. These powers, in a prevailing model of sensibility, are represented as greater than those of words. Tears, blushes, and sighs— and a range of postures and gestures—reveal conditions of feeling which can connote exceptional virtue or allow for intensified forms of communication. Feeling is above all observable, and the body through which it throbs is peculiarly excitable and responsive.[17]

Studies by George Sebastian Rousseau and David E. Shuttleton, especially, further exemplify this association; they examine the proximity of conceptions developed in literature and in medicine in the case of *Sir Charles Grandison*, a novel in which Richardson borrows theories set out by his doctor and friend George Cheyne.[18]

Describing physiological phenomena allowed writers to set forth the scope of feelings and the variety of their registers. In fact, the great majority of texts drew on the vapors chiefly to illustrate the emotional life of their characters. The author would adopt an established vocabulary and knowledge as a means of constructing a character, and rarely was a new conceptualization of the vapors proposed. In the following, I will not seek to determine how the literary corpus was influenced by the medical corpus, or to prove the concomitance of a vision of the body across disciplinary boundaries, but will focus instead on the process by which, at a certain moment in the eighteenth century, some physiological phenomena were identified in the novel as standing at the limit of propriety. How were these physiological disorders evoked and framed at the heart of the eighteenth century—what were their contradictory roles, invested with narrative stakes that altered their significations, their imaginary outlines, and their value? Each mention of such fits was the occasion to develop a particular perception: enigma, sensibility, trickery, coquetry, or fate. Some authors went beyond simply defining a character by hystericism, and instead made the fits of the body into signs of emotions, signs with which to interrogate representation itself. In the novels discussed here, physiological manifestations are interpreted as signs of sensibility and education. They are presented as a language of the body, marking turning points in the progression of feelings, and often use extreme manifestations of the phenomena as a means to question the relationship between identity and representation.

Charlotte Lennox consummately portrays the interpretation of each bodily gesture and affection with Arabella, the young heroine of *The Female Quixote* (1752, published in German in 1753, in French in 1773, and in Spanish in 1808). Arabella cannot imagine seeing and feeling outside the criteria of chivalrous novels. She interprets the gestures of those surrounding her through the code developed in the novels of Scudery and her contemporaries, thus remaining locked in a world she has constructed and unable to understand the efforts of her suitor, Glanville, to disabuse her of her illusions. A doctor finally succeeds in freeing her from her eccentric ideas, talking with her during her convalescence from an incident where she threw herself in a lake to avoid a kidnapping she believed to be imminent. A long conversation with the doctor shows her the mistaken conceptions that motivated her reaction. She then understands Glanville's constancy and rewards it by accepting his hand.

Arabella's feeling, like that of other eighteenth-century heroines, is identifiable through a series of physiological disorders. The role of the imagination, paralleling doctors' claims at the time, makes this novel a milestone in the representation of hysteric pathology.[19] It is through reading that Arabella develops her sensibility. Not only do her reactions follow the proper way to react as set out in novels, but she interprets the actions of the people around her according to what generally happens to heroines. She predicts a series of events that in fact do not transpire, seeing in a clumsy gardener a disguised aristocrat and repeatedly accusing passing horsemen of attempting to kidnap her. She reacts with the behavior appropriate to a heroine: shedding tears, trembling, and, in the most critical cases, swooning.[20]

For Arabella, the vapors are not a means to obtain specific favors, but a language in which all emotions must be expressed. They take shape in the imitation of novels, function as truths, and indeed become the heroine's only reality. They eradicate the opposition between truth and appearance, nature and culture, emotion and expression; instead, bodily reactions are direct illustrations of sensibility. Arabella cannot conceive of emotions outside their embodiment in the vapors, and she devotes herself to imitating that embodiment: experienced as the pure and innate motion of sensibility, the vapors are repeated with the diligence required by an exercise in style. In Lennox's book, physiological disorders appear with all the force of the simulacrum, yet they are not a lie.[21] They function as signs, as quotations, in the awareness of an infinite reproducibility. Lennox constructs Arabella's world as one that is neither false nor transparent: it is a world in which perceptions, reactions, and emotions are only born with the code. The heroine only seeks conformity with the code, and to question its truth is meaningless for her.

In the absence of spectators, Arabella is the spectator of her own troubles, memorizing all the variations her body undergoes and the symptoms she displays. Foreseeing the visit of men "of quality," she asks her maid to tell her life story when they arrive. The maid is puzzled, so that Arabella advises her to describe smiles, half smiles, variations in her rosiness, hesitations in her voice. For Arabella, the narration of such signs amounts to listing the marks of her delicacy and sensibility. By unfolding her emotions, she aims to prove a nobility of her heart rivaling that of other great heroines. Thus, even while the vapors stage self-abandon, they also function as an emblem, and exist to be seen and narrated.

On several occasions, Arabella speculates on the pathology as an infallible sign of a love that does not dare reveal itself to her. She is surprised by the apparent lack of symptoms in the gardener, and his sudden paleness is enough to convince her of his feelings.[22] When she visits the ailing Glanville, the doctor advises her not to speak to him for fear of fatiguing him, but she replies that her presence will save him, that she will be the best physician for a disorder that belongs more to the mind than to the body.[23] She receives the news of his improvement without astonishment, reading it as the effect of her visit.[24] Approaching the sick body through its role in literature, she rejects any other reality. Only promises or a sudden indulgence can cure a lover. In her view, pathology is a sign, and can be cured only by signs. The novel thus pushes the medical interpretation of the role of the imagination to its extreme, medicine being no longer necessary to heal, and doctors influencing Arabella merely by speaking with her.

Lennox's novel also offers the opportunity to distinguish hysterical affection from other nervous pathologies. Glanville is constantly tormented by crises of melancholia and nervousness that appear as the effect of his sense of the painful gap between two modes of reading events—his own and Arabella's—and his incapacity to communicate this to Arabella. As for Arabella, she lives in a state of complete discrepancy from others' perceptions. Protected by her illusion, she is unable to perceive this gap, and adheres to her interpretation of the signs by inserting them into a narrative in which they become an expression, a response. Once put right by the doctor, she finds that gestures are hollowed out, voided of their special signification as they drain off toward their more obvious destinations. Gestures and signs are now tautology; they have no hidden meaning. The subject appears as the origin of all interpretation, while gestures are given back their finitude.

By making the vapors the principal motif of her book, Lennox indicates a risk in the interpretation of physiological troubles—the risk of giving a free rein to narratives, which function according to their own laws and references. In the attempt to escape the finitude of the body, one may become lost in the infinity of the sign. The body is a language only because the subject claims to decrypt it, and with this claim lodges a new meaning in the gestures. The pathological attains an essential role in the circulation of the symbolic because it interrupts the order of things and makes space for mystery. Lennox shows that the interpretations of fits and disarray are nevertheless always only symbolic projections, and are all mobilized by an investment of affects. Interpreting the body of the other means first of all the possibility of staging one's own body. Arabella puts herself into play by reading the cause of the other's gestures through her own expectations: to become a heroine, she has to make the other's body signify. The physiological disorders of her own body can then become an appropriate response. All those around her become the ignorant actors of particular roles, mere cogs in a plot she is the only one to see through. She is a naïve victim of the power of signs, basing her existence in the creation of new stories, and, as a perfect actress in an order devoted to representation, she cannot exist outside it. Her repeated movement of reconstruction proceeds into a mise en abyme of representation. With Lennox, the hysteric body appears as a symbolic production that has no other purpose than the perpetuation of the order of representation.

If Lennox's novel explored the fallibility of the signs of hysteria, William Godwin deploys the unfolding of physiological disorder to reveal a truth of the subject. Godwin's first novel, *Things As They Are; or, The Adventures of Caleb Williams*, was published in 1794, one year after his study *Enquiry Concerning Political Justice*. The novel's narrator, Caleb Williams, presents his memoirs as an attempt to calm the disarray that lingers in his soul, hoping that posterity will render him the justice that his contemporaries have refused.[25] Caleb becomes curious and suspicious after he sees his master Mr. Falkland undergoing convulsions. Despite Falkland's past acquittal, Caleb believes he is guilty of murder. Disturbed by Caleb's questions and insinuations, Falkland succumbs to convulsive fits with increased frequency. When he confesses his guilt to Caleb, the power relationship between the two is transformed: on the basis of the confidence, Falkland informs Caleb that he now regards him no longer as his secretary, but as his prisoner. Caleb tries to flee, and Falkland vindictively accuses him of stealing. Caleb's eloquence cannot overcome the prejudices of class and reputation, and he is sentenced to jail. The resulting series of imprisonments, escapes, and prosecutions obliges Williams to hide, disguise himself, and lie. Falkland's withdrawal of his complaint against Caleb begins the last stage of the novel. Though Caleb's name is cleared, Falkland continues to monitor his movements and destroy his reputation wherever he goes. The tale concludes with Falkland's incrimination for murder. Falkland confesses his crimes, and Caleb's innocence is recognized once and for all, but Caleb can find no peace and is indelibly marked by his encounter.

This novel offers an exemplary case of the interpretation of a vaporous pathology.[26] The plot takes its first turn after a convulsive fit that Caleb accidentally witnesses, and in fact, Caleb's narration of Falkland immediately presents him as someone subject to vaporous fits: an aristocrat and man of letters, he seems a perfect client for Tissot or Cullen. His sensibility, his manners, and his honor make him a chivalrous character whose nerves may be overexcited at any time.[27] The reader recognizes his vulnerability through his face and movements, described as being often marked by anxiety.[28] The narrator very quickly sets out the opposition between the appearance presented by Mr. Falkland at first sight, and his real temperament, which undergoes extreme manifestations when he is alone.[29] Falkland appears as a double character, and his duplicity as the effect of an overly present body whose untimely manifestations alternate with courtesy and self-control. When surprised in the midst of a fit, he expresses anger in order to mask the violence of his disarray, which the narrator describes in terms of "symptoms." The fragility of Falkland's nerves becomes an object of interrogation for the narrator when he observes that this frailty is not accepted by Falkland himself, who tries to hide as soon as it surfaces.

Caleb positions himself as a detective. Scrutinizing Falkland's every move, he sees his disorder as the manifestation of an internal conflict, and his fits as attesting the presence of a lie. When Caleb alludes to historical figures disgraced by their crimes, judging them with an affected juvenile candor, Falkland's body reacts immediately.[30] Caleb reads in the convulsions—arising in empathy with his own words—a sign of Falkland's culpability.[31] For his social environment, Falkland's past behavior and dignified countenance function as proof of his innocence; it is only his fits that shake this composure, leaving him in a state of confusion. Crises therefore act as an avowal, while Falkland's words attempt fruitlessly to deflect the bodily confession. Caleb recounts the physiological effects of his many insinuations as clues collected in a process where the two men seek to identify the effect they have on each other's thoughts without ever asking direct questions. Throughout the novel, convulsive fits appear as the price of dissimulation, and are given as proof of Falkland's crime, preparing the ground for a future confession.[32]

Years after entrusting Caleb with his secret, Falkland tries to obtain from him a declaration of his innocence. The opposition between Caleb and Falkland is never so stark as in this scene, where Caleb rejects his master's request: "You may destroy me, but you cannot make me tremble."[33] Without supporters or fortune, Caleb has nothing to hide, and his body is characterized by his control, which he uses as a force to exploit Falkland's vulnerability. Indeed, even if Falkland were to benefit from the public declaration, his body would still betray his lies through its untimely fits.[34]

In Godwin's novel, pathology functions as the effect of the unsaid. It is interpreted as an avowal that denounces the lie in speech. The fit's confusion signals a moment of authenticity, a moment in which the body becomes the testimony and performance of its truth. Caleb postulates that the body cannot remain ignorant of its own history, and that even if the mind can live with the lie, the body will wreck every artifice. It escapes all mastery and proffers an alternate story that comes to confront the authorized version. Throughout the novel, convulsive fits stage a stifled discourse, a silence that wants to be heard. The nervous crises rapidly become the core of the plot, originating a series of discourses that put them in perspective, justify them, or elucidate them. Confession, calumny, retraction, and accusation are all consequences of the lack of an initial avowal. They are uttered to compensate for the effect of the fits, and willingly or unwillingly confirm the origin of those fits in a previous lie.

Two discourses in particular trigger turns in the plot and enable its resolution. The first articulates Falkland's truth, the second Caleb's, and they occasion a reversal in the power relationship between the two men. The first passage is an avowal: Falkland confesses his crime. His speech takes place after a series of nervous crises that announce and necessitate it. Yet it is only uttered at the moment when it has become unnecessary—when Caleb has already become convinced of Falkland's guilt. The avowal follows the silence and inaugurates a new silence, that of Caleb, who must henceforth hold his tongue. Curiously, Falkland's avowal is nothing but the affirmation of a new power, arbitrary and unlimited. He does not bestow a confidence, but asserts Caleb's culpability in having extorted the confession, and makes him responsible for the future.[35] This avowal, then, is a trap. It is the moment when Falkland takes possession of his own history, seeks to determine what can be said about it, and sets the price to be paid for knowledge. From now on, he no longer has to keep watch on himself before Caleb, but to keep watch on Caleb. His display of his own life becomes the cause to invade Caleb's life, now the depositary of his secret. No further mention is made of fits, indirectly reaffirming the previous identification of crises with what was then unsaid.

The second key speech is that of Caleb to the court, near the end of the novel. Once freed from the accusation of felony, Caleb has the power to accuse Falkland. This accusation is the last encounter between the two, followed by Falkland's death, and shatters Caleb, making him regret having incriminated his master. The nature of this last speech is paradoxical: it restores Caleb's legitimacy in the public gaze, but at the same time initiates his damning self-critique.[36] Caleb makes his speech at a point when he has ceased wanting to make it, when it has become useless for him and an object of anticipated remorse. He voices what no longer needs to be said. Initially conceived as an incrimination, the speech instead becomes an acclamation of Falkland and an account of Caleb's error. [37] What Caleb states is that while believing he was unmasking Falkland, he was in fact giving him another mask: by translating the physiological disorders into signs, he wrongly granted sole authority to those signs. Attracted by the mysterious nature of the vaporous fits, he saw all of Falkland's other expressions as lies, his calm politeness as base dissimulation. Caleb took the body to be a possible witness to truth only when in crisis; he read the pathology as truth because it disordered the representation. It is this disruption that authorizes the narrator to postulate an alternate reality: being excessive in its manifestation, it interrupts the dominant reading and asks to be read as a sign. From that moment on, everything else topples into falsehood. Only during the final speech does Caleb grasp the reduction that is produced by this dichotomy. Seeing Falkland as a murderer has prevented him from seeing him as an honest man, which he was at every moment except that of the crime. These two realities coexist. Caleb, busy converting the world into signs, has proved his own cleverness in discovering a criminal, but ultimately he has only played out certain signs against others without attaining an absolute truth.[38]

In both *The Female Quixote* and *Caleb Williams*, the relationship to the body is based on acts of deciphering. The central characters are interested only in the signification of the fits, which they invest with the value of testimony. The two novels are structured around the existence of two contradictory perceptions of the world. In *The Female Quixote*, the presence of hysterical fits signals a move to the order of the simulacrum; in *Caleb Williams*, it signals a move to the order of the truth. The characters are both obsessed with their reading of the body, and both finally come to regard their reading as a mistake. Both novels indicate how, during the eighteenth century, the body is seen as providing its spectators with meaning, and simultaneously how the perception of the body as signifying is a source of error—for if the body means something, it may give access to the truth, but equally it may mislead spectators into another interpretation of events, which, though different from that provided by speech, is not more true than that speech. The interpretation of the disorders cannot be separated from an interrogation of the possibility of an innocence of the body, its capacity for simulation or dissimulation. The body's faculty to convince and to manipulate others' emotions both displaces and confirms the order of representation: On the one hand, it leads the spectator to privilege another interpretation of the events, and thus might tear apart his beliefs about people's roles. On the other, by offering the spectator new meaning, it creates a new way to interpret—fits become the truth—and as such provides the ground for a new representation to take place.

With his epistolary novel *La Religieuse* (The Nun),[39] Denis Diderot's use of hysteric pathology succeeds in combining the body's power of truth with a power of manipulation. The narrator, Suzanne, is compelled by her parents to become a nun, and struggles to escape over a period of many years. She addresses her memoirs to the Marquis de Croismare in the hope that he will help her to find a place to settle, telling him that she started her narrative when she decided to break her vows, and ended it when she had finally fled convent life after encountering four different mother superiors. In the first convent, she complains about the superior's hypocrisy and refuses to take her vows, upon which her parents hold her captive until she agrees to return. She comes to trust the superior of the second convent and takes her vows, but when the superior dies, her replacement's authoritarian manner is rejected by Suzanne, who determines to revoke her vows again. The resulting violence against her persuades the vicar general to have her moved to a third convent, where the superior's loving attitude is the source of a new type of excess: she is induced by the superior to touch her. At the end of her narrative, Suzanne has fled with a priest whom she then has to abandon, and is now looking for some honest way to earn a living. Describing the constraints and humiliation Suzanne has undergone allows Diderot to draw attention to the reciprocal influence of the soul and the body. Physiological troubles here operate to crystallize power relationships, as different characters recount their own disorders and interpret the fits they have witnessed.

Diderot successively unfolds different types of vapors (furor, ecstasy, nymphomania): the space of the convent functions as a theater, each of whose actors and spectators can affect the others. The special interest of the work lies in making the physiological phenomenon not the feature of a single character, but something that strikes each character successively. Each convent is the setting for a new, spasmodic choreography; the narrator's meeting with each mother superior is marked by the arrival of what contemporary physicians called frenzy, catalepsy, ecstasy, and uterine furors. Diderot's descriptions run the full gamut of symptoms, of the body's possibilities: vapors, disorder, distraction, abandon, trembling, convulsions, fits, ecstasy, lack of sensation, and loss of consciousness. Each crisis is a form of response, giving voice to what cannot otherwise be expressed. The elucidation of these crises is made possible by the context, for, as Anne Vila has noted, the reader is drawn into the interpretation of the signs of sensibility.[40] The progression of the symptoms throughout the nun's life makes them a series of physiological reactions to specific circumstances, and at times a bodily emulation of what the person has witnessed, as a reproduction of signs. From compassion to piety, from the indignation of the narrator's mother to the anger of Sister Christine and her favorites, from Sister Suzanne's anguish to her disarray, the motives of these physiological disorders multiply throughout the narrative.

When Sister Suzanne finds herself obliged to stay in the convent, her body offers itself to be read as a movement of revolt. The impossibility of accepting her fate is manifested by her inactive, incapacitated, or rebellious body. The narrator thus evinces the same frenzied movements that she saw in another sister. Cast into the same dungeon, her body reproduces precisely the scene she witnessed before, illustrating doctors' conceptions of hysteric illness as contagious. For both nuns, the body exhausts itself in protest. The disorder is successfully repulsed by some, but gains hold of the bodies of those who cannot resist it. Its symptoms are reproduced in a new context, on a new occasion, in a process of imitation, as if the body's memory guided its movements.

The ecstasy overpowering the second mother superior radiates a force that wins her spectators. She becomes a charismatic model of the transport and beatitude a nun may experience. Illustrating again the medical concept of the sympathy—that is, the possible contagion of troubles between the afflicted and her spectators—Mother Moni communicates her ecstasy to Sister Suzanne without Suzanne having to experience the same piety. Suzanne's ecstasies mirror Moni's, even though she claims not to share the force of Moni's faith.

When the novel's fourth mother superior covers Suzanne with caresses, she returns them; her sojourn in the last convent becomes an occasion for discovering another form of disorder, that explained in the eighteenth century as "uterine furors."[41] The turmoil that invades her when she is with the superior demonstrates once again the influence of the woman in a state of crisis, creating a third occasion for sympathy. For the first time, the narrator herself offers a pathological interpretation of the phenomena. She refers to them as a symptomatology, indicating that her innocence prevents her from conceiving the bodily manifestation of the superior in terms of sexuality. The pathological approach allows Suzanne to avoid questioning her own behavior, and her confessor will use the same vocabulary. Speaking with him later, Suzanne cites the rapidity of events as a justification, implying she had no time to think and was merely the dumbfounded onlooker of a state she was unable to understand.[42] She describes her superior as "hare-brained,"[43] and Thérèse, the superior's companion, as "crazy."[44] The novel highlights the influence that can be exerted by a person in the grip of crisis, the fascination of hysterical access, and the ascendancy of its victim over the onlookers: Suzanne's mother obtains her daughter's promise to take the veil after a fit, the mother superior convinces her to join the convent through the aura of her ecstasy. On repeated occasions, the narrator observes the power of hysteric fits, in the shape of the emotions that her moments of transport provoke in other nuns, who watch her with fascination. The narrator shudders when she realizes the power she could have held in the convent if she had been inclined toward "hypocritical or fanatical tendencies."[45] Transport displays itself as a spectacle—and the narrator knows it, for in the midst of a crisis she remains conscious of her effect on others. Even in this moment of abandon, Suzanne is still entangled in a web of power, but far from being a powerless victim, she discovers the force of ecstasy's fascination. Though she denies having the desire to use it, the narrator confirms the possible instrumentalization of the transport and its unequaled capacity for persuasion.

Manifold levels of interpretation appear as Suzanne alternates an internal gaze with an external gaze to describe her own disorder.[46] To these must be added the other characters' interpretations of her fits and of the fits of other nuns. Across the spectrum of crises, what counts is the narration of an unfolding of signs. They are received, interpreted, accepted by some subjects and rejected by others. The richness of Diderot's novel lies in its indication of the extent to which the body is caught between conflicting interpretations and manipulations. The deciphering of the body becomes a stake of power, and its interpretation catalyzes power relationships.

The narrator understands that her crises benefit the mother superiors and the nuns, who can use them to stigmatize all rebellious nuns and present any rebel as unbridled, even possessed. However, those wanting to manipulate Sister Suzanne for their own ends find themselves faced with a new will to mastery. The nuns do not succeed despite the ingenuity of their crimes, for weak though she is, Suzanne no longer surrenders to disorders.[47] Having failed to provoke an actual crisis, the nuns hope to produce its signs when the vicar general comes to visit, with her body's dirtiness and decay: "There was a great deal of fumbling and violence over [someone who] only needed it because they had arranged it so. It was important that this priest should see me haunted, possessed or crazy."[48] Unable to control the body, they aim to control its interpretation. When Suzanne appears before the vicar general in the greatest destitution, the narrator understands that her torn clothing may be interpreted either as the effect of possession or, on the contrary, as evidence of the deprivation of her rights, depending on the angle of the gaze directed at her situation. To redirect the interpretation of this mise-en-scène is to turn the scenery inside out and expose the elements of the set (torn clothing, cords tying her arms) and the wings (Suzanne's cell). But this does not stop the nuns from trying to provoke Suzanne into a fit when the vicar general asks if she renounces Satan:

> Instead of replying I suddenly leaped forward and uttered a loud scream, and the end of his stole came off my head. He was very upset, and his companions changed colour. Some of the nuns fled and the others, who were in their stalls, left them in an uproar. He made a sign for quiet, but he kept his eyes on me, expecting some extraordinary manifestation. I reassured him: "Sir, it isn't anything; one of the nuns stuck something sharp right into me."[49]

The elaborate staging is foiled. Sister Suzanne's successful struggle not to be overwhelmed by disorder demonstrates, like her past crises, how much hinges on bodily reactions. Only a mastery of gesture will allow her to limit the power of the others and to negotiate a space of autonomy.

Later, when Suzanne joins the convent where the mother superior indulges in touching, the reader witnesses a reversal of this scene. The confessor denounces the practices of the mother superior, and Suzanne joins in his interpretation, fleeing her and crying, "Get thee behind me, Satan!"[50] But the narrator does not examine the contradiction, and it is left to the reader to retrace the power of the interpretations, which delete and deflect signs depending on who speaks and who decides on the interpretive model.

The narration of crimes and fits affirms alternately the vulnerability and the resistance of the body, its innocence and its capacity for manipulation. Convulsive access appears as the result of power relationships and constantly generates new ones. The force is dual. On the one hand, the fits can manipulate characters and weaken them. The fits move, disturb, inaugurate, a memory of the body that no discourse can fight. Frenzy, ecstasy, uterine furors, overcome the narrator through an effect of sympathy. On the other, the physiological disorders deflect head-to-head confrontation, and thus facilitate the power of the nuns to encroach on novices' wills and rights. The interpretation of bodily turmoil enables the negotiation of everything that the novices' state does not allow them to say openly. The body taken by confusion simultaneously becomes the plaything of the established nuns, and shows its defeat with the display of fits.

Diderot establishes that the convulsive body has its codes just as much as the mastered body does. Suzanne's narration presents the power of fascination that arises when bodies tear themselves away from what is expected of them. There is a fascination of the body for itself, for its resistance, its vulnerability, its movements. The convulsive body is condemned to its own loss and to the judgment of others, for as long as it refuses to be disciplined or to be forgotten. The body at once surrenders itself and stages itself. An interpretive violence threatens the body that moves out of the expectations placed on it. Suzanne also reveals that, during a crisis, the body temporarily breaks only that regime of representation to which it currently belongs. The body may attempt to undo that regime, but will only be all the more rigorously subjected to the interpretation of a will, of emotions, or of an education. Thus, when Suzanne says she has been attacked with a needle, she makes the interpretation of demonic possession impossible, and the whole scene in which the nuns appear as the agent of her shriek breaks that regime of interpretation — yet as soon as the religious interpretation of the fits has been discarded, another interpretation immediately arises. If the caresses of the final superior are seen as a pathology by Suzanne, the reader sees them in terms of sexuality; the confessor then guides Suzanne to the interpretation that her superior is possessed by the devil. One order of interpretation is broken, another sets in.

There is no way of recounting gestures as mere gestures, for to recount them is already to interpret them.

Thus, the particularity of Diderot's novel is to play with the body's passivity and its power, illustrating all the ambiguity of this dual effect. As the objects of antinomic interpretations, signs can provide the proof of innocence or possession. In this novel, the truth of the sign does not exist. There are only interpretations, which, grafted onto signs, can manipulate perception. All reasons invoked to explain the crises are presented as the result of exterior projections—yet the art of representation is precisely the art of making what one sees signify. Diderot reveals the interest of each party in directing interpretation, and obliges readers to admit that denouncing one representation will make them the prisoners of another. The spectacular dimension of hysterical affection solicits, even forces, interpretation. Vaporous affections play a pivotal role on which turns a whole set of representations. They catalyze representations without stabilizing them, and accelerate the division of the discourses charged with explaining them.

The Rise of Medical Narrative

While literary writers seized on fits as signs carrying a signification inside a narration, in the late eighteenth century physicians became increasingly interested in narrating their patients' disorder—identifying the pathology in the process of narrating it, and thereby instituting it as an object of knowledge. They also regularly invoked the imagination to explain troubles, relocating the diagnosis within a narrative understanding. But their aim was different from literary writers of the same period: by tracking fits, they were trying to keep improving medicine. For that purpose, they recorded their encounters using texts they called "observations": reports on consultations, written in the form of narratives in which the physician described his patient's symptoms, the diagnosis he made, and his therapeutic endeavors.[51] Interestingly, the French physicians who chose the nosological form for the overall structure of their treatises tended to include observations as well.[52] The length of such texts varies, but increases toward the end of the century. As historians have shown in the last twenty years, the observation format was no eighteenth-century novelty; physicians had been writing down their observations of patients long before.[53] But from the eighteenth century, writing down observations became a much more widespread activity. Observations took up an ever greater role in treatises, works devoted to consultations were published, and many new medical journals were launched, publishing large numbers of cases.[54] The *Journal des Sçavans* and *Philosophical Transac-*

tions, both inaugurated in 1665, were supplemented by journals specifically dedicated to medicine. To mention only the most important one, Nicolas Bertrand et Grasse founded the *Journal de Médecine, Chirurgie, Pharmacie* in 1754, with the goal of gathering together all kinds of observations for members of the profession. In 1773 Joseph-Jacques de Gardanne launched *Gazette de santé* (Newspaper on Health), addressing a wider public.

As we will see, in recounting their encounters with the patient as observations, doctors stimulated interpretive models of the body and reoriented an epistemological field—medicine—by affirming their own tenacity in the identification of symptoms and knowledge of each patient. Observations framed the pathology and diagnosis in the format of a narration, as stories. If narrative style lends itself to experience, it was nevertheless more than a way of keeping track of the case; it was the instrument of an elaboration of knowledge, presented in the making.[55] In these observations, the doctor does not record his conversations with patients but sets out his own role, becoming, so to speak, the hero. He appropriates the case in the very movement of describing the symptoms. In such texts, the position of the physician is dual: he is the practitioner, the successful or unsuccessful party in a series of battles against physiological troubles, and simultaneously the narrator, who presents the case after the fact and inscribes it into a particular epistemological perspective. The erasure of this gap by the narration valorizes not only the doctor-narrator's therapeutic competence, but also his capacity to grasp each specific case within a broader field of knowledge.

My assumption here is that physicians should be regarded not as the origin of discourse, but as relays caught within different stakes—determining therapeutic practices, reinforcing the distinction between the professions of physician and surgeon, constructing a medical category, establishing the authority of the discipline. Their texts reproduced and displaced the exigencies of the time. Rather than analyzing doctors' supposed "distortion" or "misrepresentation" of the body, I look at their attempts to consider symptoms in relation to the temporality of patients' biographies; and rather than focusing on the "gaze,"[56] I show how the demands of narrative appear to shape the writers' understanding of the pathology and attribution of specific diagnoses. In examining these observations, the heterogeneity of the encounters between doctors and patients will become clear—they varied widely based on age, gender, social class, and station of the patient and the site of the meeting.[57] Indeed, doctors and patients cannot be posited as two unified blocs: eighteenth-century texts demonstrate that, on the one hand, doctors do not hesitate to oppose their colleagues in their publications, and, on the other, patients developed extremely varied approaches to their own disorders.

In the eighteenth century, many other roles emerged for the physician besides that of diagnosing an illness and prescribing a therapy, including acting
as confidant, courtier, investigator, or mentor. To underpin their singularity,
doctors used their correspondence and publications as a forum to distinguish
themselves from charlatans, surgeons, and their own colleagues.[58] Even if surgery was taught as an art in France from 1743, and surgeons separated from
barbers in England in 1745 to form the Company of Surgeons, it was rare for
a doctor to speak of surgeons with respect. Once the therapeutic value of
bleeding had been questioned, the surgeon's role in the treatment of hysteric
affection was easily criticized. Surgeons were even identified with empirics, as
physicians opposed the variety of their own remedies to the surgeons' bleedings and the empirics' emetics. They regarded these violent therapies as relics
of an outdated medicine, and regularly bemoaned them in the observations
they sent to societies of medicine.[59] Toussaint Guindant, for example, complains about the ways surgeons abusively practice bleedings[60] and prescribe
purgatives:[61] "I will say en passant . . . that a Surgeon Major at the Hospital of
Chantilly bled a young woman afflicted with hysteric vapors a thousand and
twenty times over nineteen years (for it was counted) in her foot and in her
arm."[62] In Guindant's view, it is town life that spreads these techniques.

According to doctors, vaporous and nervous affections, long ignored, are
those for which recourse to a physician is both least likely and most necessary.
The publisher of the translation of Whytt's treatise on the vapors warns of an
ignorance that is bound to be perilous.[63] His worry was a common one at the
time, and indeed in the 1780s the poverty of the countryside condemned the
population to remain untouched by therapeutic evolution. Beyond their criticism of individual charlatans, surgeons, or doctors, the polemics of this period have to be understood as expressing a desire for steps toward the reform
of medicine. They insisted on the need for wider-ranging care, for a health
policy across all of France, a change that would come only with the Revolution. In fact, the move toward detailed narrations of encounters with patients
was encouraged by medical academies themselves. The Société Royale de Médecine (SRM) paid physicians to travel or move to more inaccessible parts of
the country, thus fostering the dissemination of knowledge and the collection
of a large corpus of observations.[64] By sending physicians to the provinces
and developing a wide network of correspondents, the SRM demonstrated
how eager it was to accelerate the circulation of therapeutic practices.

Other academies and medical societies solicited consultation reports and
regularly gathered to read and comment on them.[65] All these institutions saw
the invention of therapeutic practices as the chief means of making medical
progress. In the observations, which retraced the events of several days in a

case, doctors gave access to the gestation of their methods of interpretation. They wove a fabric of hypotheses and of moral and anthropological conceptions with which to clothe the body. Some saw the submission of observations to the academies as a step toward publishing a full treatise, and would accompany their observations with a letter asking for support in publication. Whether or not this strategy succeeded, physicians publishing treatises on the vapors often belonged to three or four academies, at times even a dozen, accumulating regional, national, and foreign endorsements that were listed on the title page of the works they published. The establishment of academies in the French provinces in the course of the eighteenth century favored emulation between colleagues,[66] rewarding some physicians with the title of "correspondent." The academies offered annual prizes for a dissertation on a topic announced in the *Journal de Médecine*.[67]

The SRM's 1786 competition asked contributors to distinguish between hystericism, hypochondria, and melancholia. One of the competing physicians entered anonymously, suggesting that participation could be motivated not solely by the prospect of a prize, but also by a desire to contribute to knowledge in progress. All the applicants provided numerous observations and devoted little space to theorization; their aim was rather to collect cases that would subsequently help to develop new theories about the body, in line with the new demands of medical knowledge. At the same time, the emerging provincial academies' need to garner legitimacy and the support of wealthy patrons left its mark on the type of knowledge and of writing that was to develop—for example, privileging a knowledge that could be demonstrated in public and that was likely to fascinate.

Accordingly, the academies supplemented their writing activity with practices of larger-scale diffusion. During the second half of the eighteenth century, public medical demonstrations began to proliferate, promoting new therapies. For example, the creation of an amphitheater for dissection in Rouen in 1736 enabled the renowned surgeon Claude-Nicolas Le Cat to start his anatomy lectures.[68] As for hysteric pathologies, medicine presented spectacles to increase its credibility, with doctors publicly demonstrating the powers of electricity, magnetism, and somnambulism. Despite their public insistence on their singularity, physicians in practice blurred the distinction between the charlatan, with his fairground theater, and the physician from the Faculty with his bookish knowledge.[69] They drew inspiration from the taste of the day, combining fascination, mystery, and demystification in the style of Robinson's spectacles of phantasmagoria.[70] These settings were so many occasions for physicians to meet their patients, transform the modalities of the encounter, and put their knowledge to the test.

André Marrigues's observation, read at the Academy of Rouen on October 1, 1762, is a case in point.[71] Marrigues, the master in surgery of the royal bailiwick of Versailles, relates how during his public demonstration of the therapeutic powers of electricity, a young woman volunteered for the experiment. The following day, she told him that her menses had returned after a long suspension, a problem doctors had not been able to remedy. This report attests that far from being locked in passivity or ignorance, eighteenth-century patients showed initiative, and could go almost as far as to subvert the doctor's authority: in this particular case, the young woman asked to try the experiment as if merely curious, when in fact she was looking for a therapy. Prepared to use new techniques still at an experimental stage, she selected the knowledge she wanted to access—electricity—and rejected that which she did not—a traditional knowledge, which would very possibly have consisted in questions about her sexuality or her desires. The surgeon makes her story into the validation of a new practice, the confirmation of a hypothesis, and a means to acquire knowledge. Dispossessed by the patient of his role identifying a diagnosis or the suitable therapy, he reappears in another role, to experiment and produce knowledge. Here, the body is read not in relation to a subject, a soul, but to a body-testimony. It is neutralized and becomes "a young woman." The goal is to particularize the case as little as possible. The young woman's attitude, offering herself fearlessly to the experiment, can only valorize the procedure, and the limitation of the surgeon's role, reduced to applying electricity, further reinforces the validity of the experiment: it is the electricity and not the doctor's presence that has exerted the effect. He thus extracts all particularity from the experience. And this is the stake of Marrigues's observation: to prove the usefulness of electricity and establish it as a therapeutic practice.

The physician's narratives, apparently designed to interpret the case, and the observations themselves happen in a predetermined field of enunciation; they are neither autonomous nor neutral, and are invested with high political and epistemological stakes. In these texts, the doctor's word is borne along by several different objectives: his observations attest to his therapeutic practice and his desire to ground his theories in experience, but the writing also has the function of attesting to the physician's expertise, earning colleagues' respect, and attracting new patients. On the other hand, many observations were intended to remain in manuscript form, shared only between colleagues. At times, doctors ask for advice, relating their uncertainties about one case or their sense of incompetence faced with another—supreme proof of the authenticity of a narrative. On other occasions, they choose to report on their own practice, their success and remarks, adding theoretical notes.

Examining the writing of these observations, the reader discovers how the relationship to the body is changing, how modes of narration themselves shift,
and how the vision of the physician's role evolves.

In the Shadow of a Gothic Tale

More often than not, physicians still privileged observations that presented a
very dramatic case and were likely to mystify their readers—a tendency that
took shape, as Gianna Pomata points out, in the sixteenth-century medical
historia: "The *historia* format was primarily used for registering the unexpected and the unaccountable, to open a chink in the armor of established
expectations through which the perception of novelty could steal in."[72] An
eighteenth-century example is the observation sent to the Société Royale de
Médecine on July 1, 1778, by Esprit Calvet, a doctor in Avignon famous for his
correspondence with the greatest savants of his time—which made him an
important participant in the Republic of Letters.[73] In it, Calvet writes about
a case that

> leads to impressive reflections on the prodigious effects of the imagination:
> the named Catherine Chaïse, about thirty years old, married and not preg-
> nant, of a quiet temperament and appearance, with regular menses and a
> good constitution, on May 25 witnessed the execution of a woman who was
> hanged. At the moment when she saw the woman being thrown from the lad-
> der, she felt strangled herself, and felt around her neck the same impression
> as a rope being pulled tight, the neck's muscles were extremely stiffened and
> hard, the spasm was extreme, the closed jaws opened only slightly for the en-
> trance of liquid nourishment. Nevertheless her speech was fairly free and the
> ideas very clear. It was in this state, after some remedies from neighbors, that
> she was taken to the hospital on June first. I found her with these very symp-
> toms, swallowing only with great difficulty, a hard pulse with no fever, her
> neck as firm as that of a sculpture, but all her limbs still supple. The spasm of
> the neck muscles was constant in her, with no variation, but fifteen or twenty
> times a day she felt a convulsion in these same muscles, with which she felt
> she was being pulled up as with a rope and I saw her make exactly the same
> grimace as a hanged person does at the moment of strangulation. . . . At last all
> the symptoms gradually disappeared and she left on the second of this month
> in a state of health.[74]

The patient cannot speak. The doctor moves from the narrative of the
public execution to that of the body's self-condemnation. He starts with the
description of the hanging even though he only saw his patient for the first
time a few days later, and bases his analysis on testimonies concerning the

patient's emotional life, not on the relationships or interactions between her organs or on her temperament. The narration presents everything in chronological order rather than beginning with the description of the body. Calvet proceeds to the description of symptoms only once he has presented his interpretation: an effect of the imagination. The observation is not, this time, about the deleterious effect of reading on the imagination, since it concerns a woman a priori shielded from such excess: a woman from the lower classes, married, with regular menses, not pregnant.

The body here becomes a mirror, and by miming suffering it becomes a body of suffering. "The prodigious effects of the imagination" consist in the transformation of the body effected by a mere sight. Seeing becomes feeling; seeing marks the body; seeing a condemnation is condemnation of the body to feel again and again; seeing is epidemic. The identification plays from woman to woman, from body to body, from spectacle to spectacle. The condemned and disarmed body becomes a disarming body, which the narrative localizes as the origin of the fits. Calvet sends this observation less to vindicate his treatment than to signal the effect of imagination, to indicate to physicians what they must take into account in their practice, and to insist on the need to further elucidate its powers. He complements his classic prescription with moral resources, acting first in the struggle against the symptoms and then in the identification of the pathology's origin. The imagination is constructed as a faculty that takes and gives all shapes and thus lends itself to all interpretations. The body and the imagination appear in a relationship of complete reciprocity, reacting to each other and staging all those reactions. Calvet's interpretation does not try to enforce specific therapeutic responses but instead to participate in the production of knowledge by gathering cases, seen as a step outside of abstract systems, in pursuit of an experience-based knowledge.

The physiological effects of emotion mobilized physicians' curiosity, and they hastened to send all kind of reports to academies of medicine communicating the violence of their cases. In 1777, Monsieur Denis, prison doctor in the town of Douai, sent the SRM an "observation on the cure of a universal spasm caused by fear":[75]

> The named Thérèse, daughter of the Campabem, aged 22, who for 5 years has been in domestic service, fled from her master's house early in the morning and went to town, where she was arrested on suspicion of the domestic theft that had occurred at the home of this bourgeois the night before her flight. When she was arrested, she was found in possession of a purse with money far beyond her estate, which led to the pronouncement, imprudently aloud, in her presence that she would soon be hanged. She was so stricken by her deten-

tion in prison and by this speech that she fainted, the menses that she had at the time disappeared, and she remained quite sick. The prison doctor had her bled immediately and took care of her.

A few days after July 17, . . . she was cold, pale, with nearly no pulse or respiration, unconscious, wordless, and in an apparent state of impending death. . . . At other moments, when she would feel a little more herself, her complexion would then regain color, she would be more agitated, throwing herself to one side and another, taking and losing almost nothing. I suspected a universal spasm: in this view, on the third day, around five hours of the watch, finding her again in one of these moments, when it seemed that there was only the last breath of life, having put her in a posture suited to receiving a spray of cold water on her face, I had some brought in and filled a pint pot which had a handle, and I threw all the water it contained into her face with a short shake. It made her wriggle and she was agitated in all her limbs.

I started again a few minutes later. After her agitation, her pulse rose and did not fall again.

Finally the functions came back, the menses reappeared and she was as well as her unhappy state allowed one to hope.

Cruel duty to take her away from the arms of death only to put her in those of the hangman. She was hanged a few weeks later.[76]

The crises function as the ultimate proof of guilt for the sergeants, and as proof of the sympathy between the soul and the body for the physician. Here, too, the description of the triggering event precedes that of the crises, and makes the crises the result of emotions. The sergeants' suspicion is confirmed by seeing the purse, and the body is attacked when the future sentence is mentioned. The physician shows the reader how the body anticipates its own becoming and offers the spectacle of its own condemnation. The sentence is not inscribed on the surface of the body, as in Kafka's *Penal Colony*;[77] on the contrary, it is the body that stages the sentence in advance. Far from fighting or rejecting it, the body anticipates its punishment and reacts to external intervention only with trembling.

The doctor acts neither as judge nor as lawyer. He does not speak to his patient as a guilty woman, nor indicate her theft as the source of the pathology. Instead, for him the origin of his patient's state is the utterance "aloud" and her detention. He limits his role to healing the body, and abides by this service to his patient without raising the question of her merits or faults. The only moral judgment that appears is "imprudently," which is directed at those who arrest the young woman. This moderation of judgment, even though the physician recognizes the relation between pathology and moral affection, is very important: it inscribes the manuscript at a great distance from the writings that will develop less than a generation later and continue well into the nineteenth century. Then, as I discuss later in this chapter and in chapter 6, the doctor will want to assume a moral role, not limiting his action to healing his patients but claiming to guide them morally in their action. The historical specificity of these manuscripts, in which the doctor circumscribes his role to the physical cure of his patients, will then become clear.

In many of these archival documents from the 1770s, the body is described in its capacity to guess, repeat, and anticipate. Physicians interpret affliction as the patient's reproduction of pivotal events, whether as a witness or as the person suffering their imagined consequences. The body becomes a receptacle for the emotions and the scene where emotions are played out. De la Porte's manuscript, received at the SRM between 1775 and 1779, recounts the case of a child:

A young girl who just died at the age of 10, with a petulant temperament and vaporous to the last degree is the subject of this [observation]. This child came to the world with that disposition and the susceptibility to irritability which gives birth to and characterizes nervous illnesses; even in the cradle, she was subject to convulsive movements of all kinds; the further this young daughter advanced in age, the more considerable the irritability became, and the stronger and more frequent the convulsive movements. The mother is very healthy; as for the father, he is valetudinarian and very feeble, and frequently stricken with colic resulting from his work as a painter. This child had a very good wet nurse; the character, which is the temperament's natural consequence, in this child was so vivacious that the smallest correction, even the smallest contrariety or the least resistance to her whims, her wills, would make her fall into suffocations, into convulsive movements, or into varied and astonishingly numerous convulsions in diverse parts of the body.

At the age of 6 or 7, fits sometimes of apoplexy, sometimes of catalepsy, and sometimes of epilepsy appeared on the scene. The little girl nevertheless was fresh and fat; with a face high in color, and a wine-colored birthmark spreading across her cheek up to the upper eyelid. In the fits, this stain would take on a more or less vivid crimson red, and always proportionate to the fit of anger and the vivacity of the fits, and—what is most remarkable—it wept and distilled drops of ruby red blood.

Finally at the age of 8 and 9, threats, privation, verbal and manual correction (means that we believe well suited to destroy fancies, whims, and stubbornness in children) which unfortunately only embittered her mood, worsened and multiplied the fits of this child, and were always followed by nosebleeds, a scarlet redness on the buttocks when she was threatened with the whip, and a quite abundant oozing of blood at the surface of the buttocks when they were abused. And finally, in the violence of her cries and weeping, there was a torrent of bloody tears, secreting from the two eyes.[78]

The writing sets out the case without making any judgment. Heredity appears as a possible cause, since the doctor regards the child as having been born with the disposition. But neither the mother nor the wet nurse is indicated as potentially having created a favorable terrain for the pathology. The body baffles the physician's expectations, yet the description is offered as if nothing could be more natural. Language is in the service of testimony, and the doctor does not want to take sides. He sees his role as rigorously reporting the pathology's spectacle.

The narrative aims to be vividly descriptive, and the doctor's report becomes hypotyposis. To describe the fits of a painter's daughter, the doctor selects a palette of reds: wine-colored, ruby, scarlet, blood, crimson. Crying, oozing, and then a torrent of tears invade the field of sight. The acceleration takes place in the rhythm of cries and whip strokes. Dread speaks in this narration of the body's strangeness, and according to the physician, it is the body that tells. The temporality of immediacy in the bodily reaction is soon replaced by that of bodily anticipation: the buttocks are scarlet even before being struck, the wine-colored cheek weeps bloody tears even before the eyes do. The body anticipates pain and multiplies its forms of disfiguration. It becomes cries, it becomes stains, it becomes stigmata—anger, outburst, monster. A detail is no longer simply a sign, but the clue to a threat. It becomes mark, symptom, omen, prodrome.

This observation shows the difficulties faced by the physician reporting his consultations: in the absence of a metaphysical explanation, everything takes on a supernatural aspect. The beauty of de la Porte's narration is that it aims to witness, and refuses to give free rein to hypotheses. By refusing to place itself at a savant distance, the narrative becomes an experience of seeing. The doctor does not describe what happened, but what he saw. The manuscript blurs the frontiers between fiction and medical narrative, and seems to be snatched from a Gothic novel by Ann Radcliffe or Matthew Gregory Lewis. This reported observation dazzles the reader with its chiaroscuro lighting. It attests to the difficulty of achieving a coherence of the body and of narrating what goes beyond the physician's expectations. When theorization leaves the field to observations, these observations are themselves governed by prejudices, by invisibility, by impossibility.

De la Porte's narration is not unique in the genre. A piece by Collet, read at the meeting of the SRM in August 1777, describes a twenty-five-year-old woman of a vivid and sanguine temperament, with "such an irritable fiber, that the least feeling of sadness or joy makes great changes in her."[79] The doctor recounts that this woman, who has never menstruated, has fallen victim to convulsive movements and bloody vomiting every month since she turned sixteen. Her body exhausts itself in bloody sweat: "The body enters into dampness, and suddenly the blood comes out, drop by drop, through the skin's pores, from the roots of the hair up to the tips of the feet and hands; now the forehead, now the face, the breast, the back, the arms, the thighs, everything is successively covered with blood."[80] Not only that, but

> when the sweats stop or do not increase according to nature's wishes, Mademoiselle falls into syncopes accompanied by the most frightening symptoms, her eyes turn, her teeth clench, her face becomes violet, her throat swells prodigiously, her chest rises, her lower belly collapses, her arms and hands tighten, one observes twitches in her tendons, the pulse is small and concentrated. She remains in this state sometimes more than fifteen minutes, and when she starts to come back, she tosses to and fro relentlessly, weeps, laughs, speaks incoherently, and complains of a great alteration.[81]

After indicating some attempted treatments, the doctor adds: "Ordinarily the day after the crisis, Mademoiselle feels well, and she is in a state to go back to work. Born of poor parents with many children, she occupies herself with spinning cotton."[82] He notes that bleedings were applied and that the crises disappeared for four months. But suddenly the patient

> felt a sweat of blood so abundant that very quickly her blankets, sheets, mattress, and even the bed's wood were soaked. It is the second time in two years. Following this extraordinary crisis, at the end of this same month, she noticed that a small discharge of blood had happened by the ordinary channel. It was the first time and it lasted only one day.
>
> When she is made to put her feet in lukewarm water, the water is soon stained with blood, and she sweats blood whenever she is a little warm.[83]

It is in excess that the body reveals itself. Pathology becomes an intermediary in knowing the body. What constitutes the force of these texts—one narration more extraordinary than the next—is the physician's complete openness for what he sees. He describes the fits with attentive precision, not insisting on his own role. The physiological disorder carries the body beyond all proprieties, in fright, in horror. The body seems to know no boundary, and its unanticipated vigor is also what ruins it: it breaks through its boundaries, becomes unlimited. There are colors, effluvia, secretions. The body becomes a spectacle, the doctor a spectator. The openness of the physician toward the possibilities of the body parallels the lack of limits in the body he tries to understand, and the modesty of the knowledge which looks for its keys. Everything, then, seems possible. Knowledge is humility in the face of the violence of what happens. Having relocated disorders from demonic possession to physiology, the savant is confronted with his ignorance more than ever before. References to philosophical, medical, or literary traditions cease. The question of sensibility itself has disappeared. Finding intelligibility now above anything else means knowing how to outline a case exhaustively. For the academies, all such cases are precious, since it is through them that the members hope to discover the body's functioning. When the body reveals its incredible violence and resistance, it becomes a case to analyze in terms of possibilities and limits. If in hysteric cases the crisis is not understood as the moment of the body's healing, it may at least be the moment of understanding physiology.[84] Here, to participate in the edifice of knowledge is not to draw up a system, but to approach the body from its limits—not to gather together all cases in one conception of the body, but to understand the reach of the body's possibilities. A last excerpt, this time about a physician gambling on the resistance of the body, will indicate how consideration of the body's outlandish force shaped late eighteenth-century physicians' expectations of their patients' bodies.

A young person strongly constituted, Mademoiselle***, had for 5 days been in a syncope that was taken for real death by three out of the four doctors present. All possible means to distinguish the state of death from that of life were employed without success. The mirror was applied to the mouth and it did not cloud. Jolts, odors as violent as the eau de luce and the volatile alkali flower, all were used in vain. The entire body was cold, without pulse, without respiration, the extremities were cold, in fact the patient represented in all respects the state of a real corpse. The eyes were fixed and turned upward but were not dull. The skin was healthy and fresh more than that of a corpse, which led me to think as the young physician, who still maintained that she was not dead. It had been suggested to slit her feet when I was called for. After having obtained the necessary information from these gentlemen about what had preceded this state and having examined the current symptoms, I said I did not believe her to be dead and that one must have recourse to a simple but sure means to ascertain her state. I proposed cold baths and ice. The three doctors opposed it because, they said, she was really cold and really dead. They wanted to burn her feet. These gentlemen went away without wanting to hear the reasons that I explained to the family and to the young physician who stayed with me. If, I told them, your young lady is really dead, the ice that I will put in the water will not melt. If she is not, and there is a remnant of internal warmth, the ice will melt and its melting will indicate to us what we can hope or fear. We don't run any risk if she is dead; if she is not, we must do everything to be sure of it. So we prepared a bath with ice in a tub and when the patient was put in her bath, I added seven to eight big pieces of ice, which melted in very little time. Hope returned to the entire family as well as to the young physician and to me. I added more and more pieces of ice as they melted, and after ten or eleven minutes I had the satisfaction of feeling a pulse that was also felt by the doctor who was holding her other arm. A few moments later, pulsations were noticeable and the sick girl heaved a deep sigh which was heard by all those present. Joy returned to the entire family, who were testifying their gratefulness when the patient started to howl like a wolf. She gnashed her teeth, struck them together forcefully, and finally opened her eyes and stared at us one after the other. I had her come out of her bath to put her in a warm bed after having dried her by the fire.

Having gone out with the parents, who were showering me, and the young physician who had insisted that she was not dead, with benedictions, I was sharing the joy of the family when we were called back. As we entered, the young person stretched out her arms to me, telling me she owed me her life, but that she was not cured. She told us she was suffering surprisingly of colic, which was gnawing at her lower belly. Having felt the womb's region, I found there a surprising tension with convulsive movements.[85]

An initial helplessness is followed by triumph, the prize wrested from endeavors undertaken in uncertainty. The doctor presents himself as the one who can see life where the majority see death, hinting at the physician's dependence on the support of colleagues sharing his opinion. Despite his ignorance of the body's mechanisms, the doctor does know how to persevere, and that culminates in the rekindling of its flame.

For the 1786 SRM competition, this doctor chooses a most spectacular case to speak about hystericism, and his text helps to define the role of the physician and of medicine at the time. In it, the physician appears as the figure who persists stubbornly and stakes his all, the figure who believes until the last moment in the possibility of healing the body. The knowledge that extreme cases exist permits a belief in the body's absolute resistance, and medicine is made of belief as much as of knowledge. It is in the obstinate endeavor, in the proliferation of attempts despite all doubts, that medicine can reach its victories.

Traps and Countertraps

While this attitude of observing the body without taking sides was spreading, another kind of approach developed in medical discourse, aiming to create meaning by inscribing physiological disorders in the linearity of a physiological and moral subject. In those cases, the pathology is seen not merely as a disorder of the body: the emphasis is no longer on the symptoms but on the afflicted themselves. The physician seems willing to detach himself from the force of the spectacle so as to analyze the crisis in relation to the person it affects and place it in relation to a story. The will to break with a perception of the hysteric fit as an extraordinary moment is manifested in the texts' integration of the pathology into daily life. The moral study of patients enlists their ailments as the effects of errors and bad habits.

Throughout these narrations, a relationship with the patient and the body is articulated whereby the doctor presents himself as serving his patients, whatever the power relationship may be. Doctors try to legitimate their role vis-à-vis their own patients and to inspire trust in order to discover the details that precede the crisis and the patient's general way of life. The narratives are structured by tensions and conflicts. As much as the struggle against the pathology, what occupies the doctor is the attempt to correct and educate the patient. Explaining the emergence of symptoms becomes a first step toward instilling discipline. Expectations and frustrations take shape in the elaboration of traps that are described, suggested, or set by doctors. These reports offer insights into the conditions of the exchange between doctor and patient, the mechanisms by which their mutual expectations, demands, and perceptions are framed.

In this type of interpretation, patients are characterized by their passivity. Even their will to recover is sometimes cast into question. Noël Retz, for example, starts his treatise on illnesses of the mind, published in 1786, by announcing the necessity of "convincing people tormented by these illnesses that they must busy themselves with recovering instead of insisting on remaining sick."[86] The same year, Dr. Lavallée in Anjou underlines the importance of obtaining patients' obedience. The *Mémoire sur les maladies nerveuses* (Memoir on Nervous Illnesses) that he sends to patients includes an "observation on hystericism and the stopping of the menses"[87] of a forty-five-year-old woman, a mother of twelve, which is mainly devoted to communicating the difficulty of establishing a relation of trust with the patient. Lavallée insists on his concern to "persuade her, as much as possible, that she represented her ailment to herself as infinitely more than what it was."[88] He continues:

> The tumultuous situation quickly ceded to this treatment, but as soon as the question arose of a curative procedure, frequently a stumbling block for women of this world, my attention was drawn to the fantasies of the patient, her bizarre appetites, which, once satisfied, threw her into a disorder that she capriciously and madly attributed to her diet. Frustrated, she called upon different empirics, soon making it necessary for her to return to me for my advice, given in vain to such an unreasonable woman.
>
> After many recurrences, disgusted, tired of wasting my time, my performances, and my efforts, I was forced to abandon her to her destiny. It is cer-

tain that if the patient had docilely followed the path indicated, she would have desired her timely recovery; she hobbled through five or six languishing years in the city, and went to finish her life in the countryside.[89]

The narrator draws the reader onto his side with the generalization "women of this world," removing the role of intersubjectivity. Medicine counts among the effects of the pathology the loss of desire to recover and the belief that the powerful physician's therapies are the source of the fits. The desire for healing is thus a step in the healing process; it is blocked by the pathology, for the patient does not conceive of herself outside that pathology, and interprets the attacks on her affection as attacks on her own desires. Lavallée presents himself as the voice of reason, excluding his patient from any capacity to obey him or indeed to act autonomously. Her actions are the result of her state, and vice versa. When she listens indifferently to doctors and charlatans, she both manifests her refusal to conform to the hierarchy of knowledge and shows just how sick she is. The doctor's abandonment of the patient is presented as the outcome of a vain struggle, in which Lavallée tried to cure her against her will. Having lost the desire to heal her, he condemns her and presents her life as an inexorable decline.

Hysteric illness is presented less as a whim than as what leads the patient to whims, or else as an effect of whims accepted for long enough to affect the body's balance. Given this particular challenge posed by hysteric illnesses, some doctors authorize themselves to make special demands, and even present the patient as the one who forces them to work out a series of tactics. They transform their own ignorance into a combat where they take up arms against not only the pathology, but the patients.

By describing the traps into which patients may fall if they fail to listen to him, the doctor presents himself as their ally. He reinforces his claim to authority and justifies his demand for exclusivity and for the patient's complete collaboration in the treatment. To evade the grip of the pathology, patients have to distrust their own impressions and leave it to the doctor to know. They are then appraised according to criteria of goodwill and cooperation. Pierre Chirac, adviser and First Physician to Louis XIV, does not hesitate to place conditions on healing in his correspondence with a female victim of hysteric illness: he can help her "as long as she wants to help herself, & to believe us blindly."[90]

In their narrative, doctors establish a discipline that will restore a subject's control of his or her body in order to cure hysteric affections. They endeavor to subdue patients' loathing for particular treatments or their adverse opinions. An equivalence is posited between mastery over the body and health.

The writers especially advise avoiding coffee, chocolate, and red meat, and privileging broth, milk, vegetables, and white meat. They invite their patients to walk or ride on horseback or, when that is not an option, to take a daily trip in a carriage. They also favor waking up early and spending no more than seven to eight hours in bed. The texts frequently mention patients' tendency not to follow the doctor's recommendations, constructing a doctor-patient relationship mediated by indiscipline and error: patients have not listened to their physicians, have despised common sense, have indulged in idleness and gluttony. Worse, in cases of nymphomania, they have diverted their body from natural functions toward pleasure.

Since patients' trust is hard to win, some physicians use cunning. Reports on their consultations devote much space to stratagems devised for speaking to patients and obtaining more information about their way of life and their affections. If it proves impossible to intimidate patients by convincing them of the trap set by the pathology, doctors themselves choose to beat them at their own game. Retz, for example, expands at length on a case for which he developed a masterly tactic:

> Almost always, the young lady had to be convinced through opposites; so as to cool her down, for example, I proposed to her some warming remedy; I was certain of her refusal, & of her taste for remedies opposed to those encompassed in my proposition; so that what suited her appeared to come from her alone; I appeared to resist, then reflect, and finally to give in; I had the appearance of according a victory to the patient, who had obliged me to believe her knowledgeable enough to choose her own remedies, & who more exactly was applying a remedy that she saw as her own creation, & I had the authority of an endorsement in which she glorified.
>
> When she had need of baths, she was antipathetic to this remedy, until she was persuaded that the bath had given her a fever each time she used it; this was announced in advance, as though to protect her from this remedy that she hated. When it was time to prescribe it, I wished, out loud, as though in passing, that the patient had a fever; I then searched seriously for means to procure her a fever; finally I insisted on the bath as a means of kindling a salutary fever that would favor the success of the remedies.
>
> It is through trickery of this nature that one must often capture patients' willingness to recover; one will not succeed otherwise; he who wants to assure the success of his knowledge must not neglect that which can allow him to make use of it; otherwise his care would be for naught; maladies of the mind most of all offer innumerable difficulties for doctors.[91]

Retz is actually the first physician to cast hysteric affection as belonging to the category of pathologies of the mind, and he thus introduces another reversal: it is the physician who worries about the health of his patients, whereas patients themselves care only about being right. Using trickery becomes the duty of the physician who acts for his patients' good, and this duty goes beyond the relationship of trust that they establish together. He alone must carry the responsibility for their healing, and this may lead him to act on his own initiative. In the case discussed by Retz, the patient belongs to the privileged class, and he understands that she could easily turn away from his services to choose a more indulgent physician. Cunning means eliminating the harsh appearance, and claiming instead a relationship of trust and reciprocal appreciation. Retz situates the physician's role as one of radical superiority, for the doctor possesses a knowledge that places him beyond confrontation. Avoiding a collision course with his patients, the doctor must show deference and even leniency and, when necessary, cheat them. He will find his satisfaction in healing his patients in spite of themselves.

The image of the flatterer as it appeared in the medical treatises written as dialogue or correspondence reappears in this narrative to support a claim to therapeutic necessity. At the time, this conception of the physician as seducing the patient toward regained health appears as a crucial tactic for the cure of madness, and is also promoted in published texts such as the commentary that Abbé Robin adds to his translation of Thomas Bowen's *Historical Account of the Origin, Progress, and Present State of Bethlem Hospital*:[92]

> The art of curing the mad is still in its infancy, & to possess it, one would have to expand one's research beyond the limits of Medicine: more than the physiology of the sick, one would have to study their morals, to follow them in the march of their ideas & their reasoning, distinguish the instants & occasions when the intellectual chain breaks, recognize which over-vivid feeling comes to throw them into disorder, know how to get closer to their opinions, their taste, their whims, always seeming to give in & subjugating their will; mastering their feelings, their ideas & reflections, just as the flexible Courtier inspires, directs, & governs his Master, who believes he can decide men's fate by himself. It is this unknown art which would imperceptibly bring the Maniac back to understanding ideas & reasonings that confused him & lost him, to see & remember objects that used to irritate and frighten him; while Medicine, for its part, would be busy discovering the injured parts in his physical organization & recovering them to their order & integrity.[93]

In these observations, establishing a relationship of trust is not only a way to convince the patient to accept a discipline: for the physician, it is also a means of finding out more about the patient. This conception, prevalent at the end of the ancien régime, would not disappear with the Revolution: Philippe Pinel also publicized his strategies for addressing all the various mental illnesses in his *Nosographie* and *Médicine clinique*. Obtaining confidences is the focus of many of these narratives in the late eighteenth century, where reconstructing the patient's character and way of life becomes crucial to the formulation of hypotheses on the origin of the pathology. The conviction develops that the pathology speaks the truth of the patient. At times, just a few gestures are enough for the physician, who construes them as a form of avowal. In one of the manuscripts sent to the SRM in 1786, a doctor recounts:

> Mademoiselle de***, boarder of the convent***, fell into a crisis of hysteric vapors, the convulsions of which ended in an abundant vomiting of blood that was followed by a weakness of short duration. I was sent for with great haste believing she was dead. On my arrival, after asking her how the crisis had begun and taking her small and very agitated pulse, she answered laughingly that she had no idea and pressed my hand to her breast with an expression that left me no doubt as to the cause. . . . It had been publicly announced that this mademoiselle was dying of a vomiting of blood taken for hemophilia, yet she reappeared in the world, of which she was one of the most beautiful ornaments, without one being able to see upon her face that she had been sick. This young maiden having married some time afterwards, I was called upon to treat a new crisis of vapors, but rather different from the first. This one was announced by a terrifying panic that was calmed by a few drops of mineral liqueur of opium in an infusion of yellow bedstraw and several days' use of frog broth. She was then pregnant and was free from further attacks until the confinement, which went very happily.[94]

The doctor recognizes three clues in the manifestations of the body: the crisis, her pressing his hand on her breast, and the brevity of the fit. To the doctor, the gesture is more instructive than the crisis itself, as a liberty transgressing codes of propriety. The doctor depicts himself as more modest than his patient. While she presses his hand to her breast, he does not name the cause of her malady, leaving readers to guess for themselves. His narrative closes with a pregnancy that ends the malady; the patient has blossomed into a woman through fulfillment of her role.

The diagnoses of vapors and of hysteric and hypochondriac illness take shape observation by observation. Narratives validate their applicability, and attest to their frequent use in naming patients' troubles, but the doctor endeavors to understand every case in its specificity. In each individual patient, everything is meaningful: the act of calling a doctor and the act of rejecting him, the act of intensifying therapies and the act of abandoning them before the disappearance of symptoms. Everything in the patient is a sign, not only the crisis. Physiology and pathology become the point of departure to consider balance and imbalance, desire and discipline; the doctor's narrative suspends the status of the patient between sickness and guilt. Narratives constitute and alter the imagining of the malady, moving from the extraordinary to individual fault. Silence becomes the sign of a dissimulation, and the dissimulation a proof of guilt. Silence thus intensifies fault and ineluctably makes the pathology a form of involuntary confession. At times the doctor doubts the utterances of his patient, and perseveres until he pries out a full account of the patient's actions the day before the fit. He imagines motivations for her reticence, taking her aside and repeating his interrogation until he is satisfied. One manuscript relates:

> A young woman artisan, feeling extremely hot after having danced too much, went to the window to take fresh air. Perspiration ceased after remaining at the window too long. The next day, she was attacked by hysteric vapors so intense that she lost consciousness and rolled on the ground crying and screaming with amazing force; her voice that was weak in her ordinary condition was then very loud and sonorous, which so surprised her parents that they sent for me. She was subject to hysteric fits but slight in comparison with these. . . . When I interrogated her as to what could have given rise to these fits, she told me that she had no idea and that she was subject to them. I prescribed a cold bath with the remedies of which I generally make use in cases of hysteric vapors by deprivation, but that evening a new crisis appeared, even more violent, which surprised me and persuaded me that she was concealing its true cause. Having brought the patient back, I again questioned her and it was then that she confessed to me that she had gone to the window after much dancing, but she begged me not to say anything to her mother who would prevent her from dancing again. I changed my treatment, I made her take a hot bath to recall the perspiration, and had a chicken stuffed with borage for her broth, I also gave her an infusion of yellow bedstraw to which I had added a pinch of flowers of nercau.[95] Her fits diminished noticeably and since then she has not had but two very slight crises. Some days later her menses came in abundance, and they completed her recovery without purging. The tongue having always been beautiful and the stomach performing its functions well, she did not have the slightest fit of fever this whole time.[96]

In this observation, again, the doctor starts with the cause of the diagnosis. He then moves to the symptoms, the evolution of the pathology, and the manner in which he succeeded in establishing its origin, before concluding with the recovery. The trap here takes the form of bait: the doctor makes it known that he will be unable to choose an adequate therapy until he is perfectly familiar with the circumstances of the hysteric crisis and the patient's sentiments and lifestyle. He plays on delays and relapses to obtain new confidences. Renewed crises are interpreted as repeated threats by a body that demands satisfaction. The delay before recovery—which might suggest the inefficacy of the doctor's treatment—in fact paradoxically provides him with further confirmation of his correctness, as he manipulates the postponement as evidence that something remains unspoken. Most important here is to establish a perfect correspondence between the avowal and the curative means applied. The doctor changes his treatment, moving from cold to hot and from anodyne water to herbal infusions, as the patient changes her narrative.

Not too much later, this opening of the window would come to be seen as a source of hysteria not because of a temperature shock, but because the young girl is exhibiting herself, taking the risk of forgetting herself, and putting herself into play. This change exemplifies the gap between the interpretive models of the late eighteenth and nineteenth centuries. Significantly, however, the deployment of force in the two models is largely the same: the doctor isolates himself with patients, pursues them until they have recounted everything, extorts details under cover of confidentiality, and wrests secrets from patients as though they were sins to be concealed.

Even if the positioning of the physician is at times similar in the manuscripts and the treatises, the type of writing is radically distinct from that developed in the treatises written as correspondence or autobiography. In those works, the greatest possible range of symptoms, distributed as widely as possible through the body, are described, from headache to palpitations to fainting; as many readers as possible are to recognize the symptoms, and their reiteration is what gives a sense of continuity. The treatises examined in chapter 3 evoked a mode of life just as much as a pathology: in them, physicians insist on the symptoms' fluctuating character and avoid any specific determination of the pathology. In the observations submitted to medical societies, in contrast, the cases are concentrated on a single scene, a dominant impression, an encounter. Uniqueness is always stressed. The observation consists in the selection of details turned into signs, recounted as exhaustively as is judged necessary. The catalog of symptoms and clues situates the patient outside of daily routine.

The Construction of Secrets

Throughout the observations, to cure is to find the story. At the end of the century, medicine transformed its perception of the body and in the process transformed its own role. The origin of the pathology is no longer seen as lying in an imbalance of the body—whether humoral, mechanical, chemical, or even nervous—but beyond the body, in the biography. Under the banner of hysteria, the doctor now does not merely invoke the patient's way of life; he seeks to identify exactly what the presence of the illness implies. He suspects that patients are better placed than himself to decipher the causes of their disorders, and regards their refusal to cooperate as expressing bashfulness, shame, distrust, or duplicity.

A 1786 excerpt from the archives of the SRM confirms this reading. An anonymous doctor, who sent several observations, narrates his encounter with a young female patient:

> Among her entire family, Mlle de*** passed as consumptive because of her extremely fine skin and delicate complexion, and because her cheeks were colored with a beautiful red, especially after meals. A dry cough then tormented her and she sometimes suffered rather long crises. I suspected her of this habit for some time, but as I was not her doctor, I could not find out for sure.
>
> Following a somewhat forced dance, this young lady suffered a suppressed perspiration that centered on her chest, increasing the cough that ended in purulent expectorations, offering further evidence that she was consumptive. Her doctor treated her accordingly, but as the malady progressed, I was consulted. At the time, I made certain of the tension of her muscles, and prescribed the same remedies as mentioned above without worrying too much about the expectoration, to which they were not contrary. However, I gave her no baths but ordered an infusion of poppy with syrup of hare so as to facilitate expectoration; by these means I came to calm all these incidents, which were connected with hysteric vapors only by her symptoms of whooping cough. In place of the severe diet on which she had been placed for several days I put her on a half-diet that was gentle and nourishing, I purged her during her convalescence, all the symptoms of irritation having disappeared. I rendered her a still greater service in leading her to end a penchant of whose consequences she was ignorant and which would have unfailingly led to her death. I did not dare to speak to her about it, but I took the initiative of bringing with me a treatise on masturbation that I purposely took from my pocket in front of her while initially refusing to satisfy her curiosity. I appeared to be ceding to her own solicitations when I lent her the book, but above all I demanded that she read it alone and without the knowledge of her parents, which she promised, from the next time that she saw me, she gave me the flattering name of her savior, a name which she has continued to adopt since then and honestly satisfies, she says, her gratitude.[97]

The narrative charts the patient's transition from the moment where she "passed as" what she was not, to the moment when she "honestly satisfies" her gratitude, an expression that the doctor interprets as a confession. As such, it loses all pretension to being observation, and puts the patient's case into perspective through the selection of certain elements to the exclusion of others. The cough is no sooner mentioned than it is dismissed, seen only as an effect or, more precisely, a mask. The doctor quickly directs the reader's attention elsewhere. The cough has played its role by signaling pathology and resisting the medications proposed as a cure, thus arousing the doctor's suspicions. He sees a second pathology, hystericism, behind this cough; and this second pathology is his reason to suspect onanist practices.[98] The doctor surmises, instructs, and rescues the patient from danger. He knows to wait for the right moment to cure her, seeing his role as reading beyond signs and symptoms, and guessing the story that his patient's body is telling.

The anonymous narrator recounts a trap with three stages: the stage of silence and temptation, the stage of negotiating the reading of the book, and a final return to the patient, who confirms her transformation by naming the doctor her "savior." Not only is there no confession; there is no interrogation either. The doctor himself does not dare to articulate the idea he seeks to communicate, but intervenes as a supposedly ignorant intermediary. By initially refusing to lend the book, he veils his suspicions and his plan; he pretends to be transgressing a rule that he himself establishes through his own hesitation to share the treatise. It can only be lent in an exchange where the patient is protected by ignorance and the doctor by secrecy. The lending of the book coins another secret, and the patient's interest becomes a confession of her habit. The avowal does not take place, but the patient's gratitude becomes its substitute and operates as her testimony.

In his account, the doctor devotes much less space to the description of symptoms than to unfurling the context that prepares his reading of the patient's symptoms. His narrative functions as a prolepsis, dismissing any possible objections to the cure, which functions as ultimate proof. The doctor's narrative grants preeminence to his judgment. It establishes itself as the law, a law supported by the book. Readers are assured that the narration omits no detail of significance, and that its every element contributes to the diagnosis.

In the act of repetition—it is by interrogating his patient repeatedly that the doctor establishes the existence of a secret—the body becomes an object of memory. Subjects cannot escape from themselves, and they must express what has troubled them in order to be treated. The doctor presses for answers, and it seems that therapy cannot begin without a complete confession. In search of that confession, he may try to force the avowal with another trap, using simulated agreement to extract the patient's confidence, then turning it against the patient. As in other, similar observations, he insinuates himself into his patient's good graces and claims to have understood the origin of the malady. He tells a story, encouraging her to identify with the portrait he paints. When a doctor discerns nymphomania, the tactics are even more elaborate. In the thesis he defended at the Faculty of Medicine in Montpellier, Pierre Lafage sets out the best procedure:

> The doctor called to a person in whom he suspects the existence of nymphomania, must first employ persuasion, endeavor to inspire true confidence, flatter to reach the tastes of the patient, bemoan her condition with her, encourage her, affect the most intense interest in her situation, and once he has achieved his goal, return by the same gradations on what he has said, making her see the danger she is in, and inspiring her with all the possible horror, frightening her if necessary, with scenes of the sad results of this cruel affection. All this must be supported with appropriate and long continued remedies.[99]

The doctor not only may contradict himself—he must do so. He claims complicity to win the confidence of the patient, tricks her into revealing secrets, and threatens later disasters. The doctor must incarnate roles that grant him the degree of intimacy necessary for a confession from the patient. In this sense, he mirrors the patients that medicine depicted as being in a constant movement of fluctuation and contradiction with themselves. The narrative seems to function regressively, recounting maladies to find the guilty party. From a helpless practitioner facing an indefinable pathology, the doctor seizes the authoritative role. He makes the afflicted believe that they have secrets, and constitutes them as patients in the course of uncovering their story, as if that story had absolute power over the body. The doctor thus creates a cause for the pathology—which remains unexplained at the physiological level—and a role for himself.

At the turn of the nineteenth century, the description of symptoms seems to take up less space in the observations. Other considerations dominate, such as the patients' history, their way of life, the emotions overwhelming them, multiplying the elements of the narrative that point to a moral cause for the illness. The space devoted to the therapy also diminishes in favor of the circumstances of the fit and its treatment. The doctor sees the identification of a cause for the pathology as a necessary key to recovery. Physiological disorders are no longer merely symptoms indicating a pathology, but signs of its causes, and the doctor moves away from a physiological conception of the pathology to take up the role of a detective. A new approach to hysteria is developing. In his *Recherches historiques et médicales sur l'hypocondrie* (Historical and Medical Research on Hypochondria), published in 1802, Jean-Baptiste Louyer-Villermay gives an example of this with his observation on a case that he characterizes as "simple hysteria":

A young girl, raised with the utmost care to assure complete physical and moral health and supplied with the best-planned education, contracted sweet manners and a lively sensibility.

At twelve years of age, menstruation began, became abundant and regular, and eighteen months passed without any remarkable phenomena. . . . At fifteen years of age, a concentrated love that she did not dare admit to herself, interior trouble, vain struggles to dismiss a first impression, absolute rule of this feeling over all her being and her senses; but also truly heroic courage, and a firm resolution to admit her love neither to the object of her affection nor to her parents, out of fear of distressing those to whom she was so justly dear. It was in these circumstances that she received a violent blow to the head; instant suppression of her menstrual flow, violent headaches, color in her face, rapidity and strength of pulse, general malaise.

Her age, her habitual good health, despite this moral upheaval, and the fear of cerebral events, led me to practice a bleeding of her foot, which was followed that same day by the return of menstruation; but upon seeing the object of her love, loss of consciousness, plaintive noises, involuntary cries, from time to time incomplete return of functions of the senses and understanding; soon total momentary extinction of intellectual functions, disordered contractions in her upper limbs, convulsive elevation of the chest, violent palpitations, spasmodic tightening of the neck, and feelings of strangulation, preceded or accompanied by a hysteric ball,[100] locking of the lower jaw as in tetanus, and soon convulsions of the surrounding muscles. This symptom above all caused me much concern: many times I feared that she would tear apart the principal organ of taste. . . . [The list of symptoms continues.]

I recognized the malady, and even the cause; I spoke about it with her parents, who dismissed my suspicions, and I was no more fortunate with my young patient, who protested that no moral affection had determined the events she had suffered.

To avoid an error in a similar case, one would have needed the powers of discernment of the doctor Erasistratus; and I firmly believed I was mistaken; but an incident came to change the scene, and hope of obtaining her parents' consent and seeing her wishes granted decided the girl to confess her love to this young man and to her parents. Political events delayed the desired moment and the entire convalescence; but finally, the condescension of her parents and the vow of nature fulfilled, rendered her perfect health, furnishing her with the opportunity to offer society a model of all social virtues.[101]

The patient in Louyer-Villermay's account is living a lie, and remains sick until marriage resolves the origin of her troubles. The doctor does not content himself with a physical cause (a violent blow to the head) that could have explained the young patient's malaise. He is searching for something else. Looking at his patient, he sees a character rather than a body; he is no longer interested in the physiological. He envisions a sequence where others had seen pure chance. Avowal becomes an act of mediation reconciling the patient and the pathology, as is attested by the disappearance of the malaise.[102] The confession functions as a refiguring of the pathology through narrative, for it confirms that far from being alien to the subject, whether as demon or Hydra, the illness described in the narrative *belongs* to the subject,[103] and indeed this is hysteria's raison d'être: its belonging to the subject. For this relationship to be established, the intermediary period between the patient's avowal and the fulfillment of her wishes is essential: the delay allows the narrator to align the symptoms with lack, solitude, desire.[104] At the end of the narrative, health becomes a mirror of the young woman's moral virtue.

Before any confession from the patient, the doctor already surmises, sees, and names the causes of the illness and its possible remedy. By saying he thinks he was mistaken, Louyer-Villermay demonstrates his openness toward patients and his willingness to believe them despite the variance of their opinions from his own. He also makes his diagnosis more incontestable, a move bolstered by his reference to Erasistratus of Chios. Erasistratus was a doctor of antiquity cited by Galen for having recognized love as originating the troubles of Antiochius, the son of Syrian king Seleucus I Nicator. Observing that Antiochius's pulse quickened in the presence of his stepmother, the king's wife, Erasistratus deduced the pathologization of an impossible love. Comparing himself to Erasistratus, Louyer-Villermay encourages his readers to pursue the analogy and compare his patient with Antiochius. By recounting the denials that he initially faced, he gives his fellow doctors an example of what they might expect to encounter. He describes himself as being powerless as long as she refuses to cooperate, and notes the risk of being accused of incompetence; his undaunted stance is ultimately rewarded with success. Such a reading cannot but tempt other doctors to privilege their own hypotheses despite a lack of evidence.

Louyer-Villermay begins his narrative by making his patient's feelings of love the context for the initial crisis, as though he has already grasped every detail of her story. The pathology is presented to be read in the framework of love, from which symptoms cannot be separated by the reader. The retrospective mode of narration permits the doctor to rewrite the patient's story and reinterpret its twists and turns. He can insist from the outset on the symptoms and circumstances necessary to his diagnosis, disposing his readers to concede the accuracy of his theories. He even uses the verb "contract" to speak of his patient's "sweet manners" and "lively sensibility," as if he were already discussing the pathology. A disjunction emerges: the progression of the narrative does not correspond to the progression of his meetings with the patient or his therapies, and certainly not to a progressive understanding of her case. As such, the narrative does not run parallel to the progression of medical knowledge. Rather than the conclusion of his encounters with the patient, the diagnosis is the point of departure from which the entirety of the story is read.

The text's temporality is also of note. Louyer-Villermay's example is built around the evocation of an ample space of time, and obeys a diversity of rhythms. Certain periods are quickly sketched while others are written out in detail, allowing the doctor to describe the states of his patient's soul. The account of pathological crises takes the shape of a list of symptoms without any further reference to temporality. At first sight, this bare list of symptoms might seem to permit Louyer-Villermay to consider the illness on its own terms, for by describing it as a simple enumeration of symptoms without stages or progression, he avoids evoking its specific development in the subject. However, in fact, Louyer-Villermay keeps readers informed of the evolution of the malady's context, and can thus make the illness a reflection of his patient's interior life. He can elect not to theorize hysteria as an affection with its own progression, and can dismiss the possibility that the aggravation of her condition is strictly physiological in nature. In the writing of his narrative, Louyer-Villermay makes hysteria a pathology without any momentum of its own; it becomes a simple reaction, disappearing as soon as it is no longer provoked by external circumstances.

Far from being just a series of identifiable symptoms, the pathology is presented as inseparable from the patient who both provokes and suffers it. The doctor introduces the patient as an agent in a teleological link between biography and physiology. Hysteria becomes necessary and inevitable. The criteria for the identification of the pathology operate through a movement of retrospective implication, so that the patient's biography is constructed like the plot of a tale. This turn to her past appears to draw its legitimacy from the absence of other possible causes, subsequently validated by the identification of an originating shock, disturbance, or struggle. The terms of the narrative compel patients to regard themselves as agents in the development of symptoms that affirm things they wish to hide or refute. The narrative thus becomes a means to reassess and reconceive their biography as a risk to their body.

The specific character of hysteria seems to arise not so much from the crises themselves, but from what provokes them. The reader's attention is redirected away from the symptoms, away from the physiological affections, and onto moral affections. It seems to be the pathology's lack of linearity that leads physicians to grant more space to biography, as a way of introducing continuity. The doctor does not so much translate the transformation of the body as make the patient's biography the key to the body's belonging to the subject. The ruptures originated by the sick body allow the doctor to reconstruct the subject as story; as its interpreter, the doctor reasserts the sovereignty of the story over the body, of the gaze over the act, of writing over the sensitive manifestation.

✶

Many medical narrations of this period described the spectacle of the symptoms to account for the intensity of the hysteric illness. The anonymous doctor and Louyer-Villermay, in contrast, downplay the fit itself to bring to life its before and after. It is possible that these "observations" were invented. But it matters little whether they are genuine or fake: what the texts demonstrate is the particular role of narration in the theoretical elaboration of the pathology at the end of the eighteenth century. The deployment of a narrative model, as witnessed here in Louyer-Villermay's work, is meant to convince the reader of hysteria's specificity—to open a window onto the patient's affection. The observation genre, then, makes it possible to contextualize the pathology and to weave a relation between symptoms and untheorized information gathered about the patient. It is a means to inscribe temporality as a key to the coherence of the pathology. Even if the association between the behavior and the affliction cannot be articulated explicitly, the building of a narrative assigns the pathology's features to the patient. The relationship between narration and conceptualization is dual: the description functions as explanation, and writing up the case is the moment when the physician constructs the pathology as arising from the patient's experience. While the observations appear to offer them as mere examples, these cases present an opportunity for implicit interpretations that affect the perception of the pathology—the narrative format allows the doctor to express what cannot be stated without proof, and takes the place of logical linking. An ideological sedimentation takes place in the gathering of these observations. The act of narrating is an act of acquiring knowledge.

If in the first half of the century the physician sought to explain the pathology by examining its crises, the new importance attached to narrative now locates the cause somewhere else. The aim is now less to find out which organs are affected, which symptoms appear: it is to piece together everything that preceded the access. Far from being "observation" as practiced in this era of the birth of the clinic,[105] the exposition makes no attempt to reproduce the moment of examining the patient. It becomes instead the unrelenting unfolding of a plot. The physician as narrator identifies the turning points, the emotional shocks, the secrets.

What distinguishes such texts from earlier approaches to hysteric passion or erotic melancholia two centuries before—for example, Louyer-Villermay's approach from Jacques Ferrand's in his 1623 work on erotic melancholia?[106] At first sight, very little: the description of symptoms, the afflicted (young women from all social classes), the causes (love, desire, social constraints), repeat themselves. Ferrand himself recounted a story similar to that of Erasistratus, and made reference to him. In some respects, there seems to be a repetition without progress, merely interweaving the ancient theories with new literary references. But the *writing* changes. At the beginning of the seventeenth century, what mattered was to handle as many cases and references as possible, without lingering on any of them, in the attempt to arrive at a general discourse. Ferrand is interested in amorous practices, which he presents as having extreme and ridiculous traits, as impetuous motions. He focuses on the obstinate character of love and makes pathology its natural conclusion. At the beginning of the nineteenth century—for example, with Louyer-Villermay—what matters is the detail of each singular case as an opportunity to understand the relationship between the body and the soul.

What develops in much medical writing over the late eighteenth century is not so much the "gaze" discussed by Foucault as a form of narrative intelligence. The narrative moves away from the spectacular approach to the disorders and instead reconstructs each of its steps, inserting continuities into what had long been presented as discontinuous. The temporal dimension of the narrative acts as an intermediary in the collection of symptoms: it allows the configuration of the diagnosis. Narrating here means creating a link between affection and pathology. It supplies the necessary mediation to move back and forth from case to pathology, thinking the pathology from a case and the case from the pathology.

In order to take the vaporous crisis outside the realm of the extraordinary, it must be seen as a *reaction* by the body; being a reaction, it can be given a meaning. The aftermath of the fit inscribes itself in continuity, so that it too can function as an explanation. Removing the demonic character from symptoms does not suffice—the doctor must align them within a linearity of the sensitive body. Nothing can assure the coherence of this linearity better than the story of the subject. The doctor has come full circle: he bears witness to a body that is testifying, a body that not only authorizes its testimony but precedes, demands, summons, and institutes it. The hypothesis of the body's coherence can thus be turned into an ascription of subjects to their bodies, assigning those bodies the responsibility of explaining what has disturbed them. Patients approach their own bodies in search of a story and provide doctors with a portrayal of their maladies that is suited to the construction of a story.

In such writings, the idea of a body-as-truth—later fundamental to psychoanalysis—emerges. The pact between the doctor and patient is presented as a pact of truth (truth of the body and symptoms, truth of narration, truth of diagnosis), but it often occurs through a series of violations: the lie of the symptoms, the lie in the account of circumstances that preceded the disorders, the lie applied to identify the cause of the pathology, the lie applied to cure. In these lies, new relationships are woven in a kind of reciprocal contamination: traps and countertraps, procedures of simulation and dissimulation, and tactics of persuasion and denial. The doctor appears to authorize himself by means of the lying body, which permits him to tell lies of his own in search of knowledge. But the lie arises out of desire for a story, and rather than being seen as a suffering body, the body is presented as a witnessing body that stages that story. The doctor witnesses the body's declarations and becomes the interpreter of its language. He institutes it as that which contains truth, a truth transmitted in the course of its suffering. The patients' narrative becomes a narrative of confession, associating the desire for therapy with a departure from dissimulation. It is thus by passing through a series of lies that doctor and patient postulate the truth of the pathological body.

There is some continuity between such narrative and the conceptualizations of hysteria through metaphors as discussed in chapter 2, for both instigate a game of hide-and-seek. If hysteria once mobilized the doctor's attention as a pathology obscured by symptoms borrowed from other afflictions, by the end of the century it is a pathology tied to revelation. What was a masked pathology now becomes a mask, a mask that reveals itself as such. The manifest aspect of the pathology—its evident presence, its violence— becomes the sign of a secret or a lie and capable of being deciphered. Hysteria no longer hides itself; it reveals. Hysteria is no longer an unpredictable and inopportune outburst; it is a result.

The shift suggests the necessity for a new science such as psychoanalysis to emerge, a science that would establish the role of the physician in this discursive exchange. And while many historians have seen the birth of psychoanalysis in the therapeutic practices of Mesmer and Puységur,[107] the texts examined here show that it is narrative that provided the fertile ground for a new understanding of hysteria, orienting the body toward what psychoanalysis would later invent.

The writing of observations and their new importance at the end of the eighteenth century confirm the transformation of the relationship between medical discourse and its object. The move to narration allows a new approach to the pathology, which has become a signifying modality. What develops is a different relationship with knowledge and language. Commotion is supplanted by the satisfaction bestowed by an explanation. Narrativization bends the biographical to the demands of the pathology's explanation. Language takes up a new role as reformulating, understanding, intelligence. There is a seizure of power by language, a move to appropriate the body. With the narrations of the end of the century, the doctor claims the superiority of the one who knows how to analyze his patients' interiority, and how to evaluate bodily disorders as signifying links. Language makes it possible to reduce the body's distraction to a series of effects; the narrative configuration makes it possible to think the diagnosis, to justify it, and to make it evident. The exercise of narrative affirms itself not only as the identification of the diagnosis, but as access to its intelligibility.

The reader may be surprised at the similarities in the use of narrative across the differences in writers' objectives, whether to divert, for literary works, or to cure, for medical ones. However, the definition of these objectives was not always as narrow and specific as it is today. In the eighteenth century, both literary and medical narratives aimed to move their readers and to produce knowledge. Both shaped understandings of the body and perceptions of its desirable and undesirable states. Questions of authority would later deepen the rift between the spheres of competence of leisure and of science, favoring the development of specific writing practices as means of affirming medical and physiological truths. But in the period I address in this chapter, novelists and physicians shared a desire to move away from abstract systems and obscure metaphysics, and joined in giving accounts of the constant variations of the body. Through its changing condition, they delineated the effects of sensibility and the powers of physiology. Literary and medical narratives provided them with styles that could render the plenitude of convulsive disorders and their unexplained components without recourse to mechanistic or strictly physiological approaches. Both kinds of writing dealt with the question of the body's transparency or opacity, and established the possibility of a continuous correspondence (or, in eighteenth-century terms, sympathy) between body and soul, the body being at times the illustration of the torments of sensibility, at times their provocation. Postulating such a strong connection seemed to explain both the vulnerability of subjects and the profusion of elements likely to affect them. We can venture to say that for late eighteenth-century writers and physicians, the later aim of objectivity would have completely missed the point of comprehending the body, omitting to trace the wide array of circumstances that exacerbated sensibility and favored bodily disorders. It was by avoiding uniform cases that these writers hoped to reach universal truths. Indeed, they specifically favored the most extreme and spectacular, putting their trust in the exposition of the multiple layers that such cases and stories would provide. Certainly, all these narratives assigned a role to gender, age, education, and way of life. But the writers did not claim to select data according to preestablished frames, and prided themselves on providing the most exhaustive description possible. Each scene, each observation, was a piece in the collection, whose absence would undermine the understanding of a character's emotions or a patient's pathology. Narrative functioned as the exercise and proof of attentiveness to the character's or patient's sensibility and thus, in turn, to the reader's own sensibility, paving the way for his progressive understanding of the affection. These disorders were comprehensible only if they could be shared, and narratives were to instruct the reader by moving him.

6

Adopting Roles and Redefining Medicine

Above all, the hysteric is someone who has a story, an histoire, and whose story is told by science. Hysteria is no longer a question of the wandering womb; it is a question of the wandering story, and of whether that story belongs to the hysteric, the doctor, the historian, or the critic.

ELAINE SHOWALTER[1]

In France between 1760 and 1820, in the context of discontent with the ancien régime and the birth of the new nation, hysteria was increasingly instrumentalized to represent a form of moral degeneracy. This final chapter asks how medical writings fashioned a role for the doctor as a savior in the battle against this pathology. It spotlights some of the medical theories and practices around hysteria, which continued to diversify. Of the multitude of approaches circulating in this period, some espoused the role of the nerves—important examples are Joseph-Louis Roger from Montpellier, with his concept of a nervous fluid that could be affected by music,[2] and Thomas Trotter from Edinburgh, with his famous treatise on the nervous temperament.[3] Inquiry into nerve physiology and the brain also began at this time, notably with the work of Thomas Brown and Charles Bell, paving the way for even more radical changes in the interpretation of the body. Many other doctors, in contrast, adopted a uterine conception of the illness; they located its origin in the uterus rather than seeing it as the result of a dysfunction. This diversity in approaches to hysteria underlines the fractured and disputed fortunes of the term, inextricable from the diversity of the era's political and epistemological stakes. Across these different theories, however, one thing recurred: the construction of a new status for the doctor, emphasizing his role and power.

As Jacques Léonard writes, "The authority of doctors expanded, their prestige increased *before* their efficacy was confirmed. Their power was relied upon *before* their knowledge had been proved."[4] The invention of hysteria well exemplifies this dynamic, contributing as it did to the establishment of doctors' ambitions in the late eighteenth century. Whereas the medical narratives presented in the previous chapter focused on the identification and

justification of the diagnosis in specific cases, the texts selected here emphasize the image of the doctor in his national role, with responsibilities around
both the pathology's cure and its prevention. This period has already been
studied in detail by historians addressing, especially, the numerous changes
in medicine's status and role in the state, but my own interest is in the part
played by discursive practices and rhetoric in promulgating a vision of medical activity as something superior, and on the use of hysteria as an exemplary pathology within such discourses. To that end, the chapter looks at
three contemporaneous frames in which the therapist was associated with
hysteria: a spiritual frame, a magnetic or somnambulic frame, and a moral
and social frame. Despite the diversity of these constructions, a new figure of
the doctor—alongside an increasing stigmatization of hysteria—takes shape
through their polemics, diatribes, narratives, and memoirs.

The chapter begins by examining Pierre Pomme's religiously framed polemic on hysteria. Needless to say, polemics between doctors were not an
eighteenth-century invention. Medical works had long been a site of furious
debate, and the late seventeenth century was marked, especially, by a violent
dispute between two celebrated doctors to the king, Raymond de Vieussens
and Pierre Chirac. The contribution of Pierre Pomme's work was to place
polemics at the center of debates on therapy. Before Pomme, polemical works
were conceived of as a territory annexed to knowledge, a space of purely rhetorical dispute: Vieussens and Chirac argued about plagiarism and immoral-
ity, and Vieussens reported that he had long refused to enter into this debate
because it took him away from his real work.[5] For Pomme, however, the use
of rhetoric is inseparable from the invention and formulation of knowledge,
from the invocation of metaphysical powers for medicine.[6] The same texts articulate both his battle against adversaries and his medical convictions, forg-
ing a new role for the doctor as a healer who acts beyond all expectations and
cannot be reduced to his knowledge of medicine or physiology.

I investigate the frame of magnetism through writings on the practices of
Franz Anton Mesmer and Jacques-Henri-Désiré Petetin. Like Pomme's, these
works propose a specific configuration of the practice of medicine that implies a new interaction between the physician and his patient, an interaction
favoring metaphysical projections; this time, magnetism is the practice that
endows the healer with extraordinary powers. Polemics, sarcastic broadsides,
and narratives contribute to the creation of a mythic role for the doctor, who
is suddenly gifted with mysterious powers. The result is a quite new scenario
for the therapeutic act. Petetin, in particular, plays strategically with a reversal
of the relationship between doctor and patient, thus opening up new ways of
envisioning their respective status.

Yet at the very same time a third configuration was at work, addressing hysteria within a moral and social frame. Inscribing the physician and medicine into the heart of ordinary daily life, these writings also reinterpreted the French Revolution and the new nation as the results of particular pathologies. This approach in no way undermined the status of the physician; on the contrary, it presented his work as most necessary in the new nation. At stake was now the physician's moral responsibility and a new conception of the role of medicine, this time dedicated to prevention. The shift from therapy to prevention led to a new focus for medicine: women. The existing gender distinction in the identification of the pathology was further developed, and hysteria's interpretation as a uterine illness was implemented in a new way.

In the simultaneous development of these various visions, a new status for the physician was asserted. As this chapter will show, early nineteenth-century changes in the discourse on women and hysteria, far from affecting women alone,[7] were part of much broader shifts—shifts in which medicine was reframed and doctors in France deliberated on how to position themselves vis-à-vis the new government. In this process, discourses on women that had been circulating in the 1770s and especially the 1780s were reinvested into a new idea of doctors' roles. Once again, the same notions circulated in new contexts; the same objects were deployed in pursuit of new stakes.

To Mystify or to Demystify? Establishing the Role of the Therapist

Pierre Pomme has long since disappeared from the annals of medicine's celebrated doctors.[8] Yet in the eighteenth century, he enjoyed unprecedented success following the publication of *Traité des affections vaporeuses des deux sexes, ou maladies nerveuses, vulgairement appelées maux de nerfs* (Treatise on Vaporous Affections of the Two Sexes; or, Nervous Illnesses Vulgarly Called Nervous Disorders). Born in Arles, Pomme studied in Montpellier and became one of the doctors of Louis XV. As early as 1767, he was awarded the titles of Consulting Doctor to the King and to the Grand Falconer, titles he would retain until the Revolution. They allowed him to practice medicine in Paris and Versailles without a further academic qualification from the Parisian Faculty of Medicine. On June 30, 1803, he was received at the Société Académique des Sciences in Paris, and in 1808 he introduced himself as a member of the Sociétés Académiques of Vaucluse and Marseille and a correspondent of the Académie Celtique and the Société de Médecine in Avignon. Initially published in 1760, his treatise was reissued with additional material in 1765 and 1767, then in two volumes in 1769, with further additions in 1771, and in quarto pages accompanied by an engraving of the author in 1782. A new

edition published in the revolutionary year VII was expanded to three volumes in 1804. The work was also published in Italian in 1765,[9] in German in
1775,[10] in English in 1777,[11] and in three Spanish editions before the century's
close. These incessant republications were met with sustained criticism. By
1763, an anonymous work had already appeared discussing Pomme's treatise
with fierce sarcasm. Pomme entered into dialogue with this kind of critique,
and even encouraged it. He parried it in a polemic published in the *Journal
de Médecine* shortly thereafter, and began publishing the attacks in new editions of his treatise that responded to his rivals and added new accounts of
successful cures. By giving a forum to his adversaries, he acted as though the
controversy were a matter of public interest, and positioned the treatment of
the vapors at the center of medical practice.

As became clear earlier in this book, the storm of debate around the vapors
was fueled in part by their fashionability, and over the course of the century a
deluge of medical treatises was devoted to the malady. The diagnosis encompassed a host of symptoms including headaches, weeping, convulsions, and
fainting. Doctors' explanations of the illness also differed widely, variously
identifying its origin in the womb, brain, fibers, nerves, worms, sensibility,
and imagination. However, Pomme's *Traité des affections vaporeuses des deux
sexes* presents readers with a new approach to the pathology. For Pomme, all
conditions of suffering, convulsion, and paralysis are caused by a hardening
of fibers arising from a dryness of the body. They may additionally be an
effect of popular tonic remedies such as quinine, but all are caused by hysteric or hypochondriac maladies.[12] To treat such conditions, he recommends
taking cold baths that may last between two and ten hours each day (here
echoing what Daniel Roche calls the late eighteenth-century "apologia" for
water[13]), moderate bleedings, and liters of chicken, veal, or frog broth.

Pomme's descriptions of his treatments give shape to a phenomenology of the therapeutic act, but his treatise is even more striking in terms of
its writing. Unlike many of his colleagues, Pomme does not linger on the
findings of renowned doctors, nor propound a detailed theorization of the
malady and its causes. Instead, his work is devoted to the narration of individual treatments carried out by himself—a focus on the heroic healer that
imitates the practice of charlatans and empirics. Observations make up almost the entirety of the *Traité*, with rare moments of theoretical reflection
subsumed in the case narratives. Pomme constantly contrasts his successes
with the fruitless treatments of his colleagues.[14] His accounts of his cases use
evocative imagery. The focus on results rather than on comprehension of the
illnesses seems to target prospective patients at least as much as fellow doctors. Pomme displaces the emphasis from the malady itself onto his patented

therapies, making skillful use of the surrounding polemics to dramatize his discourse. The dissemination of Pomme's treatise was accelerated by the controversy surrounding it, and it eventually attracted the interest of Voltaire. Voltaire references Pomme in his *Questions sur l'Encyclopédie*, casting him as an exorcist for modern times. For the philosopher, Pomme's treatment of hysteric and hypochondriac affections exemplifies the struggle against fanaticism and ignorance. The term "exorcist" evokes a set of images in which the status of both Pomme himself and the discipline of medicine are at stake. The figurative schemes developed by Pomme and Voltaire in their ensuing private correspondence, as will be seen, helped give rise to a myth of therapy, as Pomme deployed rhetoric to create an image of his role as a doctor while pursuing his own strategic aims.

Voltaire's *Questions sur l'Encyclopédie*, published in nine volumes between 1770 and 1772, accords Pomme a position of prestige in the entry "The Great Exorcist":

> The possessed, the maniacs, the exorcised, "or more exactly" those made sick by their womb or chlorosis, hypochondriacs, epileptics, cataleptics, all are cured by the emollients of Mr. Pomme, the Great Exorcist.
>
> The vaporous, the epileptics, women tormented by their uterus are always said to be the victims of malignant spirits, evil demons, and the vengeance of the gods. We have seen these pains called the sacred sickness, and the priests of antiquity were given responsibility for the maladies, since the doctors of the time were completely ignorant.
>
> When symptoms were very complicated, it was said that one had many demons in the body, a demon of furor, one of lust, one of contraction, one of tautness, one of bedazzlement, one of deafness; and surely, the exorcist himself had demons of absurdity and mischief.[15]

Voltaire bypasses more celebrated doctors such as Willis, Sydenham, Lange, and Brisseau, describing Pomme as armed with an original array of techniques: bathing, broths, and bleeding. Because Pomme achieves the primary goal of moving away from superstition, he is offered as a symbol of the Enlightenment. His work contributes to a new construction of knowledge through its affirmation that hysteric maladies are physiological and rooted in the body. The past naming of demons is rejected and replaced by a theory on the mechanics of different parts of the body, with Beelzebub and Astaroth relegated to an illusionary world.

Retelling the story of Saint Paulin's treatment of a possessed man, Voltaire lifts an episode from religious history and reduces it to the level of anecdote. The hagiography is divested of the authority of the religious tradition on which it relies. Voltaire continues his attack on superstition by recounting

the floating test, for a long time used by religious authorities on those suspected of demonic possession:

> It often happened that epileptics had withered fibers and muscles that weighed
> less than an equivalent volume of water, and that they floated when put in a
> bath. People cried: Miracle! People said: They are possessed or a witch; and
> went for holy water or an executioner. It served as unquestionable proof ei-
> ther that the devil controlled their bodies, or that they had given themselves
> over to him. In the former case they were exorcised, and in the latter they were
> burned.[16]

Voltaire rereads the floating test in light of Pomme's experiences bathing his
patients. Pomme's medical treatise states that a loosening of tissue leads the
body to weigh more and sink to the bottom of the tub.[17] Floating becomes
an indication of the hardening of the nerves, signaling hysteric or hypochondriac pathology.

After reading "The Great Exorcist," Pomme responded to Voltaire with
a letter carrying an entirely different interpretation of his own significance.
Seizing on Voltaire's use of religious terms only to negate their meaning, he
also begins to cast his work in a theological framework, but undoes the irony
of Voltaire's writings:

> Yes, Sir, I heal the demoniacs, I return sight to the blind; I make the lame walk;
> I resuscitate the dead; and I proclaim to the entire world the simple manner
> in which I perform these miracles; however, I have not yet succeeded in cur-
> ing the hard-headed, the fanatics, and the doctors devoid of faith; it is on this
> miracle that I now must work. . . . I had lost hope completely; but I now feel
> my courage returning to me after reading your encyclopedia; when I see you
> celebrating my miracles, you give me the desire to return to my work. I will
> take up my quill again to attack the perverse sect that is the enemy of truth
> and all that is good.[18]

While Voltaire associates Pomme's therapies with exorcisms as a part of
his attack on religion, Pomme views the connection as a mark of prestige. The
doctor appropriates the theological imagining by reinvigorating Voltaire's
metaphor with added details and images. Not limiting himself to the figure
of the exorcist, Pomme evokes the image of Christ, healing the lame and the
blind. And instead of using the term "miracle" from a critical distance, he restores the signification of the word and recovers its power. Voltaire's affected
gullibility derisively derails the meaning of the scriptures, dissolving its pretension to truth while emphasizing its history. Pomme, in contrast, imports
the original significations of the terms without their context, repeating them

in a grandiloquent style proper to the narration of miracles. A true subversion of religious discourse emerges, and one can well imagine that Pomme's response must have amused Voltaire. Pomme's identification with religion functions to reduce theology to a world devoid of transcendence. On June 27, 1771, Voltaire writes to Pomme from his exile in Ferney:

> Madame Racle, who lives in my wilderness and has long been possessed by the same demon as the woman with the issue of blood, has not yet been healed by your liquids; but such demons take time to be chased away, and I still believe you to be a very good exorcist.
>
> I know that in your path you will encounter scribes and Pharisees who will endeavor to disparage your miracles; but no matter what they say, your kingdom remains in this world. As for me, I am possessed by a demon that makes my eyes as red as the touring festivals in the almanacs and almost completely deprives me of my sight; yet I will make myself read all that you write against the enemies of your doctrine with the greatest pleasure. I have faith in your gospel, though the men of our time remain difficult to persuade.[19]

Voltaire alludes to the miracle of a woman whose hemorrhaging ceased after she touched Christ's garments, and goes on to spin a web of metaphors.[20] The terms "miracle," "Pharisee," "demon," and "gospel" extend the parallels established in *Questions sur l'Encyclopédie*. Pomme responds immediately, maintaining his identification with Christ:

> What must be done to convert the unbelievers? . . . Miracles! Yet those that Jesus Christ performed before Israel angered the Jews, who crucified him rather than converting. I almost suffered the same fate in Paris and Lyon, as everyone knew at the time. Like these hardened Jews, doctors will never convert; or will, like the Jews, not convert until the end of the world.[21]

Far from questioning Voltaire's use of theology, Pomme continues to interpret religious associations as a mark of prestige, and his apocalyptic tone transcends the usual polemics against colleagues. He brandishes religious imagery to valorize his practices. His stubbornness becomes devotion, his writings gospel. Metaphors allow an aura to arise around Pomme, through signs he borrows from religious scripture. He begins to present himself as a savior who brings the dead back to life, confronting masses of infidels who aim to stop him at any cost. Every circumstance becomes a pretext for the employment of religious language. In the 1782 edition of his treatise, Pomme states that after his departure from Paris a public notice in Amiens reported that he had become an expatriate, while a bulletin in Bordeaux declared him dead. The doctor includes these reports in the preface to the next edition of his work, in which Voltaire is also cited.[22] Pomme also relates Voltaire's comparison in a letter to the Marquis de Grimaldi, published as part of the sixth edition of his treatise:

> The news you have heard is not without foundation; for I am in fact mad; how could I not be, I who take it upon myself to correct the doctors, preaching morals contrary to their beliefs; for them, this is a new religion, and you know what that cost the apostles who preached the religion of Christ. As such, one must suppose either that my zeal is truly apostolic or that I am mad. Voltaire wrote to me from Ferney, asking my advice on the illness of Madame Racle, attacked by a loss of blood that he compared to the sufferer of Jerusalem. He said "That difficult though he is to persuade, he believes in my gospel; but that in my path I would encounter scribes and Pharisees who would dispute my miracles; and that my kingdom was not of this world." Voltaire knew, as I do, that *nulla est invidia supra invidia medicorum.*[23]

In the letter Voltaire addressed to Pomme, he affirmed that Pomme's kingdom *was* of this world, distancing medical practices from the notion of metaphysical power. Pomme rewrites Voltaire's phrase to read that his kingdom is *not* of this world, proclaiming a boundless authority. He uses Voltaire to make himself appear a divinely chosen healer and to grant himself the very powers that the philosopher denounced. Rather than demystifying the mechanics of the convulsive body, Pomme mythologizes his own medical practice. With the influence of the church in decline, the religious lexicon becomes an open repertoire of metaphors, surrounding medicine with the air of authority it had previously lacked.

Voltaire's final mention of Pomme appears in a letter to a friend describing a "ferment" agitating Paris and the countryside. Reporting that two of his nephews are competing for a seat in the Parisian parliament, Voltaire writes: "It makes me laugh, and I laugh because I do not believe that our national illness is fatal. These symptoms are vertigos that ought to be cured by Mr. Pomme."[24] For Voltaire, the reference to Pomme serves to relativize and defuse tensions. By presenting political disputes as forms of pathology, Voltaire dismisses their importance. He casts them as uncontrollable movements that do not merit serious attention and are more aptly described with the language of pathology than that of reason. Presenting them as a malady suggests a temporary reaction that could never cast the French political system into question.[25]

Pomme, in contrast, from this point on constantly invokes theological references. Presenting himself as a savior, he crafts a new image for medicine. Following his exchange with Voltaire, Pomme recalls and rewrites past successes for new editions of his treatise. Two predominant strategies come to characterize his works: the use of metaphors, and the writing of repeated new versions of patient observations as emblems for his therapies.

The doctor's weaving of metaphors becomes increasingly hyperbolic as his work gravitates toward the genre of the epideictic. The most extravagant of images, such as martyrdom and resurrection, are repeated at regular intervals. The sixth edition of Pomme's treatise, in particular, piles up such terms: "What it costs to speak the truth! And when one has the courage to declare oneself an apostle of the truth, should one not expect to become its martyr?"[26] Pomme reinterprets and reframes his actions through religious imagining, and presents his entire practice through the prism of religion. He speaks of converting rather than convincing, his patients are lame rather than feverish, and he sees paralysis rather than numbness. He recounts the suffering of those who did not heed his advice in a kind of catalog of martyrs[27] and likens the pharmacists and doctors who oppose him to an impious sect. Pomme rescues dying patients, proclaims compassion for doctors who oppose his doctrine,[28] and pronounces a mea culpa when recalling the patients he was unable to help earlier in his career.[29] Pomme writes that he is stopped during his travels by sufferers on the roadside begging for his care,[30] and leaves his door open to "paupers."[31] He speaks of his own words as "sacred truths."[32] He repeatedly emphasizes the loyalty of his patients—they become his faithful.

Pomme continued to publish until he was almost eighty years old, citing his longevity as evidence of the righteousness of his struggle. In that period of his life, he writes that he is comforted by the attention his treatises have received: "Did I preach in the wilderness, like John the Baptist? No, thank God, no. My anathema against the abuse of quinine woke the savants from their slumber regarding the most interesting aspect of clinical medicine."[33] Throughout his ever-expanding repertoire of references, Pomme uses signs of compassion and divine calling, and signs of competence and recognition. These signs stretch across descriptions of individual cases, weaving a mythology of therapeutic deeds. As part of this movement of expansion, Pomme appropriates the history of religion, religion's lexicon, and its role in society. He consecrates his work through the act of writing. Pomme attributes a transcendent value to his medical practice, hypostatizing medical discourse by raising it to the status of truth and faith.

The use of rhetorical figures is supplemented by narrative strategies in the rewriting of cases. His illustrations operate as evidence persuading readers to support him, and the treatment of a patient named Lazarus Vidal offers a unique showcase for Pomme's talents as a writer. Here is the first version, published in the 1765 edition of his treatise, before his exchange with Voltaire.

It appears as a footnote and stresses Pomme's professionalism in employing a variety of methods to save a patient:

> Lazarus Vidal, a native of the village of La Baume in the principality of Orange, eighteen years of age, with a dry and slight temperament, was brought to the Hospital on July 9, 1763, by men of the countryside, who found him lying under a tree without feeling or movement, such that they believed he was dead, I found this patient at the hospital at the hour of my visit. His pulse was very slow and concentrated, his jaw was so stiff and immobile that it was impossible to make him drink a single drop of water; this indicated that he required vesicatories. The next day, July 10, there was no change. On July 11 he remained the same. On July 12 leeches and irritating enemas were applied, they were applied again on July 13, still without success, & on July 14 he developed a fever. Such a desperate case allowed for all sorts of experiments. Ice was applied to the patient's head.[34]

In this first version, Pomme writes a detailed progression of the malady. He demonstrates the benefit of his intervention by describing the initial hopelessness of the case. He presents his perseverance as a testament to his professionalism, recounting a succession of experiments as though illness could never withstand such medical devotion. But the tone and meaning change drastically when the case is retold in 1771, befitting a case ideal for embellishment in the terms suggested by Pomme's exchanges with Voltaire. Here is the tale of Lazarus as published in *Nouveau recueil des pièces publiées pour l'instruction du procès que le traitement des vapeurs a fait naître parmi les médecins*:

> A poor beggar (Lazarus Vidal of Baume . . .) was found dead under a tree, exposed to the scorching heat of the sun. He was sent to the Arles hospital and placed in a coffin. They were about to bury him when I arrived for my visit to the hospital: I received the information necessary to understand his sad death; I examined the corpse with the surgeon of the establishment. He found the man's jaw to be in convulsion since it was stiff and immobile; from which I concluded that Lazarus was not dead, and that like the Lazarus of the Gospel, he could be brought back to life. The corpse was thus plunged into a cold bath in my presence, and returned to life.[35]

In this second version, Pomme no longer concentrates on the description of the patient's condition or the treatments applied, but instead on staging the doctor's personal intervention. Pomme opts to forget the five days of fruitless efforts, and the patient is now instantly healed upon Pomme's initial treatment. What was a perilous lesson in 1765 becomes an immediate success in 1771. There is a clearly dramatic flair to the second account of the story, with the young patient lying in a coffin upon Pomme's arrival. The doctor is not treating a living patient; he is examining a corpse. The cure thus resembles a miracle, and the scene is made more suggestive by recasting the "native of the village of La Baume" as "a poor beggar." With this rewriting, Pomme positions himself as a savior, and unequivocally evokes the image of Christ.

A third retelling of the episode goes even further. After a purely oratorical reference to his methods, Pomme's 1804 *Supplément au traité des affections vaporeuses des deux sexes, ou maladies nerveuses* repeats his list of preferred treatments. These therapies are reviewed summarily, emphasizing Pomme's essential role in their application. To substantiate the value of bathing, Pomme again cites the case of Lazarus:

Must one cite the mute? Must one cite the lame? Must one resuscitate a dead man? Lazarus Vidal of La Beaume . . . was found dead under a tree, exposed to the power of the sun; he is brought to the hospital in Arles where I was working at the time; he is covered with a shroud and they are about to bury him, this corpse is plunged into a cold bath in my presence and that of . . . the dean and the entire faculty of the hospital; he comes back to life.[36]

In this third version, Pomme employs the rhetorical device of enallage, using grammatical inconsistency in his choice of tenses to add urgency to the scene. The account takes the form of a hypotypose, with the cure seeming to take place before the reader's very eyes.[37] The scenographic quality of this narrative adds an element of performance to Pomme's medical intervention. The use of an ordinary bath without additional prescriptions focuses attention on Pomme himself, and he now acts alone, watched by the other members of the faculty. Pomme's writing seeks the power of immediate presence. His literary and rhetorical strategies enable a constant refashioning of therapies that are themselves not particularly unusual. At the same time, by placing water at the center of his treatments, Pomme lends them to a symbolic reading. Water bestows new life, and new life is procured by the presence of Pomme as much as by the bathing itself.

When Pomme cites Voltaire to underpin his religious imagery, he plays a trick on the philosopher, using his name, his authority, and his celebrity while reversing the implication of his reading. In taking him at his word, he catches him in a trap. With great élan, then, Pomme blends biblical references, invocations of philosophical authority, and espousals of his therapies.

In Pomme's writing, a textual strategy heralds a new practice of medicine, one constructed through naming and citation. Pomme's work is characteristic of eighteenth-century medicine in claiming to renew the practice of medicine and using published polemics to build his reputation as a doctor. It would be possible to interpret these trends as signs of the weakness of eighteenth-century medicine, the discipline resorting to literary strategies in an attempt to make outdated practices appear new. Yet there is more at play in Pomme's writings. He understands that he needs to speculate on the public perception of medicine and of the physician in order to affirm the authority of his practice. A minor dispute between colleagues would fall short of inaugurating a new vision of medicine, but Pomme can assert his standing through continuous extrapolation of his role as a doctor. Controversy is placed at the very heart of the text, giving it its raison d'être—or, to speak in eighteenth-century terms, its nerves. Pomme's extravagance functions to confirm his authority as doctor. To convey the enormous stakes involved in his writing, he adopts a tone that is both lyrical and severe. Pomme's treatise hurries readers into taking sides and making judgments. He demands a partisan reading. Through his publications and polemics, Pomme draws on the growing status of medicine to cast himself in a unique role. When other doctors reject his methods, they are only proving his singularity.

In concentrating his energy on the narration of treatments rather than descriptions of illness, Pomme emphasizes his experience in daily encounters with patients, countering the image of doctors as savants lost amid antiquated books and obscure references. Unlike many of his colleagues, he does not advance abstract theories on the origin of maladies supported by only a handful of examples.[38] In addition, his attention to the success of his treatments allows him to profess devotion to the health of his patients, and in return, he pressures them to actively endorse his claims. Rather than combating illness alone, Pomme also appears in a struggle against his colleagues and their remedies—his fervor indicates his desire for action more than reflection. His role is depicted as being to transform rather than to observe. His patients are no longer seen as victims of their bodies, nor are their illnesses viewed as judgments on the excesses of their lifestyles. In Pomme's writings, patients are depicted as martyrs who have suffered at the hands of callous and greedy doctors. Pomme maps the discipline of medicine as a battlefield, but this war is waged in the space of writing, in the game of citations, and on the field of language. By writing himself into a Christ-like image, Pomme aligns the therapeutic with the moral, the quotidian with the spiritual. The prestige of his personal practice lifts him above the customary position of a doctor in society. No longer is he a servant to aristocrats, catering to their every caprice. Through his writings Pomme ascribes to himself a dual mission to save the sick and enlighten his colleagues. In the course of his hyperbolizing retellings, he emerges as a healer apart, his portrait reflected in the mirror of each account.

The historicization of hysteria has traditionally led to a narrative in which the perception of the body first moved from the extraordinary to the ordinary—from an understanding in terms of theological manifestations to one framed in terms of pathological disorders.[39] The mystery of transcendence was supplanted by the problem of physiology, a problem for which an early solution was expected. Certainly, chapter 4 suggested a continuity between Willis and La Mettrie or Le Camus (to name just a few) in reinterpreting the role of physiology and consequently questioning the ascendancy of the soul. In this context, physicians, whether they had a specifically materialist agenda or not, appeared as advocates of the faculties of the body. In chapters 4 and 5 we saw physicians in the 1780s describing the phenomena they have witnessed as being physiological, yet outside of their realm of knowledge. But cases such as Pomme's indicate other, even diametrically opposed, perspectives. In these discourses, the physician presents himself as invested with immense powers of healing.[40] A similar emphasis on therapeutic success also marks the work of Franz Anton Mesmer and Jacques-Henri-Désiré Petetin.

Magnetism, Parodies, and Mystification: The Art of Framing a Therapeutic Practice

In Pomme's work, the focus of the extraordinary spectacle of the body shifted from the mysteries of physiological powers to those of moral and spiritual powers. The locus of fascination changed, and it would change even further with Franz Anton Mesmer in the case of magnetism, and with Armand Marie-Jacques de Chastenet de Puységur and Jacques-Henri-Désiré Petetin for somnambulism. In their approaches, a perfect knowledge of the body is no longer deemed necessary: what matters is knowledge of the physician's powers over the body. Mesmer and Petetin return to the dimension of spectacle, but this time the performance is directed by the physician.

Mesmer's therapy was based on what he called animal magnetism. In propositions 1 and 2 of his *Aphorisms*, he defined this as "a mutual influence between celestial bodies, the Earth & Animated Bodies,"[41] an influence that worked by means of "a fluid universally spread, & continued so that it suffers no void, and whose subtlety allows for no comparison, & which, by nature, is able to receive, propagate, & communicate all impressions of movement."[42] Mesmer describes animal magnetism as a flux of energy that circulates from one body to another. His three works, published between 1779 and 1785, are very abstract, describing neither symptoms nor the organism. The therapeutic practice is theorized as the use of a magnetic flux to restore the magnetic balance.[43] These ideas were not new, many dissertations having been published on the role of magnets since the fourteenth century.[44] In the seventeenth century Jan Baptist van Helmont was deeply interested in the use of magnetism,[45] and in 1750 François Planque published a report sent by a patient frequently afflicted with trembling who has been advised to carry magnets.[46] The patient sees this as the solution to his fits, also insisting on the magnets' preventive effects as demonstrated by his suffering on the occasions when he forgot to carry them. All these ideas had been developed further throughout the eighteenth century, especially in the shape of possible therapies with electricity. As Jessica Riskin argues, "Mesmer did not depart from the standard wisdom by relating health to the regulation of imponderable fluids in the body; he merely stated a commonly held belief among natural philosophers."[47] But Mesmer took full advantage of the power with which the therapy endowed him. He tied it into a much longer-standing therapeutic tradition in France, as Laurence Brockliss and Colin Jones have stressed, noting that Mesmer "would conclude business by embracing his acolytes with the words, 'Go forth, touch, cure.' The thaumaturgical strain in such a formulation recalled the ancient refrain with which the kings of France touched those affected with the king's evil."[48]

By focusing on the use of the category, and not the history of an illness, my reading differs considerably from the way that historians have associated hysteria and magnetism in past decades. In her 1965 study of the history of hysteria, for example, Ilza Veith presents Mesmer as the precursor of psychoanalysis. Veith regards the therapeutic practice of magnetism as a first experimentation with hypnosis, and Mesmer as a figure who could cure hysteria even without giving a name to the forces he was putting into play.[49] A similar reading was endorsed by Henri F. Ellenberger in 1970 in his seminal *The Discovery of the Unconscious*, though unlike Veith he singled out one of Mesmer's followers, Puységur, who created a cure based on somnambulism.[50] However, precursors exist only in predetermined histories; this book, in contrast, argues that agents' roles and hierarchies change depending on the particular interpretive stakes. This divergence of historical reading is exemplified in the association of Mesmer and Puységur with hysteria. Neither of them sought to cure specifically those patients said to be hysteric, nor did they talk about hysteria or mental illness or illness of the mind. In fact, they were not interested in giving a name to their patients' disorders or distinguishing between pathologies at all, but solely in curing them. The only exception to this may be Mesmer's 1779 distinction between the nervous illnesses, which he engaged to heal immediately, and the others, which he said could be healed "mediately"[51]—but he did not explain what characterized or separated these.

In fact it was Mesmer's therapeutic practice, and not his theoretical con-

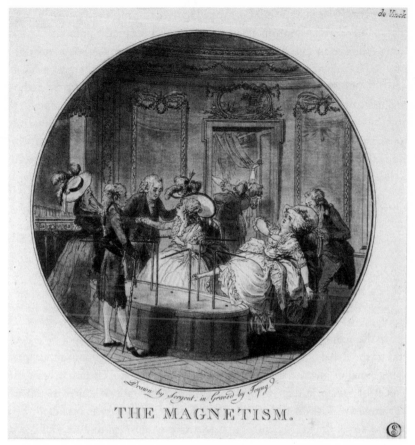

THE MAGNETISM.

FIGURE 12. Laurent Guyot, "The Magnetism" (Paris: Tilliard, 1784). Image courtesy of the Bibliothèque Nationale de France.

ceptions, that led to the association of magnetism with hysteric affections (see fig. 12). Many vaporous patients were said to have tried magnetism and famous hysteric patients such as the Princesse de Lamballe to have joined in. Mesmer first used magnets by wearing them as he touched his patients, sitting in front of them, their knees between his, one hand on their belly, one on the chest. The touch, the gaze, and the listening were part of a single sequence, and his practice remained similar when he renounced the use of magnets. When inundated by patients, he started using a vessel known as a *baquet* or "tub," in which they would sit together on a magnetized bench, holding ropes that had been magnetized in advance. At times he also had musicians play, especially the glass harmonica, to enhance the crises. Convulsions provoked by magnetism also required a private space, named the "chamber of the crises," to which he or his assistants would take the patients one by one once the crises began. The therapy itself provoked hysterical symptoms: a state of crisis that mainly consisted in convulsions, possibly rolling on the floor, and was at times followed by evacuations (tears, vomiting, excrements). Mesmer would proceed to palpate the patients at magnetic points to restore their calm. Magnetism made the hysteric crisis the therapy. The crisis itself was provoked by magnetic means in order to accelerate the pathology and the cure. The therapy thus rested on a hierarchy in which patients would abandon themselves to magnetic flux mastered by the doctor. An advocate of magnetism who studied and practiced with Mesmer until their falling out, Charles d'Eslon, writes:

> From there it follows that if he undertakes to cure a madman, he will only cure him by occasioning fits of madness. The vaporous will have vaporous fits, epileptics fits of epilepsy, &c.
>
> The great advantage of Animal Magnetism thus consists in accelerating crises with no danger, for example, we may suppose that a crisis happening in nine days with Nature left to its own devices will be obtained in nine hours with the help of Animal Magnetism.[52]

In the eighteenth century, Mesmer's association with the vapors was central to the debate on magnetism, and was the nub of responses either querying, magnifying, or sneering at his healing powers. D'Eslon stresses the aspect of spectacle that would later come to discredit Mesmer's practices: "From six o'clock in the morning until night, his house is assaulted: it is the theater of the most bizarre spectacle. One laughs, another cries: this one yawns, & the other shouts. Vapors, convulsions, delirium, & weaknesses come to ornament the scene together or in turn. He can never expect to have a free seat."[53] Mesmer's detractors use the image of the hysteric to stigmatize his practice and associate it with simulation and whim, associations underpinned in etchings and pamphlets that claim hysteric and magnetic healers are no more than the accomplices of a puppet master.[54] In these attacks, hysteria appears as a simulation of symptoms, while magnetism is the simulation of a cure; hysteric affections become a sexual exaltation, and magnetism the response to it.

At the end of the eighteenth century and the beginning of the nineteenth, partisans of animal magnetism themselves began to associate Mesmer's therapeutic practice with hystericism, producing a different narrative of natural secret powers. His adepts present him not as the first discoverer of magnetism, but as an important landmark in its use. The link between magnetism and hysteria recurs in an early nineteenth-century journal, *Bibliothèque du magnétisme animal,* devoted to the cause of magnetism: "Animal magnetism, we dare to say, is as ancient as the world; we find it in all eras and all the points of the earth; not with the modern name that it bears today, but under the veil of a mysterious science."[55] In order to give credit to the powers of magnetism, the author equates the experiences of Mesmer's patients with that of the sibyls, the similarity being that their physiological state leads to a crisis:

> Hystericism most often produced this prophetic state. One also observes that we qualify as virgins those who bear the name of sibyls. One should not believe that they kept their virginity because of an obligation imposed by divinity. What would that virginity have mattered to Apollo, to whom they were dedicated? The conduct attributed to him sufficiently proves he cared little about it. But priests might have a great interest in keeping them away from marriage, which is the natural remedy for hystericism.[56]

In line with many early nineteenth-century theories on hysteria, this member of the Société du Magnétisme identifies a lack of sexual relations as the cause of the insensible state in which women would utter prophecies. According to him, young women are driven into a state of hysteria so as to be presented to the people as sibyls. On one level, Mesmer is presented as being in touch with archaic forces, and his cure as located in a genealogy of the supernatural.[57] On another, hysteria is no longer only a pathology; it becomes the direct effect of virginity and of suggestion, a state that favors inspiration, exaltation, predictions. The writer also identifies the same magnetic forces in Christian miracles, in which he recognizes "the instinctive faculty of Man."[58] In this case, then, Mesmer is associated with hysteria not to discredit him but to place magnetism in an ancient tradition that Mesmer is capable of renewing. In this context, the doctor and the patient are caught up in a new relationship of power, where the traditional hierarchy is displaced and a new order instigated in which the diagnosis is unimportant: the focus is entirely on the physician's capacity to activate, then tame, the powers of his patient.

Other narrations soon counterbalanced the aridity of Mesmer's own publications, as if medicine could not make do only with theorems but had to inscribe the practice in commentaries and stories. Many of Mesmer's contemporaries applied his techniques and praised the possibilities opened up by these new therapeutic routes for patients whom they described as afflicted by somnambulism or hysteria. Hour by hour, they wrote down the cries, suffocations, and tremblings they were witnessing, often without the aim of publishing the accounts. Given the recent editions of several of these manuscripts, accompanied by nuanced analysis of the diversity of conceptions of the body on which they draw,[59] I will here look only briefly at one of the doctors in this tradition, Petetin, who did publish his observations.

Petetin identified the study of "the discovery of physical and moral phenomena that are veiled by hysteric Catalepsy, and that are its inevitable effects."[60] He treated his patients while they were in a state of what he called somnambulic catalepsy. These patients, "crisiacs" (*crisiaques*), are characterized by their capacity to be conscious of their surroundings while in a somnambulic state. His treatment utilized electricity, drawing off sparks with an electric machine he called an "excitator." It was applied with the patient lying down or in a bath[61] in order to dissipate catalepsy or modify it into "perfect catalepsy."[62] On other occasions, he used magnets to redirect magnetic flux. Here, Mesmer's therapeutic theatricality is rewritten: the orchestra, the stage, and the wings—the chamber of crises—planned by Mesmer are reinterpreted by Petetin as a chamber spectacle, and he invites the family and friends to the room where the sick woman is performing. The doctor visits his patient daily when she is afflicted by a crisis, at the moment she has predicted during her previous crisis. He often engages in several experiments before publishing them.[63] Extraordinary narratives are recounted: he explains how he communicates with his patient through her stomach, describing how she recognizes and names objects placed on her belly while she is in a cataleptic state. He notes how she describes all the objects around her and hears what he whispers behind one hand while touching her belly with the other. She predicts the time of her fits with exactitude. During the consultations, she becomes an oracle who speaks very clearly and divines who is around her, what the guests are wearing and thinking about, and at one point guesses the contents of the letter the doctor brings.[64] The doctor repeatedly plays with this effect of spectacle in his narrative, and reinforces the analogy by calling her the "prophetess."[65] But her dramatic new faculties are destined to be revoked: as the patient explains, it is only by losing these extraordinary faculties that she will regain sight and hearing through the organs intended to perform them.

The singularity of writing on these phenomena lies in its aim of making the reader marvel, playing with the irony of inversion: the patient becomes the one who knows, while the healthy person experiences his ignorance and inability to perform such actions. Petetin offers his patients the means to create their spectacle, so that indulgence and complicity between the patient and the doctor here reach their greatest extent. In hysteric pathology, the patient appears as the director of the play—yet the hierarchy does not disappear, for the exchange demonstrates the doctor's nimbleness in mastering knowledge, while the patient's state is exploited like a sideshow. In one of his last works, Puységur would radically criticize this therapeutic practice in reference to Mesmer: "After having obtained from a few cataleptic, hysteric, or somnambulic women effects all the more satisfying for these doctors' self-esteem, they always agreed with their own previous theory. They merely gave us systems."[66] And indeed, the theoretical construction of the diagnosis of hysteria and the therapeutic practice around it were means for physicians to affirm their own status at the end of the eighteenth century.

The differences between Pomme's, Mesmer's, and Petetin's writing practices indicate that the construction of hysteria was not a simple evolution of the understanding of bodily disorders, moving smoothly from a narrative of spectacular manifestations to a regularization of the fits as a spectrum of phenomena capable of explanation. Instead, many other dynamics, such as the resurgence of a fascination with the extraordinary and the search for a metaphysical force, were also involved. No teleological agenda toward a rationalized approach to the body can be identified in the history of the category; diversity was the most striking feature in the production of this knowledge. The rest of this chapter addresses the refashioning of traditional medical discourse (as opposed to magnetism) as a moral force in a corrupted country at the time of the French Revolution, doctors' investment in a strategic role for themselves in relation to the Republic, and the impact of this now-expanded role on constructions of women after the Revolution.

Strategies of Legitimation and Definitions of the Patient to Come

Whereas in the Enlightenment period the discourses of doctors and men of letters about the vapors coincided around the themes of emotions, imagination, and fibers and nerves, and sometimes seemed almost interchangeable in terms of style, at the time of the French Revolution the stakes of medical writing moved further away from those of its literary counterpart. The literary and medical texts considered in chapters 3, 4, and 5 often talk about the body in similar ways, both mainly concerned with the relationship between body and soul, but now starker divisions emerged between writers. As Dena Goodman has shown, the Republic of Letters was independent from royal institutions, was transnational, and framed its intellectual pursuits as being, although at times political, separate from the goals of the state.[67] In contrast, the doctors of the new political Republic envisioned a role as members of the state, fulfilling a national mission in their writing. French doctors might write about the immoderation of the Parisian lifestyle and the court, they might write about the "English" malady, but they did so in pursuit of universal conceptions of human nature or of modernity and its excesses. Doctors publishing after the French Revolution claimed a national value for their discourse, since it was committed to the future of the nation and the development of the full potentialities of its citizens.

In this section, I seek to analyze how physicians deployed the revolutionary discourse to secure their connections and sculpt their role in the nation,
a role they had only begun to claim since the creation of the Société Royale
de Médecine. I ask how forms of these discourses circulated, and how their
dissemination allowed perspectives that secured the status of the physician
to become predominant. Conceptions of hysteria in this period bear strong
traces of the political changes in France—in particular, there was a complete
reorientation of the category that made the womb, and the role of women,
crucial to a new future.

Two short letters sent to Samuel-Auguste Tissot in 1792 will serve as a
backdrop for this discussion. In July 1792, Gabriel Bègue wrote to Tissot:

> When I had the honor of consulting you a few months ago, to remind you of
> a more accurate idea of my illness, I sent you the copy of a letter you had writ-
> ten me in '72, twenty years ago, in which you advised me to take quinquina
> and cold baths and told me that my illness was a nervous hypochondria. I
> have certainly received, during the space of thirty years, a dozen letters from
> you. In the last one, of the month of April, you told me that my illness was the
> same as twenty years ago and that you could not give me other advice than
> what you gave me then.
>
> I would thus be happy if you could give me a simple declaration that I am
> afflicted with a nervous illness or with a nervous hypochondria for which I
> have consulted you. Anyway, whichever form you find appropriate for this
> certificate, I would be infinitely obliged to you.[68]

Curiously, in 1792 the diagnosis of a nervous illness could procure salvation for aristocrats. It promised to gain them access to a nursing home, the
director of which could refuse to give them up upon demand, thus protecting
them from the revolutionaries' attacks.[69] Another letter from an aristocrat,
dated late September 1792, supplements this first picture:

> I come as an old acquaintance to ask you for an essential service. To try to pre-
> serve my mother's fortune and ours, it would be necessary, Sir, for you to have
> the goodness to provide us with an attestation in the correct form according
> to which my mother Madame the Countess of Fcey arrived in Lausanne on
> July 20, 1789. At the time, she lodged the first night in the inn of the Golden
> Lion and later at the Spa of Bourras. Having always had in you, Sir, the greatest
> trust, the purpose of her trip was to consult you. I believe it is also necessary
> to put in the certificate that my Mother occupied her apartment at the Baths
> in Bouvian until the month of September 1789, and that she left to go to Basle.
>
> Dates so little favorable to us attest, I hope, my sincerity. We flatter our-
> selves to obtain an exception, since my Mother left this town in the month of
> June 1789 to come to you, Sir. If she did not arrive until July 1, it is because she

stopped on the way at our estate in Reugney to take milk and baths, and then went on her way on July 19, before being informed (in the mountains, where she was sick) of the events that happened in Paris on July 14.

I remember perfectly, Sir, your care during my childhood and my small-pox. . . . Time is very short for the service I take the liberty of asking you.[70]

The memory of childhood, left vague and gratuitous, is designed to awaken the doctor's sympathy, and contrasts with the great precision provided for the month of July 1789. The style of the letter, and its being written at a time of the most violent massacres, leave no doubt as to its motivations. The aim was to escape the Terror and to safeguard a property under threat after the flight of its owners. The diagnosis of vaporous and nervous illness appears here as something capable of saving those who initially bore it as a derogatory and stigmatizing label. The doctor finds himself credited with a new authority, bestowed by knowledge, which gives his word the value of competence and neutrality.

The two letters emblematize the reversal that this section attempts to trace: with the Revolution, doctors who previously paid court to aristocrats could suddenly lay claim to a new power. This was not the power to save by therapeutic means, but to certify in the name of their status. If the edicts of 1793 closed the academies and schools of medicine, opening access to medicine to all comers, the reorganization and centralization of their profession would soon give doctors a new authority, paralleling developments in the other natural sciences.[71]

At this time the medical profession was undergoing numerous changes, which have been studied by many historians over the last forty years. Despite the disputes between the Foucauldian reading and the Roy Porter tradition in British and North American scholarship, there is broad agreement that the turn of the nineteenth century marked a major change in the medical sphere. Foucault discussed the emergence of the "clinic," and in his wake many historians have focused on the epistemological change in the hospital.[72] Thus, Othmar Keel, while refuting Foucault's privileging of nosology in research on the hospital, agrees that Europe as a whole saw a rise in teaching in the hospital's spaces.[73] Others have shown that this development not only affected the space of the hospital and the methods of teaching, but altered the vision of medicine itself as a profession. Dora Weiner has analyzed the planning of health care as a national objective in the wake of the Revolution, and Ann F. La Berge has described the increasing role of hygiene, which made health a political priority in both cities and the countryside.[74]

Accompanying and crosscutting these transformations of actual medical practice and education, however, were changes in the discursive envisioning of medical roles. In the case presented here, they were driven by the theorization of a medical category: doctors' constructions of hysteria allowed them to produce claims for a new role. What interests me here is how the rhetorical refashioning of the position of the physician, from a court physician to a citizen physician,[75] accompanied the transformation of hysteria from a pathology described as a pathology of aristocrats—whatever the actual uses of the diagnosis among all classes may have been—to a pathology presented as threatening the nation as a whole (through the vulnerable constitution of women). This shift in meaning can be investigated by analyzing the discursive practices of a selection of physicians between 1770 and 1820, for physicians' assertions of the national importance of health measures and the crucial role of medicine were based not primarily on medical innovations, but rather on their elaboration of a new rhetoric around these ideas.

Until the French Revolution, doctors had little authority and were often perceived as incompetent. For example, it was extremely difficult for them to persuade the public to accept the vaccine against smallpox, despite having statistical evidence that it was a brilliant and efficient invention. Even after the Duke of Orleans had his children vaccinated, the wider public refused to follow suit. If doctors' lack of credibility changed after the Revolution, it was not, or not only, because of an invention—there was no major therapeutic innovation in those years. The authority that doctors gained was a moral authority, at a moment when the existing moral authority, that of the church, was under fire. As the church lost ground, a space became free that physicians could invest with their discourse of importance to the lives of the mother and the citizen-patient.

The ambition to write for the nation and thereby contribute to fortifying the French people thus galvanized the demands placed on the writing of medicine around the time of the Revolution. One of the resulting shifts is the subject of Foucault's *Birth of the Clinic*, which shows the emergence of a type of writing—the nosology—going hand in hand with the new importance of the hospital to medical knowledge. Foucault traces the interdependence of gaze and language in the process of observation. In the attempt, at the turn of the century, to move away from the medical conceptions that had previously framed the approach to patients (temperaments or humors, iatrochemistry, mechanics physiology, nerves), a new vocabulary and a new syntax were to be implemented, language that remained on the level of the visible without further implications beyond what could be seen. Foucault's identification of a method of interrogating patients serves to demonstrate the scientific evolution of a practice and its development in the public space, that of the hospital.

But the method Foucault studies was not the only writing practice aris-ing in the period. In this section, I trace the emergence of a different kind of discourse, not observational but rhetorical, one designed to reach patients in their home—a moral and social discourse that affirmed a new place for physicians in society and also involved the definition of the patient to come. I will focus here on two debates that developed before the Revolution as hys-teric and vaporous affections, on the one hand, and the magnetic cure, on the other, were made into emblems of civilization's excesses. My discussion aims not to retrospectively reconstruct the revolutionary climate, but to un-derstand the selection in the 1770s and 1780s of seminal themes that then take on a new scope after the Revolution; not to present doctors as revolutionaries or even to situate their participation in the Revolution, but to analyze their reinvestment of existing modes of describing the body of the nation—that is, the discussion that follows analyzes forms of writing and interpretive models through the prism of pathology, as forms of thinking about the social body that are inscribed in actuality.[76]

The eighteenth century was a period marked by criticism of the con-temporary. This vision was anticipated notably by the seventeenth-century writer Nicolas Venette (1633–98) in *La Génération de l'homme; ou, Le Ta-bleau de l'amour conjugal, divisé en quatre parties* (The Generation of Man; or a Depiction of Conjugal Love, in Four Parts) a work, frequently reprinted during the eighteenth century, that accused mankind of decadence: "In a word, since [mankind] is luxurious, it is also stricken with weakness of the nerves, gout, stupidity, & an infinity of other illnesses that overwhelm it."[77] As Sean M. Quinlan emphasizes, in 1750s France "socially engaged doctors injected themselves into a public debate about morality and social class . . . in which contemporaries discussed how they could reform what were perceived to be debauched upper-class mores."[78] To give one example, in 1766 Raulin produced a long harangue on society's degeneracy in the preface to his *Traité des fleurs blanches avec la méthode de les guérir* (Treatise on White Flowers with the Method to Cure Them). To reinforce his claim, he presents it not as a judgment, but as an affirmation that arises from nature itself: "The hu-man species degenerates, this is the general cry of Nature today."[79] The terms "degenerate," "debility," and "decline," which appear every two pages, give further emphasis to his point.[80] The interpretation of the emergence of mo-dernity is allied with moral considerations as a weakening of physical forces. Doctors proclaim the urgency of regaining the health of society in the name of a golden age, which some see as still surviving in the countryside. They make reference to the great value of the Scythian women, whose martial vir-tues kept vapors at bay.[81]

In late eighteenth-century writings, these vaporous pathologies came to exemplify the loosening of morals. They were hypostatized as signs of mankind's decline, and as civilization's mark and price.[82] In the United States, Benjamin Rush adopted the same discourse, regarding the hysteric pathology's diffusion throughout all classes as the effect of modernity.[83] In fact, naming the vapors was often the pretext for a critique of modernity and for expressing fears about the future. In the foreword of his *Nouveau Traité des vapeurs* (New Treatise on Vapors) of 1770, Pressavin proclaims the dangers of the "humidifying" therapies most often used against vapors, which lead the male body to lose the strength and resistance that properly distinguishes it from the female: "Woe to mankind if this prejudice [the use of such medicine against vapors] extends its rule to the common people; no more plowmen, no more artisans or soldiers, because they will soon be deprived of the force and vigor necessary to their work."[84] The diagnosis as threat spreads from one book to the next, gathering pace as it goes. The polemical dissertation of Montpellier physician Hugues Maret,[85] *Mémoire dans lequel on cherche à déterminer quelle influence les mœurs des Français ont sur la santé* (Memoir in Which One Seeks to Determine What Influence the Morals of the French Have on Health), which won the Academy of Amiens prize in 1771,[86] makes the corruption of morals the origin of the proliferation of hysteric and hypochondriac affections. Maret sees an intrinsic relation between these pathologies and the eighteenth century:[87]

> Without the confusion our morals bring to our soul, without the disorders that they occasion in the functions of the body, we would not be weakened by a great number of illnesses that destroy us or make our existence painful; we would not have lost our gaiety; we would not groan under the blows of an infinity of convulsive illnesses; we would not be devoured by worries; & vapors would still be little known to us.[88]

Maret's work traces the moral deterioration back to François I and appeals to the French to limit their excesses. His moralizing soon takes on an apocalyptic tone. Willing to make a diagnosis of society's body, Maret places himself in a position of exteriority:

> Oh my fellow citizens, you see the source of your evils, it is for you to dry up that source. The natural desire for well-being misled you; your care to avoid everything that could be necessary to your health exposed you unceasingly to the risk of losing it. Pleasure escapes you because you seek it too much. You have polished your surface only at the price of your being. You have finally reached the target you were striving to reach. Turn around; go back to the point where your ancestors were in the middle of the seventeenth century. You

will seem less agreeable, but your company will be surer. Some of your defects
will remain, but you will have few vices. You will even be able, avoiding part
of their excesses, to shirk part of the evils they were prey to . . . return to the
bosom of your families, from which the love of pleasure takes you away too
often, & avoid, by this conduct, the evils the future is preparing for you & and
that the ill education of your children makes inevitable.[89]

Maret attempts to seduce his readers by talking about their personal happiness, and he insists that he is their very own mouthpiece.

De l'Influence des affections de l'âme dans les maladies nerveuses des femmes,
avec le traitement qui convient à ces maladies (On the Influence of the Affections of the Soul in Nervous Illnesses, with the Treatment Which Is Suitable to These Maladies) was published in 1781 by Edme-Pierre Chauvot de
Beauchêne, who graduated in Montpellier and was physician to the king's
brother. His paraphrase of Maret's ideas is not as virulent in tone, but it too
outlines a new role for the physician:

The doctor's task is fulfilled when he has found the means to reestablish the
balance between moral affections & organic movements in the proportion of
their mutually affecting faculties, & following the order fixed by nature, he
can only reach this point by using moral assistance & material agents. . . . It
is necessary to regulate the distribution of one's time . . . ; at first leave a few
hours to madness, diminish them imperceptibly to give more to reason, music, conversation, a few instants to reading, at moments when the delirium is
suspended; everything, in fact, right down to the color of one's chamber, can
influence one's cure.[90]

In this passage, a change in the therapy promises a greater role for the physician, who is there not merely to treat pathology but to reinstate a moral order
of nature.[91] The physician locates his task in the regulation of the activities of
citizens even before they become his patients, framing his role not in the gaze
but in persuasive recommendations, admonitions, and warnings. This type
of rhetoric, with its criticism of the aristocratic way of life, takes full account
of revolutionary discourse.

In the 1780s, opposition to the progress of civilization and its evil effects
on human nature hardened and shifted, with a new responsibility accorded
to mothers. The concept of civilization taking man further away from a primitive innocence was partially superseded by the demands of a new nation.
The prerevolutionary discourse, built on the myth of a golden age, that told
aristocratic women to prevent vaporous fits by going to bed early, spending
time in the countryside, and not wearing corsets was replaced by a programmatic discourse in which doctors presented their role as a comprehensive and

crucial one: the full understanding of human nature. As Elizabeth Williams puts it, "Medical authority derived not merely from the technical, legal, or bureaucratic advances of medicine but primarily from its success in reconstituting itself as an anthropological endeavor, as an embracive 'science of man.'"[92] Though doctors did not attain such a central role in practice, this change in the positioning of their discourse nevertheless matters, because it enabled them to present assumptions that contributed to the construction of everyday perceptions about patients, about women, about citizens. Conditioning the relationship between doctors and patients, those assumptions created certain expectations, and as a result also potentially distrust and frustration, since they made particular demands on the state and the population (here especially women, but also men, as fathers and spouses).

Within the diagnosis—shared by doctors, pamphlet writers, and agitators—of modernity and the progress of civilization as a danger to the body, and increasingly also as a danger to morality, hysteria became one of the focal points of physicians' rhetoric—as an exemplary pathology allowing them to formulate the necessity of their role within the nation. The pathology was constructed to project them into a strategic political position, and was presented according to their aspirations to change their status in society. While other developments in the field were no doubt more directly instrumental in changing the status of medicine,[93] doctors used hysteria to inaugurate a type of discourse, spread a kind of fear, assert the need for their monitoring services, and conceptualize a link between the medical, the political, and the moral in the new French nation, by associating hysteria with young females and, as such, with motherhood. Hysteria exemplified the elements that doctors wanted to highlight and in relation to which they could fashion an active role for themselves.

Setting these constructions in the context of the great debate on magnetism, raging at exactly the same time, casts further light on the changes that resulted in the mutation of the doctor's role. Mesmer's success, promising to heal all illnesses, drove medicine into a state of alert. In response, members of the Faculty presented magnetism as a dangerous activity that favored the imagination's excess through sexual attraction and visual contagion. Against this they set the moral rigor of classical medicine, insisting that even if the therapy sometimes fails to heal, it at least always preserves good order. It is thus in the attack on magnetism that doctors assert their role in the protection of morals. The public move to avert the danger of such fashions implicitly affirms the importance of medicine and its connection with the state, as can be seen in Nicolas Bergasse's interventions in the debate on magnetism and Mesmer that arose in the 1780s.[94]

That decade saw magnetism's advocates and enemies tear each other apart, each trying to make the other a sign of the corruption of the day. In *Mesmerism and the End of the Enlightenment in France*, Robert Darnton interprets the polemics on magnetism in the context of the coming Revolution as a locus for political insurrection—as a means to criticize the established order and to radicalize discussions on morals and the corruption of nations. Darnton makes Mesmer a landmark of the intellectual agitation of the pre-revolutionary period, noting that proponents of Mesmer's theories likened the order of medicine to the monarchic order.[95] He focuses on advocates of magnetism critical of the monarchy, such as Antoine Servan, who declares: "[You] maintain ceaselessly the most complete despotism of which man is capable . . . you become absolute sovereigns over the sick common people."[96] For Darnton, Mesmer's effect was to catalyze a whole wave of political agitation. However, rather than analyzing the texts in terms of the political events of the Revolution that followed, I would like to consider instead how the ardent defense of or attacks on magnetism offered opportunities to develop a particular discourse on the nation's health. That discourse would reappear after the Revolution, when its themes were picked up again by classically trained doctors without reference to the proponents of mesmerism, demonstrating once again the circulation of the theme of hysteric pathology in different contexts with different political agendas.

Trained as a lawyer and later active in politics, Bergasse fervently advocated magnetism in two of his works, which propose a new conception of medicine. The first of these, written in 1781, is framed in an ironic mode.[97] Piling up exaggerated praise for classical medicine, the author pretends to take up a position against magnetism:

> Then, is it not one of the principles of a healthy legislation, a legislation whose only goal must be to civilize men, to be on guard that no innovation be made in Medicine that could deprive it of its abuses? If, by chance, animal Magnetism existed, if by means of that singular discovery, one could, as I do not doubt, replace this science which we improperly call the Art of curing, the more useful Art of preserving; what revolution, I ask you, Sir, should we expect? . . . While a more robust organization would remind us of independence, when with a different constitution we would need other morals, because we would have different ways of being and enjoying, how could we stand the yoke of the institutions that govern us today? . . .
>
> Thus, Sir, there is an essential relationship between legislation, morals, & a People's Medicine; thus the more a People is civilized, the more it is important to maintain, as a constant means of civilization, all the prejudices which can render Medicine respectable: thus among us, the Body of Physicians is

a political Body, whose destiny is linked with that of the State & whose existence is absolutely essential to its prosperity; thus, in the social order, we
absolutely need illnesses, medicines, & Laws; & the distributors of medicines
and illnesses may influence the habits of a Nation as well as the depositaries
of the Laws.[98]

Bergasse pushes the adversary's thesis far enough to prove its fallibility. A
note endows the reader with the role of drawing the necessary conclusions:
"There would be more than one response to all this: but something must be
left to be done by the Reader's sagacity. Comparing what I have just said, to
all that is objected against me, he will untangle without difficulty on which
side lives the abuse of facts & the false use of reasoning."[99] Bergasse's strategy
is to move precisely those who would seek arguments against magnetism, win
them as readers, and convince them, almost without their knowing it, of the
opposite cause. He tries to guide the reader to react independently against the
Faculty's positions.

At issue here are no longer aristocratic vapors, but those of the people,
and a characterization of the social body's political role: the vapors are what
keeps the people in its place, and as such they sustain the cohesion of the
social body. In this sense, they are framed as the trace of conservative forces
throughout society. If the aristocratic vapors resulted from luxury, the cause
of their more widespread counterpart lies somewhat similarly in depraved
morals, but ones that have now gained hold over the lower classes.

Bergasse's second work, *Considérations sur le magnétisme animal ou sur
la théorie du monde et des êtres organisés d'après les principes de M. Mesmer*
(Considerations on Animal Magnetism or on the Theory of the World and
of Organized Beings, Based upon the Principles of M. Mesmer), published in
1784, does not apply irony but defends the interests of magnetism directly.[100]
Here Bergasse presents magnetism's partisans as those capable of renewing
the order of morals and purifying the people of the corruption that has taken
hold: "There is a time in the progress of civilization when our morals necessarily lose their power. It is the time when our too numerous arts have
too much increased our enjoyments; then our sensibility is too divided &
becomes too uncertain between the impressions it feels, for a single one to
predominate with any duration."[101] Pathologies appear here as what makes
possible the permanence of authority, holding people in servitude and preventing any thoughts of a better life. Bergasse describes the body of physicians as a body connected with the government, and the pains of the third
estate as a political tool to keep it from ambition; health, he argues, would

bring the risk of insurrection. The goal of protecting medicine is to protect an authoritarian government:

> We hardly belong to nature any longer. All these affections, all these passions that society gives us, even these prejudices, these opinions, these customs to which it subjugates us; these laws of all kinds with which by compelling the impetuosity of our propensities, society almost always depraves them; these arts that are its work & that, by bringing too numerous enjoyments to the soul, fatigue, alter in a thousand ways our sensibility even more than they develop and satisfy it; these torments of hope & fear, these ordinary scourges shared by all men who lead a social life; these habits either false or too deep that are given by ambition, sadness, over-long disquiet, contention of the mind, whatever its object might be; that diversity in the manner of living, according to everyone's affluence, needs, and whims, all these things must work in a thousand ways on human organizations, & after an interval of time more or less long, rob them, in great part, of their initial character.[102]

Bergasse's interest in magnetism is accompanied by a conservative politics in a time of revolution. Indeed, it is precisely in the name of a past age of innocence that he declares physicians to be useless. As a politician, he extols mesmerism in terms of a correspondence between the health of the people and the health of recovered morals. Inspired by eighteenth-century philosophical writings that opposed a delinquent society at home to an age of innocence to be found in conquered lands, Bergasse sees the diffusion of magnetism as offering a possible return to the golden age. He writes:

> In medicine, the art of preserving is to the art of curing what in legislation the art of preventing crimes is to the easy art of punishing them. Unfortunately, in legislation the art of preventing crimes and in Medicine the art of preservation have so far been two almost ignored arts.
>
> Medicine and Legislation, as they exist today, resemble each other: the object of each seems only to be to appease symptoms; but finding out why man is sick or wicked, in a better order of things diverting him from the causes that produce vice or pain in him, this is what they are far from being able to do.[103]

Medicine is here presented as a form of government of the body that could with equal ease keep it in either a state of oppression or one of vitality that would favor insurrection. Whereas Maret's discourse aimed to blame only morals, in this case the quarrels over magnetism offer an opportunity to present medicine as the accomplice of a hierarchized society. The medical body is held responsible, accused of complicity with a state in which inequality rules.

Once court society was dismantled after the Revolution, the discourse on morals did not disappear. On the contrary, it took on almost hyperbolic terms. The idea of a law of nature governing morals, set out by Bergasse, was adopted by the anatomist Marc-Antoine Petit (1766–1811), surgeon major at Lyon's main hospital, the Hotel-Dieu, in his "Discours sur l'influence de la Révolution Française sur la santé publique" (Discourse on the Influence of the French Revolution on Public Health). Read in 1796, the address summarizes the effect of the Revolution on the French people and their physiology, conceiving of two types of movements as participating in the equilibrium and agitation of the human species. The first comes from individuals and modifies itself throughout their life in relation to their desire, their will, and their organism. The second is alien to them and carries them off course by constantly varying its energy and orientations. Petit compares these movements to powerful levers that lift and move the population both morally and physiologically. Unseen in quiet times, they take on an unexpected force in times of conflict. These motions then give birth to the most powerful passions, tearing individuals apart and bringing upheaval in all classes of society. Petit enlists the theories of the vitalist doctor of Montpellier, Paul-Joseph Barthez,[104] giving them a political turn they were far from possessing before:

> All the facts I have just assembled, gentlemen, prove, I believe evidently, that there is a class of illnesses that the revolution was bound to ease or cure. . . . It has thus, so to say, put into action the method that BARTHEZ named a disruptive one, and crushed, by its salutary jolts, the hysteric and hypochondriac diathesis.[105] It has thus corrected this nervous idiosyncrasy, too mobile, and too heightened by our former feebleness, giving more energy to springs previously more delicate and too weak: thus chronic illnesses which came from the inertia of these organs are much less frequent than before; and this relationship has diminished the death rate.[106]

Promoted above the laws of nature, the Revolution galvanizes minds and calls the social order into question. It is from divergences of opinion that revolutions are born. Even better, they are therapeutic:

> Revolutions, especially those by which a people soars up toward freedom, must thus give its temperament a new caliber: the illnesses that come from a slowness of circulation, from a thickening of the lymph in the absence of movement, from the debility of the nervous fiber must then be rarely observed; because the birth of an illness needs to be favored by the participation of all the causes that develop it, and in these circumstances that participation is almost none. By the same token, already established affections must weaken gradually in times of revolution, and even heal radically; just as in the beautiful days of spring one sees the illnesses born of winter vanish by degrees.[107]

According to Petit, the hysteric affection is thus destined to vanish now that the people's minds are clear. His text makes hysteric and hypochondriac illnesses exemplary cases that prove the salutary power of the Revolution. The Revolution appears as the vivifying force that annihilates putrid odors and lascivious flaccidity. It drains off the generalized fermentation that condemned women to uterine furors. The Revolution operates as an astringent; once the first pains are over, it hardens the body and purifies the soul:

> We busy ourselves to modify [the laws] or destroy them: all passions are at play; all souls are exalted; sensibility doubles their force; energy is everywhere; and every man becomes indignant at the very idea of an injustice. It is then that opinion, more powerful than laws, rises above them and masters them; it is then that opinion is truly named queen of the world. . . . Among these diverse frictions, all minds are electrified; thought is born greater and more proud; men capable of everything proliferate. Genius shines where one thought ignorance was to be seen; and human reason, drenched in new fire, seems destined only to conceive great things.[108]

When a generalized pathology is the index of dissatisfaction and corruption, the health of the nation demonstrates the virtue of a political body; the exaltation of all social classes attests a new vitality. Through the use of metaphors, pathology becomes the expression of political trouble, but far from being only a vice of the political body, it also becomes the possibility of recovery. Political thought, as radical and emancipatory activity, is at once trouble and therapy. It is the disorder of a social body becoming alien to what is expected of it; it is a therapy of the social body that examines society's pains in the bright light of the political arena. Petit's address, written three years after the end of the Revolution, and after the wave of revolutionary pamphlets, rewrites these diagnoses yet again, and strikingly refers once again to hysteric and hypochondriac affections as pathologies exemplifying the ancien régime.

In contrast, M.-S.-N. Guillon-Pastel, Philippe Pinel, and other physicians associated with the "ideologues"[109] read the revolutionary years as a form of hysteria: the pathology explained all the revolutionary excesses whose negative effects were to be corrected by the new republic. Increasingly, the association of the body with morality was presented as an inevitable one. In 1800, Guillon-Pastel, a member of the newly created Société des Observateurs de l'Homme,[110] described the opening events of the Revolution as having been followed by "hysteric, nervous illnesses, accompanied by convulsive spasms, spreading at the speed of epidemics, with a very particular energy,"[111] and thus wrote the Revolution as a physiological disorder. In the aftermath of the Revolution, physicians along with men of letters celebrated the nation's return to health. Goldstein has noted that Philippe Pinel and Pierre-Jean-Georges Cabanis initially viewed the new republican political constitution as a "chance to diminish the incidence of insanity and all other disorders of the mind,"[112] and many other physicians proclaimed the disappearance of physical troubles under the aegis of a new liberty, equality, and brotherhood.[113] As early as 1792, François-Xavier Lanthénas, a doctor from Paris, writes, "Freedom, whose influence is immediate and certain on people's health and happiness, must bring the necessary changes to medicine."[114] He praises the benefits to come—the effect of instruction, popular societies, fraternity, and improved morals.[115] These conceptions, which mostly developed in the circle of the ideologues, played a great part in fashioning the agenda of doctors, but are not discussed further here, having been studied in detail by, most notably, Elizabeth Williams, Martin S. Staum, and Jean-Luc Chappey.[116]

Whether hysteria was considered a precondition of the Revolution or its consequence, writers did not cease to instrumentalize it as a way of explaining the violence of those years. If they assert the imminent disappearance of pathology, this does not prevent them from giving it a crucial role in the discourse of a national health. The concept of hysteric affection, coined to name the pathology of a depraved aristocracy detached from the interests of its people, is easily transformed into the notion that health can be rejuvenated by revolutionary ardor. With this move, physicians of the period were attributing a new role to medicine: the context of a new nation presented them with an effective argument for prevention, and they espoused the superiority of preservation to cure.[117] The circulation of revolutionary ideas thus simultaneously nourished medical discourse: these discourses were relocalized, yet retained the same force, drawing their virulence from the state of national emergency. At stake was the invigoration of the nation's social body.

In its guise as prevention, medicine becomes the means by which to guard against possible social disorder and secure a moral order in the burgeoning nation. Citizen Pierre Sue, ex-secretary general of the SRM and later professor and librarian at the Paris School of Medicine, points this out in the first paragraph of his speech at the fifth public meeting of the Société de Médecine, on the 22 Pluviôse of the year VIII. The address introduces a general survey of legal medicine that he will publish in 1799:

It is not only in their illnesses that the physician lends men a helping hand. Warning of, relieving, or curing these physical disorders: this is doubtlessly one of the greatest roles of the man of the art. But his knowledge often has an application no less useful in the moral order, and in the field of justice every day one encounters circumstances in which the concordance of the laws with the principles of physiology becomes absolutely necessary.[118]

The intertwining of moral order, laws, and medicine is still the preferred theme, but the declamatory language of polemics, lampoons, and memoranda now cedes to a theoretical style of writing in medical treatises. In such texts, physicians seek not to offer appropriate therapies for particular pathologies, but to invest in the image of the health of the national body, the equilibrium of society.

Cabanis, a member of the Sénat-Conservateur and associated with the Institut National, the School of Medicine, and the Société de Médecine in Paris, ratifies this type of position in the preface to his work *Du Degré de certitude de la médecine* (Of the Degree of Certainty in Medicine), published in 1798:

> If medicine did not have, in the evils it aims to relieve and cure, a direct goal of usefulness, it would still deserve great attention as the basis of all good rational philosophy. In fact, it alone can help us know the laws of the living machine, the regular course of sensibility in the healthy state, the modifications that this faculty can undergo during the state of illness; it shows naked the physical man, of whom the moral man is only a part or—if one prefers—another face. The physician does not only see ideas and passions born from physical sensibility, he also sees, so to speak, how they come about, at least what favors or hinders their formation, and it is always in some organic state that he finds the solution to each problem.[119]

Cabanis later adds: "Medicine provides one more essential service. Like the other physical sciences, like the other arts that lean on the delicate observation of nature, it tends directly to dissipate all the phantoms that fascinate and torment imaginations."[120] Without discussing Cabanis in detail here,[121] it is interesting to note how large and new a field of action he opens for medicine. Situating the foundations of imagination and sensibility in physiology assures the importance of the physician's role. Cabanis writes medicine as equal with other arts; for a profession that was long despised, this equation is an unprecedented promotion. The word "certainty" attached to medicine in the title, even if to talk about degrees of it, means placing medicine in a perspective of progress.

Sue's and Cabanis's contemporary Alexandre Bacher, a coeditor of the *Journal de Médecine, Chirurgie, Pharmacie* since 1776, begins his work *De la Médecine considérée politiquement* (On Medicine Considered Politically), published in year II, by quoting Rousseau: "Science that instructs and Medicine that heals are very good, without a doubt; but science that deceives and medicine that kills are bad: to learn to distinguish them is then the crux of the question."[122] For Bacher, there is no better way of confronting such skepticism than preventive medicine. Doctors can assert their authority with reference to preventive discourses even if their success is disputable at the level of the cures themselves. This line of argument presents the opportunity for medicine to unite with politics and become not only the ally but the governess of public order.

A more general comparison between hysteria and insanity may be instructive here. L. S. Jacyna comments that nobody had the monopoly on lunacy in the early eighteenth century, and that "it was only late in the century that the medical profession began to claim a special competence in this area and to attempt to exclude other groups from a share of this growing empire."[123] In a similar way, in the late eighteenth century hysteria was only just becoming an object of study over which doctors asserted their authority. The emergence of a moral discourse in that process might be regarded as paralleling the shift happening around insanity, where a moral therapy was gradually introduced, its best-known foundation being the myth of Pinel removing the chains of the insane.[124] Commenting on Tuke, and on a widespread image at the time of inmates taking tea in an insane asylum, Andrew Scull has called that shift a "domestication of madness."[125] But looking more closely, it becomes clear that the shift from the concept of the vapors to that of hysteria in the late eighteenth century describes a different trajectory, moving from a conduct deemed an exceptional pathology that required the patient to stay at home to a common physiological risk threatening all citizens, one that—though not until the second half of the nineteenth century—would become a pathology to be dealt with in hospitals, with increased use of coercive therapeutic practices and a further intensification of its most spectacular features in the popular and literary imagination.

In both cases, the immediate postrevolutionary period saw a shift in understandings of medicine's role as part of the redefinition of the needs of the nation. Medical doctors now presented a particular definition of "hysteria" well tailored to their aspirations: hysteria was no longer an illness of the aristocratic sensibility, but an illness of the potential mother.

Investing in Women

As Quinlan has argued, in the revolutionary period physicians "cast themselves as the true defenders of civic and moral virtue, saying they best understood the present and future health needs of the French nation. In this discourse, they often made one particular group scapegoats: women."[126] Quinlan attributes this to the fear of depopulation, a regular theme at the time despite demographic growth.[127] Others relate it to the French revolutionaries' ideal of equality, which led to the need to reinstate some hierarchy—now a gendered one. Nicole Edelman, Martin S. Staum, Londa Schiebinger, Thomas Laqueur, and Ludmilla Jordanova,[128] for example, claim that in the wake of the French Revolution women were portrayed as vulnerable so that they could be defined by the role of mother and be required to stay at home to raise their children. This argument is supported by the absence of such a change in English treatises at the time: in England there was no such sudden surge of publications on women's illnesses, and the texts published in the period are not concerned in the same way to establish the singularity of women's pathologies or weave a web of references to confirm their moral responsibility.[129] Complementing these observations, and building on, especially, Jordanova's and Williams's and Daston and Pomata's groundwork for a history of the construction of sexual difference,[130] this section focuses on the emergence of a type of discourse, typical of France in these years, in which the physician links his role to a higher duty, a national one, by developing certain assumptions about women. I trace the radicalization of a discourse on hysteria, its focalization on women and hygiene in the name of a new nation, and the ways in which a new place for medicine was negotiated within the framework of that discourse.[131]

The 1770s in France already saw numerous treatises on women, such as that by Pierre Roussel,[132] and on onanism, such as that by D. T. de Bienville.[133] But Roussel restricts himself to distinguishing the contours of women's fragility and sensibility, and does not discuss any need for a programmatic discipline. In postrevolutionary France, after several reforms in the 1770s and 1780s, the state's intervention in medical care and social policy underwent its first radical development with the Convention's plan for assistance to all expectant mothers from the sixth month of pregnancy,[134] presenting the health of the citizens and even the unborn as the state's responsibility. As a whole, as La Berge contends, the revolutionary idea of health as "a natural right to which all citizens were entitled," positing "that if governments were instituted to protect natural rights, then public health was a duty of the state,"[135] played a pivotal role in shaping and institutionalizing public health in the subsequent years. Though this and many other of the Convention's schemes remained unfulfilled for lack of funds, they were circulated in each district and introduced a new tone.[136] The time had come to think about the future of human species: either in terms of *mégalanthropogénésie* (mega-anthropogenesis), as for Louis-Joseph-Marie Robert,[137] or, more frequently, by raising the question of the health and lifestyle of mothers, thus shifting the discourse on hysteric affection in a new direction.

The tenor is set by one of the first works in this frame, Pierre Boyveau-

Laffecteur's *Essai sur les maladies physiques et morales des femmes* (Essay on Women's Physical and Moral Illnesses), published in 1794. The essay's tone is one of deference, of consideration, even homage. The first lines run: "Talking about women, I speak to women: I strive only for their approval; curing them or, at least, soothing their pains and consoling them is my goal and my reward."[138] Similarly to Hunauld a few decades before, Boyveau-Laffecteur justifies the language of his work: "This plan imposes on me the law of making universal, or if one prefers familiar, the language of medicine."[139] He opposes the language of erudition to the "enigmatic idiom of oracles":[140] "If I made the reading of this book difficult to the [female] sex, I could not call women back to nature, and not capturing their attention, I would not deserve to be their interpreter."[141] His proposed "simple but decent style"[142] invests language itself with moral value. The doctor insists on his "disinterestedness,"[143] and, even if his work is aimed at uneducated eyes, he does not fail to immediately supply the necessary citations to prove his legitimacy. He mentions forty-seven physicians going back to antiquity,[144] listing numerous obscure doctors and a selection of renowned contemporaries, at times to recommend their work, such as La Mettrie's on hysteric catalepsy,[145] and at times to vigorously denigrate it, such as Le Camus's medical novel offering beauty advice to women.[146]

The narrator speaks as a citizen, and this is how he introduces himself on the title page of his work, embellishing the label with a note of his national responsibilities: "Supplier of the Navy Hospitals, charged by the government with treating Venereal illnesses judged to be incurable by Mercury."[147] Woman is thought from the point of view of her dependence on the physician. Boyveau-Laffecteur ties this knot more tightly using metaphors borrowed from medical language: he describes his role as "bringing back to nature a sex that the contagion of example leads astray,"[148] to which he opposes the physician's capacity to be guided by his benevolent aspirations. He defines the benefit of women as "the only goal of my works and my sweetest reward."[149] With the metaphor of contagion, women's customs appear to proceed as an illness; doctors' acts are motivated by sentimental gratification.

Between 1794 and 1802, the Catholic religion was banned, and the figure of the doctor took on the position previously occupied by the clergyman. The association is reinforced in Boyveau-Laffecteur's text through the intermingling of moral and pathological references, the doctor presenting himself as necessary to the good behavior of the family. His concern to define women's being becomes a way of claiming that medicine is the social binder. Espousing the theme of moral corruption in modern civilization, he regularly recurs to the image of a golden age, and the use of the phrases "call back," "bring back," and "give back" underlines the idea of a necessary return. His aim is not merely to cure, but to "give back a spouse to her husband and a mother to her children, to strengthen the links which chain wellborn beings to the social order."[150] Consequently, the role of the doctor is redefined: "The moral causes are those upon which the enlightened physician must focus."[151] The doctor is the one who guards the family boundaries, ensuring order and the consent of each individual to his or her role. To secure social cohesion, he must warn of illnesses and advise the nation to be on the alert against the dangers posed by the era. Boyveau-Laffecteur goes as far as to say: "A woman with regular menses is yet not protected from all the dangers threatening her constitution: and these all too well-founded fears lead me to provide more details."[152] Constant vigilance is required when the doctor bears a moral responsibility for woman's health. Noting the importance of "the balance between physical forces and moral forces,"[153] Boyveau-Laffecteur reiterates this point in some detail:

> The reasoned reading of this chapter leads us to conclude that in general there are no innate illnesses in the sex, that those inherited from their fathers are cured without difficulty, from the cradle to the age of puberty: having arrived at this time, a woman, apart from cases of contagion and venereal diseases, is only sick when she moves away from nature. It depends on her to come back by the peaceful study of her crises, simple remedies, and the balance between moral and physical forces that ensures the vigor of temperament.[154]

Woman must study her body under the watchful guidance of the physician, for the inseparable relationship between soul and body means that "if organs act upon the soul, the soul reacts upon the organs."[155] Boyveau-Laffecteur prescribes a rule of conduct to help women learn to combat their emotions: "For women, it is only one step from deep sadness to exaggerated joy: everyone knows what devastating effects on the animal organization this feeling, destined to be the expression of our happiness, can have when it takes hold of the faculties of our soul too quickly; it makes the smile of nature degenerate into convulsions."[156] Boyveau-Laffecteur enlists strong metaphors, and his language becomes increasingly trenchant: "Capital cities are a tomb for asthmatic, hysteric, vaporous, and hypochondriac women."[157]

In writings of this kind, physicians redeploy medicine in support of women's education. The particular stakes that motivate this restructuring in France at this specific time may help explain why such a reorganization was not yet happening in England: the defense of a nation fragilized by internal troubles, and the desire to export the republican model. The foundations of a new society are to be laid, which requires devotion to the children still to be born. And a hierarchy is to be inscribed in this new nation, a hierarchy seen as necessary in a society whose class system seems to have disappeared. Finally, these writers carve out an additional role for the physician: still powerless in the face of the pathology, he is powerful in his prescriptions of the duties of women. Approaches to the female body are entirely remodeled as a result.

Whereas in the eighteenth century the dominant explanation of sensibility was as an effect of the fibers, encouraged by education and the mode of life, in the nineteenth century, for the doctor Julien Joseph Virey (one of the founders of the Academy of Medicine in 1830) it becomes "the secret influence of the sexual organ."[158] The division between male and female takes the place of the previous division between the aristocracy and the people. A new hierarchy partitions the population, associating the brain with man and thus distinguishing him from woman with her uterus. At a time when some were proposing universal suffrage, giving women the same rights as men, others chose to affirm an originary inequality. Writing by medical men on women's education, hygiene, discipline, and morals proliferated.

These physicians worked to fashion a new role for medicine in prevention, and particularly the prevention of hysteria—now seen as a double danger, both physiological and moral. While previously generation had been presented as keeping all risk of hysteria at bay, hysteria is now seen as posing a threat to generation.[159] For Jean-Baptiste Louyer-Villermay, writing in 1816, "It is certainly possible that hysterical people may be sterile all their life."[160] In *Traité des maladies des femmes enceintes* (Treatise on Pregnant Women's Illnesses) of 1799, doctors Baignères and Perral report Petit's lessons held in the year VII, noting twice that a mother's convulsions can provoke the death of the child.[161] The same year, Pierre Géraud's *Aperçu sur l'hygiène des femmes enceintes* (Brief Survey of Pregnant Women's Hygiene), presented to the Medical Faculty in Montpellier, addresses the danger of women who "have exchanged their primitive and natural constitution for pleasures, feebleness, hypochondria, and vapors."[162] Géraud casts vapors as a choice that the physician must counter in the name of a society to come, and concludes that "if women's way of life were closer to the one that nature seems to prescribe to all animals; if they accustomed themselves early to exercise, to bad weather, and to different kinds of alimentation, they would tolerate their gestation as easily as all other natural functions and their life or health would no longer be endangered."[163] The notion of nature has here become wholly moral. For writers like Diderot throughout the eighteenth century, the myth of the golden age was a way to combat austere morality and make physiology the preferred guide for reflection on morals; at the turn of the nineteenth century, nature begins to be assigned a moral value as a way of classifying individuals. It is no longer a synonym for the freedom of morals, but a rating applied to each person. Nature's principal function must govern what each person does and how, and women's role is located in generation.

The parallel with animals now takes hold as a favored device. Virey's *De la Femme sous ses rapports physiologique, moral et littéraire* (On Woman, from the Point of View of Her Physiological, Moral, and Literary Relations) regularly uses animals as a model for feminine physiology, comparing animal and human cadavers and establishing parallels in childbirth. The absence of convulsive pathologies makes animals' maternal functions and orgasm a suitable model for women.[164] Hysteria is thought of as the manifestation of sexual desire, and the consequences (onanism, secret desires, jealousy, excessive piety, or strange behavior) are depicted as dreadful. Virey recommends marriage. For cases where this is not an option,[165] he suggests castration as a possible way of combating hysteria and nymphomania—again offering parallels with animals—but concedes that it is not always a successful therapy. What Ambroise Paré and his followers had regarded as a possible means of healing the suffocation of the womb, friction of the genital parts, is now presented as a moral and physiological threat. It is a threat that arises only in women, resulting from their sexual development during puberty: "Sexual constitution must be understood as the cause of such differences, the vital force that develops the superior organs in the man's body and the inferior organs in the woman's. In the former there is a tendency to superiority, to elevation; in the latter, one observes a contrary impulse; life blossoms toward the head in men, it concentrates toward the womb in women."[166]

The physician Gabriel Jouard returns to these ideas in 1804, this time to attack the idea that a woman could be the victim of nervous affections. He reserves nervosity to men, associating it with the brain: "The nervous system is the great motor, the unique regulator of sensible & insensible action, the evident & occult, the deep & superficial, the voluntary & involuntary action of all organs. The brain itself [is] the meeting point of all nervous impressions."[167] Jouard opposes this to hysteria—women's illnesses, which originate in the womb: "The uterus, despite the bizarre opinion that makes of it another nervous center or second brain, receives only very few nerves . . . and yet which organ presents more of those extraordinary phenomena that are supposedly nervous?"[168] The division between brain and uterus both illustrates and perpetuates the division between the sexes. Pathology reinforces it, condemning women to organic affections while men are the victims of affections emanating from the nervous system.

Such claims make the theorization of hysteria part of the construction of a differentiation between the sexes. Jouard goes further: "It is necessary to shake prejudices and extricate the question from all the figurative aspects lent it by the unhinged imagination of a few heads too *little nervous*, examining it in the positive either to put paid to it or to render it its proper acceptance, & I think that it then will be applied more rarely to women."[169] Jouard radicalizes his colleagues' conceptions of the role of women, again making the pathology a consequence of women's part in generation.[170] He localizes its cause in the blood system, and concludes: "As it is only for the accomplishment of such surprising phenomena that this singular variety or portion of the human species seems to have received existence, is it surprising that the diverse particular functions, in which she has been given a share for the accomplishment of this design, influence those other functions that serve to maintain her individual life & come to complexify it?"[171] Puberty and menopause are associated with pathology and become a means of understanding it.

Jouard thus creates equivalences between pathology and gender that classify hysteric disorders as a female characteristic. If for Jean Astruc,[172] professor and physician to the king and the Duke of Orleans, writing in the 1740s–1760s, it is because the womb is full of nerves that it can be the seat of hysteric fits,[173] for Jouard the womb's paucity of nerves means hysteria is not a nervous illness. By this point, the womb has been fully established by substantial parts of the medical profession as the origin of the pathology. Hysteria has moved from being a nervous pathology in which the womb could play a role to being a pathology of generation.

In the same period, hysteria is the object of a striking series of rewritings in the various publications of Philippe Pinel.[174] Pinel first presents it in the category of "vesania or alienation of the mind" (*vésanies ou aliénation de l'esprit*) to which women are particularly prey because of their sensibility, the energy of their affection, and the vivacity of their imagination. He indicates marriage as the therapy,[175] picking up the recommendation in the Hippocratic corpus for combating uterine suffocation—an approach that had been rejected by most physicians in the eighteenth century. By 1801, Pinel has already ceased to view hysteria as a mental illness: the pathology is absent from his *Traité médico-philosophique sur l'aliénation mentale, ou la manie* (Medico-Philosophical Treatise on Mental Alienation, or Mania), even though at this time he was the chief physician of the Salpétrière women's hospital. Neither does it appear in the revised version, now in three volumes, published in 1809.[176] Only his *Médecine clinique*, published in 1802, allocates it a chapter.[177] There, hysteria is associated with menstruation, the cause to which Pinel attributes the majority of women's illnesses. In his new and enlarged edition of the *Nosographie philosophique*, published in 1802–3, hysteria appears with a new classification in order V, "neuroses of generation."[178]

Pinel enlists Cullen's newly coined term to talk about illnesses of the mind, but gives it a new twist by associating it with generation in the subdivision "genital neuroses of women."[179] The definition of hysteria is located just after that of nymphomania, ratifying the division between mental illness and the physiological or sexual illnesses; similar symptoms (trembling, convulsions, migraines) are assigned to two different categories constructed as antinomic. On three occasions Pinel repeats the importance of marriage. He states that hysteria must be looked at in terms of its unique character, and this character is now defined as resulting from its origin in the uterus.

All these theories are far from new, but the tone has changed, and the precise contours of the affirmation are different. From the beginning of the seventeenth to the beginning of the nineteenth century, hysteria is associated with animality, then with excessive sensibility, then with an unbridled way of life, then with a sexual organ. If at the turn of the nineteenth century physicians take up the theorization of the womb, it is not the womb as presented in the seventeenth century. The seventeenth-century womb endangered women because it collected blood and semen that fermented and could become infected if the menses were delayed. Only a pregnant woman seemed protected from such irregularities. The role of the womb was thus a dual one, as the seat of infection and the crucible of generation. When the womb reappears at the end of the eighteenth century, it has been washed clean of its association with waste—but it has become a source of trouble on its own account, and has been promoted to the central organ of the female body. Woman's biological reality operates as a threat. The womb is no longer filled up with menstrual blood and semen endangering the body's equilibrium, but its gaping openness is just as frightening, and from this void it governs the body.

Even if their language at times seems less virulent than that of the seventeenth century,[180] such early nineteenth-century theorizations stigmatize women far more. By this time, the body as a whole has been valorized, having in the eighteenth century become the site of the soul's expression. It has been rethought through an education of the feelings, and interest in educating it has grown. In the course of the century, individuals have appropriated their own bodies. But by the same token, the pathological body now implicates the subject. It designates weak individuals, their excesses, their morals. Gender is here inscribed in all points of the body, and pathology itself comes to participate in this assignation of the subject to its sex. In these treatises on women, hysteria strikes the female body and thus marks everything that distinguishes this body from the male one. In 1817, the Paris physician Capuron writes of hysteria: "It attacks equally the girl who lives in chastity and the one with dissolute morals; the wife whose desires are satisfied and the widow who observes continence; the woman whose menses run periodically every month and the one whose discharge happens to suppress itself, or disappears completely around the critical age or the return."[181] It is not only the lexicon (with terms such as suffocation of the womb, fits of the mother, uterine furors) that undergoes a metamorphosis, but also the point of view from which woman is seen. If previously the importance of the woman and the man in generation was the subject of discussions that tried to assign the principal role to men, it is no longer controversial to grant preeminence to women.[182] This very preeminence provides the rationale for a discourse that aims to control her way of life. In very broad strokes: in the seventeenth century, woman plays a minor role in generation; in the eighteenth century, she becomes responsible for generation; and at the beginning of the nineteenth century, she is no longer distinguishable from this role.[183]

A portrait appears here of woman as a victim of her uncontrollable womb. Her body is both an indicator and a trial of her virtue. The womb has the power to affect the female moral and intellectual functions. It not only assures the pathological identity of woman, but also determines her identity in times of health. Treatises on women establish an equation in which woman is potential hysteria, and hysteria is actualized woman. They interpret pathology as a revelation of the female body: not merely as a crisis that suddenly interrupts the body's balance, but as signaling an inherent problem. Hysteria becomes woman's truth; woman is assigned to her bodily role and yet constantly threatens to fail in that role. This threat necessitates the intervention of the doctor, whose part is now to regulate the life of his patient, fixing limits and prohibitions. An array of medical works points out the importance of a successful education; the patient becomes responsible for her troubles. The idea of feminine manners and sensibility has disappeared in these writings, dethroned by the bodily reality of the womb.

Most striking is the doctors' repeated move to associate what they consider a physiological necessity, sexual intercourse, with the most rigorous morality. After a description of hysterical symptoms in which he sees a danger of death, P. Bauduit states in his 1822 medical thesis, "Marriage dissipates so many symptoms because one submits to the commandment to be fruitful and multiply; because one settles in the honorable station of being a mother."[184] The reciprocity between physiological health and social role is the new key to social order.[185] That "Medical Considerations on Marriage" can be the title of a medical thesis says much about the pact now formed between medicine and public order. Bauduit underlines his argument with the vividly recounted story of a "wellborn girl" who strays into prostitution and only after her eventual marriage "conducts herself in the most regular way."[186]

In this account, a woman's conduct is entirely subjugated to the order of the body.[187] A hysterical body results in action that contravenes the rules of the moral order. The most regular conduct will be that of the body become healthy. Doctors make their appearance as the priests and guides of morality. Their role is defined not by prescribing therapies, but by establishing a discipline to prevent the need for cure. Marriage becomes both a social and a hygienic necessity, instituting a physiological balance by means of a measured sexuality, cleanliness, a self-contained body, exercise without excess—a list of the required conditions to become a mother. Only by knowing how to anticipate the body's troubles can women avoid the otherwise inevitable rule of disorder.

To this extent, doctors working with hysteria could and did attempt to secure their role through social and philosophical arguments: a morality that ignores the laws of nature, they claimed, must lead to perdition. Only an enlightened approach, with full knowledge of the body, could foresee its desires and control them in time. In these imaginings, the doctor is a figure who makes possible the cohesion of society, and he offers a new image of medicine to that end. Far from limiting himself to the perimeters of the hospital, he develops forms of writing that aim to move his fellow citizens and cement his role in society as a whole.

★

In this context, it is perhaps unsurprising that Madame de Staël's mention of the vapors in her 1806 novel, *Corinne*, is a derogatory one.[188] Describing a man that her stepmother has chosen for her, Corinne explains, "He was a good man, incapable of causing suffering; yet if I had talked to him about the numberless sorrows that can torment an active and terrible soul, he would have regarded me as a vaporous person, and would have simply advised me to go horse riding and take the air."[189] Madame de Staël constructs her heroine as independent of all physiological or sexual needs. It is the force of Corinne's imagination alone that causes the swelling of her sensibility, an imagination here defined outside of physiological conceptions. By insisting that the cause of her troubles lies in the mind, Corinne locates them beyond the grasp of doctors and of all those who believe in the vapors. It is the men incapable of understanding her genius and its associated afflictions who see its effects as an illness. Breaking radically with the idea that women could be under the influence of their uterus, Madame de Staël here opposes Diderot. In fact, Madame de Staël's novel might be read as a response to Diderot's theory of women's genius based on a physiological analysis of their body: she offers a diametrically opposite description of a woman whose genius cannot be located in the needs

and impulses of a womb, or in physiology. From this point of view, the turn
of the nineteenth century could be interpreted as opening up Diderot's thesis,
which forks into two opposite claims: a position like Madame de Staël's that
accepts the idea of genius but refuses to make the womb an element in understanding women, and that of the majority of doctors, associating women with
their womb but denying them the creativity of genius.[190]

While it might seem at this point that in medical discourse hysteria was
exclusively defined by an origin in the uterus, the 1820s would witness the
reemergence of the brain as an origin, centuries after Le Pois,[191] with physicians such as Georget in France or Hosack in England,[192] and a strengthening
of the conception of the role of nerves in pathologies that included hysteria.[193]
Many works now excluded hysteria from their lists of "nervous" pathologies,
as if they meant to refocus the study of the disorder outside of an inquiry into
nerves.[194] The spectrum of different interpretations and investments was,
then, a broad one. But between the doctors of the mid-eighteenth century
and Petit or Boyveau-Laffecteur at the century's close, something fundamental has changed. These doctors' aim is no longer to convince aristocrats that
they will find their doctor an attentive listener, nor to speak in a submissive
and seductive voice, interpolating salon wit into medical treatises. Rather,
they speak to the female citizen, probably bourgeois, who is reflecting on the
education she wants for her children; to the married man and father, who is
concerned for the future of his family and wants to see the signs of bloom-
ing health.[195] In contrast to the accounts that accord prime importance to
Xavier Bichat's writings at the turn of the nineteenth century, particularly his
Recherches physiologiques sur la vie et la mort (Physiological Research on Life
and Death)[196] with its discussion of hundreds of dissections, I am interested
in another type of rhetoric emerging at the time—one privileging not the
description of anatomy but the evaluation of the role of morals, education,
and hygiene. Within that rhetoric, shifts in the physician's place in society are
accompanied by shifts in the tone and terms of his message. These physicians
claim the authority of the possessor of knowledge, addressing women without depending on them. They must be seen in the context of the construction
of a new social body that is accompanied by a whole set of transformations: in
their writings, medicine takes up the role of educating the citizen in hygienic,
moral, and social terms, intervening in every level of society. The medical
policing that developed in the eighteenth century (attentive to garbage, excrements in the streets, or cemeteries) is now supplanted by a politicized
medicine.[197]

This chapter has shown that alongside the rise of a form of writing, described by Foucault, that aspired to be a strict registration of symptoms and

depended on the visibility of the body, a further type of writing arose as well. In it, the physician places himself in the spotlight, invoking patients he claims to have known rather than to have observed. These medical texts adopt the format of narrative, implying a narrator and a community of human experience. This type of medical enunciation is founded on a common present, referring to a past to be fled or rediscovered and a future to be built. The measured and dry words of nosologies are left aside by these writers. Operating through comparisons and in the name of lived experience, their writing produces a knowledge of human tendencies that has the force of law. They extrapolate this line of interpretation to define women by the risk of hysteria, embracing both the physical condition and the moral point of view and making their concatenation the basis of the physician's authority. In the process, physicians focusing on women and nervous illnesses position themselves as if the needs of the new nation had moved their status from a position of inferiority to one of superiority—a change, however, that was effective less in practice than in discourse. That discourse is instituted not from the distanced position of universal knowledge, but from the proximity of a figure who claims to understand men, women, mothers, and youngsters, who has lived enough to know their weaknesses, and who speaks from the knowledge he has gathered directly from his patients. It is the claim of an empirical knowledge, based on shared experience, that he uses to postulate his authority over his patients, and a doctor in this mold comes close to the pastor in his words and tone of voice, as if aspiring to guide the new society like a shepherd. The novelty of a French nation only stimulates such doctors' will to take on responsibility for the body and the soul. The role they forge portrays the physician as crucial for the state, for he is the one to procure the preconditions for a healthy national mind and body. Invoking the too easily hysteric body, the physician prescribes a discipline. More than that, he claims the folds of the soul as his own territory.

The treatises I have presented here by no means settled the interpretation of the pathology. On the contrary, its medical denominations and explanations continued to proliferate, as Goldstein notes: "At midcentury, hysteria and a bevy of other newly and loosely defined nervous conditions— *sureexcitation nerveuse, nervosisme, névropathie*—which typically included some hysterical symptoms and which, like hysteria, 'fell short' of insanity, had come to the fore in the medical community."[198] During the nineteenth century, just as in the eighteenth, the diagnosis would recur in literature in many different forms as well,[199] and Janet Beizer discusses its use as a metaphor in political writings and pamphlets through the case of the Paris Commune.[200] The development of criminology, psychology, and psychiatry and the invention of psychoanalysis were further occasions for the term to be invested with new meanings, associating it with crime, kleptomania, sexuality, prostitution, and much more. A new set of discourses and investments arose, at times referring to older conceptualizations of hysteria when that suited their respective purposes. Even the symptoms changed, privileging the opisthotonic or arched posture, which had very rarely been reported by doctors until the 1780s.[201] Interestingly, even in the later nineteenth century, when hospitals were admitting patients diagnosed with hysteria, consensus on the pathology remained elusive: its name was sometimes contested, and although it was generally seen as an illness, some still considered it mere simulation. Dictionaries and nosologies continued to revise its definition, even if only to reiterate the impossibility of listing all the symptoms it encompassed.[202] But however broad the spectrum of its signs and investments, from the beginning of the nineteenth century hysteria was established as a category.

Conclusion

Tracing the transformations of the category of hysteria between approximately 1670 and 1820 and asking what was at stake in writing that diagnosis, this book makes three key contributions to the history of eighteenth-century medicine. Firstly, it measures the rupture that the appearance of the term "hysteria" produced in the enunciation of knowledge, examining how specific meanings were created in this process and to what ends. It emerges that religiously oriented conceptions of convulsions made up only a very small part of the writing on hysteric pathology. The affection was widely discussed in medical contexts, and was also a common reference in fiction, public political debate, and even philosophical writing. Men and women of letters described vaporous fits in order to exemplify the relationship between body and mind beyond strictly medical concerns, and portrayed physiological phenomena alternately as a code, a truth, or a manipulation, depending on their particular narrative requirements. Hysteria was not seen as a female malady in this period; until the French Revolution, it was presented as an aristocratic pathology.

Secondly, as a study of the genealogy of a medical category, *On Hysteria* adds to our understanding of the emergence of a discipline, specifically how the role and status of medicine became established in French society during the late eighteenth century. Unlike existing studies in this area, most of which look at institutions or the medical gaze, this book addresses the formation of a new image of the physician-courtier and the later transition from the doctor attempting to emulate aristocratic ways to the doctor-citizen. After the Revolution, with the new focus on the health of the nation's body, the stakes of hysteric illness shifted: it was no longer a diagnosis belonging to the upper classes, but facilitated a medical discourse on women and their national responsibilities—thereby helping to establish the public role of medicine.

Thirdly, this work proposes new methodological options for the history of medicine, and more broadly for the history of science, by addressing the written formats and rhetorical strategies used to present the category—as opposed to describing successive definitions and interpretations. I focus on political and epistemological investments in the category and the constant transformation of its connotations and implications. By attending carefully to writing strategies, webs of citation, and the circulation of rhetorical figures, this book presents a history of science as a history of knowledge in the making. It finds both coherence in the invention of this crucial medical category and constant breaches of that coherence through misinterpretation, recontextualized citation, and the strategic redeployment of narratives.

In its analysis of the enunciation of medical knowledge on hysteria, *On Hysteria* has taken texts as events. Proposing the infinitely changing character of medical knowledge, it has delineated the scope of the writing practices that permitted the emergence and diffusion of that knowledge. To create meaning is to create a status from which to speak, and to negotiate significations in a regime of values and references. Such meaning must be shared, taken forth, discussed, and contradicted. It must signify outside the sphere of the specialists, and describing the process of a medical category's construction implies considering how actors move within wider webs of meaning. The use of a new category demands special attention to invoking lines of tradition, citation, and metaphors in order to ensure its success. Those who speak position themselves by speaking—not so much revealing some more or less obscure motivation as perpetuating systems of representation and accreditation and creating the possibility for something new, an event in the field of knowledge.

Constructing a pathology is thus more than indicating a list of symptoms, assigning a cause, proposing a therapy. It is mobilized by many motives: the affirmation of competence in relation to troubles, a novel approach to the body, the classification of patients, the demarcation of the scope of medical activity. And even when a diagnosis becomes established, its name changes depending on the stakes it bears at a particular time. In this study, therefore, the category of hysteria emerges as a means to configure knowledge and theorize it, a means arising from the new demands of a historical period.

After 1800, the deployment of the term "hysteria" brings a new configuration of knowledge into existence: it is language that assigns and delimits roles and pathologies. The articulation of the diagnosis contributes to constructing difference between the genders. Producing a new organization of knowledge, the establishment of the category proceeds by the citation of cases, types, categories of people, social classes, and sexes. The act of denomination cuts through the volume of knowledge and reorganizes it, and this new cut gives shape to the diagnosis and guides its perception, obliterating a whole series of other categories that will from now on be judged irrelevant.

In terms of the category's genealogy, the evolution of the conception of hysteria is inseparable from the evolution of the ways that physicians write their role in society. In the seventeenth century, the theorization of hysterical affections offered the opportunity to build a therapeutic space out of strictly physiological interpretations of troubles, now seen as symptoms. The physician used potions and bleedings to achieve the balance of the body. If the theoretician advised his colleagues to reassure their patients, often highly distressed, or to convince them of the need to change their lifestyle, this was secondary; what was important was to gain the patient's assent to follow the treatment and be cured.

When the eighteenth-century physician wants to heal vapors, he elaborates a very different relationship with the pathology and his patient. He puts multiple and ambivalent troubles into words, reading as symptoms what others see as an unfolding of sensibility. When Hunauld meticulously describes through Sophie's voice the effects of the emotions' variations, when Raulin warns against taking an ironic or belittling view of the vapors, they are rewriting emotions as a space in which pathology is at play. In turn, sensibility becomes a necessary detour to think the physiological. At a time when the practice of diary writing and correspondence is developing, the physician expands his sphere of competence and seeks forms of medical writing suited to accounting for what will come to be read as hysteria. Doctors and the lettered elite report on fits, and invest the affection with multilayered meanings. They choose to write on "hysteria" for all the paths it allows them to pursue: the reciprocal influence between the body and the soul, imagination, sensibility, sympathy, genius, and more. Their formulation of troubles or symptoms illustrates particular momentums, trajectories, crystallizations of ideas.

As this detour via sensibility became established, the role of medicine expanded. In this regard, On Hysteria has shown that reading only Sauvages, Cullen, and Pinel, with their clearly "medical" inventories, would miss the singularity and scope of the medicine of the time. The nosology is only one of the medical writing genres of the eighteenth century. Far from being fixated on diagnoses and lists of symptoms, physicians in this period instigate forms of writing that allow them to maintain a gray zone between pathology and health; they mingle emotions, ways of life, myths, and symptoms in a single move. Eighteenth-century medicine's singularity lies precisely in its capacity to use writing practices as a way of fashioning a new role for itself at the core of society.

When the question of the nation succeeds the cult of delicacy in the late eighteenth and early nineteenth centuries, theorizing hysteria in medical texts becomes a means to locate dangers at the very site of childbirth: at a time when the importance of procreation is affirmed with new urgency, hysteria is again thought in relation to sexuality. Such theories are, however, far removed from the ones propounded by Ambroise Paré and André du Laurens in the late sixteenth century. In the early nineteenth century, localizing hysteria in the womb is no longer a way of thinking the physiological vagaries that dispossess woman of her will; hysteria is the mark of the responsibility borne by woman's body. Morals are henceforth necessary to social health and the formation of a political body of valorous citizens. Imagined as threat, hysteria offers opportunities to redefine the role of medicine. At a time when society is thought in terms of its future, the pathology localizes that future in the uterus. And what greater role could medicine have than to place itself in close proximity to each individual, even those in perfect health—foreseeing perils and conjuring up dangers?

In France and in England, similar writing practices contributed to the conceptualization of hysteric illnesses in the seventeenth and most of the eighteenth century. They flourished in novels, correspondence, medical autobiographies, and observations as well as medical treatises, and circulated a similar range of interpretations. Pamphlets and dissertations written during the crisis of the Convulsionaries in France in many ways mirrored the crisis of the French Prophets in England. In France, however, the Revolution gave rise to further uses for the cultural imagining of hysteria and, more importantly, a new context for fashioning medical practice. The linkage of hysteria and women took on a specific dimension in the phraseology of a newborn republic.

After the end point of this book, the publication of Louyer-Villermay's definition of hysteria in 1818, the category would undergo further reincarnations in France. As the nineteenth century unfolded, hysteria continued to attract the interest of both literary writers and scientists. In the literary realm, Flaubert, Huysmans, Zola, Baudelaire, Jules and Edmond de Goncourt, and Maupassant—to name only some of the most famous—endlessly elaborated the features of the hysteric, whether male or female, artist, alcoholic, worker, or bourgeois. In hospitals and psychological laboratories, hysteria would become an object of study for neurology and experimental psychology with Charcot and the Salpêtrière clinic, the School of Nancy, and Pierre Janet at the turn of the twentieth century, while Sigmund Freud created a new discipline in the course of his study of the affection.

The methodology proposed by this book implies that the variations in the interpretation of a pathology must be read as something other than progress in the knowledge of the body; rather, they are the intersections of epistemological demands and political necessities, power relationships between competences, strategies of instrumentalizing values and imaginings. They enable social ascent for physicians and increasing recognition for medicine. This is not to accuse physicians of opportunism, but to understand that the transformation of the writing of the body cannot be separated from a redefinition of medicine. There is a reciprocal movement in these mutations: new ways of thinking lead the physician to aspire to new roles, and new roles suggest new ways of thinking about the pathology. The history of knowledge happens in the entanglement of epistemological and political stakes. For categories and conceptions to emerge, be accepted, and circulate at a given moment, they need to fit within a political horizon, which they challenge and are stimulated by—whether in agreement or contradiction with the direct context. As these horizons change, the frames of possible knowledge change as well, offering the possibility of new understandings or the return of older, long-neglected ones. It is the study of the various and possibly conflicting frameworks for conceptualizing a medical category that can further our understanding of the ways knowledge develops. In different spaces (whether private, such as the salon, or institutional, such as the academies, polite society, and the university) and with different ways of imagining an audience (according to social class, activity, gender, cultural interests), different writing strategies arose and in turn led to different enunciations of the category.

This book, in other words, goes beyond the question of the category of hysteria itself to offer methodological options for the historian of science. Hysteria is an exemplary case of the determination of a medical category. It offers the opportunity to assess the extent to which the elaboration of knowledge produces differences rather than reflecting them. We see doctors sculpting patients' expectations and medicine's roles by offering modes of reading the body's troubles, by reading those troubles in terms of symptoms, and by inserting them into the linearity of organic or moral causes.

In describing an example of a knowledge process, *On Hysteria* has highlighted the many paradoxes encountered there, and claimed that they are crucial to the understanding of the history of science. It has attempted to seize on fumbling attempts, gaps, oversights, lacunae, and occasional resurgences of concepts and imaginings; it has sought to identify investments of meaning, to see how they operate, and to analyze how certain texts outwit, undermine, or thwart others. In this approach, history of science is not the history of an idea, but the history of divergences erased to secure a retrospective construction. Unthought differences, detours, inflexions, and misunderstandings all take their turn in highlighting the priorities that will determine a pathology.

In the paradoxical movements of knowledge, certain constellations are undone, but new terms appear; metaphors and references endure, but undertake new roles. The animal is no longer thought of as a reality of the body, but it is used as an image for thinking the body. The metaphor of Proteus crosses divergent understandings and recurs over many centuries as a figure around which the ideas of a period could crystallize. The image of the miracle is enlisted outside of a religious debate, to conceptualize the powers of the physician. Older images, references, and quotations are grafted onto the new discourse. They are called on for their force of ascription, and allow full importance to what cannot be theorized in a strict regime of signification—thus, enunciations and stakes change, but the same words reappear. If signifieds are displaced and motivations fade, the return of particular words ensures a continuity in the formation of knowledge. In this movement of repetition, knowledge is consolidated: it is presented as a revised vision now come to maturity. As theories and imaginings travel from text to text, the resulting mirroring effects make it possible to claim a coherence of knowledge that goes beyond the fluctuations of medical approaches.

It is precisely the enormous diversity of approaches in the eighteenth century that makes it necessary to repeat common references as a means of legitimating their various claims. It is the very gap between the stakes that necessitates points of encounter, allowing the discourse to anchor itself more strongly; it is the infinitely problematic and uncertain character of these texts that makes them apply authoritative references and a logic of proof. Far from impeding the evolution of knowledge, the insertion of such long-standing references offers fresh possibilities of meaning, the renewal of perceptions, the movement from one category to another. The historian of science's role, in that light, is to track how misinterpretation, reading against the grain, or reading between the lines can configure the constant redefinition of knowledge.

It will be clear from this that taking the measure of the multiplicity of political and epistemological stakes does not mean inscribing the physicians' strategies with determinism. To trace repetitions, metaphors, or references is by no means to establish a linearity in knowledge, but indicates instead its extraordinariness, its fundamentally unprecedented dimension, its singularity. Guided by the emergence of constantly redefined priorities and writing themselves into constantly new webs of accreditation, discourses rework their approaches in relation to the demands of the time—whether to obey those demands, nuance them, contradict them, or completely redefine them.

This book has attempted to acknowledge both language's power of ascription and its extraordinary power to displace. How can one make meaning with the body? How does one legitimate meaning? How is that meaning perpetuated or dismissed? How does one text defeat the assumptions of another? Each writing participates in meaning, modifies it, and at times, diverts it. As they actualize meaning and inscribe it into a new distribution of knowledge, texts simultaneously redefine themselves within it, against it. The analysis of particular enunciations has shown how knowledge escapes those who construct and use it, and how its evolution depends on fragmentary and conflicted processes. Counter to an approach that sees knowledge only as the product of conditions of possibility at a certain time, this book hopes to have indicated that it runs in all directions. Knowledge is written in the unsaid and in the excess; in the rewriting, the detours and false starts, it takes on new shape. And it is in the interplay of discourses that knowledge is repositioned. Its force of ascription, thwarted, attaches itself to new stakes or new objects. It is in the excess of meaning, in misunderstanding, that knowledge always takes its zigzag course.

Acknowledgments

It is a pleasure to thank those who supported this project in its very early stages: Evelyn Ackerman, Michel Beaujour, Vincent Crapanzano, Anne Deneys-Tunney, and Arlette Farge. I later received inspiring input from Antoine de Baecque, Peter Cryle, Jan Goldstein, Katherine Park, Catriona Seth, Anne Vila, Caroline Warman, and Elizabeth Williams. Excerpts of this work have been presented at various conferences and seminars: my thanks to the organizers for these opportunities to discuss my ideas. The study first took shape in French, and I would probably never have undertaken an English version myself without the generous encouragement of David Walker. Kate Sturge helped me negotiate the tensions arising from the difficult dialogue between the French and the English versions. I would also like to thank Marie Kloos, Jean Terrier, and most especially Thu-Tra Dang for their collaboration in the last phase of the book, and Marian Rogers for her final editing.

The breadth of this book's scope depended on reliable access to rare sources. For their unfailing assistance, I am eager to thank the librarians of the following institutions: Académie Nationale de Médecine in Paris, Bibliothèque Interuniversitaire de Santé, Bibliothèque Historique de l'Assistance Publique, Max Planck Institute for the History of Science, Bibliothèque Nationale de France, Bibliothèque Universitaire de Médecine de Montpellier, Archives Municipales d'Arles, Archives Départementales de Vaucluse, Archives of the Académie des Sciences, Belles-Lettres et Arts de Rouen, Bibliothèque Cantonale et Universitaire-Lausanne, Bibliothèque Publique et Universitaire de Genève, the Oskar Diethelm Library, the New York Public Library, the New York Academy of Medicine, the Moody Medical Library of the University of Texas Medical Branch, and the Francis A. Countway Library at Harvard University.

Sincere thanks for financial support are due to the Société Internationale d'Étude des Femmes de l'Ancien Régime, the New York Academy of Medicine, the Graduate Center of the City University of New York, and the Francis A. Countway Library at Harvard University. The Melbern G. Glasscock Center for Humanities Research at Texas A&M University kindly funded a research trip, and two fruitful residencies were made possible by the Institute for Cultural Inquiry, Berlin and the Institute for Medical Humanities at the University of Texas Medical Branch in Galveston. This invaluable help enabled me to explore my research with both scholars and students, and gave me the time and space to develop my thoughts. The Max Planck Institute for the History of Science in Berlin provided me with the best possible environment to conclude the study and prepare the English version. Thank you for deciding to support this work!

My deepest gratitude is to my parents and to Brian Price, who supported this research with all his energy. I hope the book itself will express my sense of indebtedness to all the artists and scholars, near and far, who have inspired it.

Parts of chapter 1 were published in an earlier form as "De la dénomination d'une maladie à son assignation: L'hystérie et la différence sexuelle face à la pathologie entre 1750 et 1820," in *Les discours sur l'égalité/l'inégalité des sexes de 1750 aux lendemains de la Révolution*, ed. Nicole Pellegrin and Martine Sonnet (Saint-Étienne: Presses de l'Université de Saint-Étienne, 2012); and "'. . . Vous en guérirez et tout sera dit': De la circonspection dans l'énonciation de la maladie au XVIIIe siècle (Suisse, France, Angleterre)," in *Qu'est-ce qu'un bon patient? Qu'est-ce qu'un bon médecin?*, ed. Claire Crignon and Marie Gaille (Paris: Seli Arslan, 2010). Portions of other chapters appeared in earlier forms as "Capturer l'indéfinissable: Métaphores et récits sur l'hystérie dans les écrits médicaux français et anglais entre 1650 et 1800," *Annales: Histoire, Sciences Sociales* 1 (2010): 63–85; "L'art de vaporiser à propos, pourparlers entre un médecin et une marquise vaporeuse," *Dix-huitième siècle* 39 (2007): 505–19; "Tension and Narrative: Autobiographies of Illness and Therapeutic Legitimacy in Eighteenth-Century French and English Medical Works," in *Tension/Spannung*, ed. Christoph Holzhey (Berlin: Turia & Kant, 2010); "Ruse and Reappropriation in *Philosophy of the Vapours, or Reasoned Letters of an Attractive Woman* by C. J. de B. de Paumerelle," *French Studies* 65 (2011): 174–87; "S/he Who Traps Last Traps Best: The Diagnosis of Vapours and the Writing of Observations in Late Eighteenth Century France," in *The Body and Its Images in Eighteenth-Century Europe/Le corps et ses images dans l'Europe du dix-huitième siècle*, ed. Sabine Arnaud and Helge Jordheim (Paris: Champion, 2012); and "Medical Writing, Philosophical Encounters, and Professional Strategy: The Defiance of Pierre Pomme," *Gesnerus—Swiss Journal of the History of Medicine and Science* 66 (2009): 218–36. The research also appeared in French, in a significantly different form, as *L'invention de l'hystérie au temps des Lumières (1670–1820)* (Paris: Éditions EHESS, 2013). I am grateful to all the publishers involved for their permission to publish revised versions of this material.

Notes

Preface

1. This book focuses on France and, secondarily, England.

2. See Irigaray (1985); Cixous (1979); Cixous and Clément (1986); Lacan (2006); Montrelay (1977). Evans traced this history (1991).

Introduction

1. See Canguilhem (1966); (1989).

2. Daston and Galison (2007), (2008).

3. James (1743–45). Denis Diderot immediately undertook a French translation with the help of François-Vincent Toussaint and Marc-Antoine Eidous; see Diderot et al. (1746–48). Until then, the standard reference dictionaries for medical terms were Nicot's *Thresor de la langue francoyse tant ancienne que moderne*; *Dictionnaire de l'Académie française*; Bayle's *Dictionnaire historique et critique*; and Furetière's *Dictionaire universel contenant généralement tous les mots françois tant vieux que modernes, et les termes de toutes les sciences et des arts*.

4. On the role of language in the writing of scientific knowledge: Anderson (1984); Rheinberger (1992); Dear (1991); Campbell and Benson (1996); Licoppe (1996); Brandt (2005); Daston (1998), 20–38; as well as the role of reading practices: Chartier (1987); Chartier and Paire (2003); Darnton (1990); Daston (2004). The methodology developed here also intersects with *Begriffsgeschichte* (history of categories): Koselleck (1979); Daston and Galison (2007); Rheinberger (1985); Rheinberger and Müller-Wille (2009).

5. Michel Foucault's work has been fundamental to the approach adopted in this book, especially in relation to his questioning of the conditions that made possible the medical gaze, the investment of moral values, and the role of literary imaginings in the construction of diagnoses. Foucault addresses hysteria in three separate passages that, despite their brevity, remain fertile and incisive. In *Mental Illness and Psychology*, Foucault unfolds the stakes involved in the diagnosis of hysteria while analyzing the encounter between the church and medicine in the seventeenth century; he further describes the transition from a theological interpretation of hysteria to a physiological interpretation of the convulsive body. In *The Birth of the Clinic*, Foucault quickly returns to hysteria in order to trace the moral and political stakes involved in the perception of this pathology in the eighteenth century; and, finally, he presents hysteria as

the representative illness of the eighteenth century. This last thesis has not been endorsed or even discussed since, as if historians had agreed to ignore it, convinced instead that it was the following century that was the "century of hysteria." Yet in *Madness and Civilization* Foucault pinpoints the radical transformation undergone by the category of hysteria in the eighteenth century. First interpreted as a surge of sensations running across the body, hysteria was not, at the beginning of the century, viewed as a form of madness. According to Foucault, a series of discourses at the end of the eighteenth century focused on the effects of excessive sensibility and made the subject responsible for his own fits. Thus, what had once been interpreted as a physiological disorder became an irregularity of sensibility that could be linked to madness. Foucault repeatedly returned to the category of hysteria (2006, 2001).

6. Foucault (1973), 140.

Chapter One

1. Jacques Rancière, *The Flesh of Words: The Politics of Writing*, translated by Charlotte Mandell (1998; Stanford, Calif.: Stanford University Press, 2004), 11.

2. This definition followed the publication of two works by the same author on the same category: Louyer-Villermay (1802), (1806).

3. Louyer-Villermay, "Hystérie" (1818), 226–72; quote at 226. The Dictionnaire des sciences médicales, published between 1812 and 1822, gathered articles from the medical luminaries of the time, notably Alembert, Pinel, Esquirol, and Laennec. The spelling and punctuation of the original are maintained in all quotes. All translations are mine unless otherwise noted.

4. Louyer-Villermay, "Hystérie" (1818), 226.

5. Ibid., 230–31.

6. Louyer-Villermay (1816), 161.

7. See Rigoli (2001); Weiner (2002), (1980); Pigeaud (2001).

8. Vigarous (1801), 447.

9. Pinel (1803), 103.

10. In the subsequent editions of Pinel's nosography, the copula "is" was replaced with the verb "to seem" ("comme l'indique son nom, paroît être la matrice"), thus qualifying Pinel's assertion. But if this revision seemed to open the possibility for other localizations of the illness, the rest of the text remained unchanged, leaving no space to question the assertion. Pinel (1807), 285; (1818), 292.

11. On the change in scientific terminologies following the French Revolution, see Corsi (2005), 223–45.

12. See Veith (1973), 38.

13. The adjective *hystérique* was used as early as 1568 in France to speak about disorders of the womb. Its use was later extended to patients suffering from such disorders. In 1615, the adjective "hysterical" emerged in England and Scotland, soon followed by "hystericks," a noun used to describe remedies against it in 1649, and by another adjectival form, "hysteric," in 1657. See *Thresor de la langue francoyse*: Nicot (1606).

14. Kaara L. Peterson (2006) analyzes the use of such terms by a man, ultimately the king, to speak about his own body.

15. For Micale, there "seems to be no precedent in medical history for Shakespeare's image of the pathogenic and peripatetic womb rising through the body cavity to the throat, either as a literary physical event in the male body or as a metaphor for emotional and existential anxiety in men." Micale (1995), 16.

16. The first four mentions of the term "hysteria" in English closely follow each other; see note 119 in this chapter.

17. Both Étienne Trillat (1986) and Helen King (1998) have exposed repeated misreadings of the Hippocratic Corpus, in which the term "hysteria" does not in fact occur. Rather, it was Émile Littré who introduced the term in his translation of the corpus, interpreting it in the light of nineteenth-century medical theory. Trillat writes: "An ambiguity remains concerning what suffocates." He judges that it could be either the woman or the womb itself. Trillat (1986), 14–15.

18. Rosenberg and Golden (1992), xiv.

19. See Green (2001), 83; Paracelse (1950), chap. 1, secs. 2 and 3, p. 51.

20. See Soranus of Ephesus (1988), 1:26–30.

21. See Grafenberg (1584).

22. See Falloppio (1578), 120.

23. See especially the third discourse of Guibelet (1603).

24. See in particular Fulgose (1581).

25. See Ferrand (1627).

26. See, for example, Crooke (1615).

27. King (1998); see also King (2004), (2007). Louyer-Villermay's bibliography for Panckoucke's dictionary listed twenty-five works in Latin on the suffocation of the womb.

28. Paré (1585); facsimile ed. (1976), 189; Paré and Johnson (1634).

29. Today's equivalent of "generation" would be "procreation." In order to avoid anachronisms, the present study will use "generation" throughout.

30. On descriptions of hysteric illnesses in case records from seventeenth-century English manuscripts, see Katherine Williams (1990).

31. For detailed accounts of the role of menstruation, see Howie and Shail (2005).

32. Liébault (1598); (1651), 44.

33. Liébault (1651), 417. See Park (2006); Mandressi (2003).

34. The author speaks of *spiracles occultes.* Paré (1976), 189.

35. Du Laurens (1646), 337.

36. Ibid.

37. Berriot-Salvadore (1993); Maclean (1980); Fissell (2004).

38. Du Laurens (1646), 337.

39. Stolberg in Howie and Shail (2005), 90–101.

40. Park (2006); Lord (1999).

41. McClive (2005), 77. McClive suggests that the term was used orally at an earlier stage, notably in Astruc's lectures. Astruc's treatise was first translated into English in 1762.

42. See Pseudo-Aristote (1668), 17 bis; Mauriceau (1694), 46; Liébault (1651), 416.

43. On melancholia, see Klibansky, Panofsky, Saxl (1964); Crignon-De Oliveira (2006); Saad (2006); Gowlands (2006).

44. "They are called perturbations of the Mynde, after M. Cicerot but Galen termeth them *Pathemara vl affectus animi,* that is the affections, or sodayn mocions of the minde, changed or altered some cause, from the ryght way of reason into some passion: and these motions of the spirites, must be as well consydered, and diligently observed of the Phisician, in the tymes of sickness, as any other common syckenes." Bullein (1579), 23.

45. Liébault (1651), 422. As Foucault emphasizes in *Madness and Civilization,* these pathologies were not seen as a male illness and a female illness. A coupling of hypochondric and hysteric illness would be theorized only in the eighteenth century. Foucault (1961), 296–99.

46. Joly (1609).

47. La Framboisière (1613), 666.

48. Crooke (1615), 252.

49. Berriot-Salvadore (1995); Laurinda Dixon (1995).

50. Bartholin fils (1647), 188.

51. Varandée (1619, 1666), 179.

52. Mauriceau and Chamberlen (1672); Mauriceau (1694).

53. Ettmüller (1698), 114. He adds that "stinking & malign vapors, coming out of the womb and moving up in the form of fumes, invade the machine of our body, which is all porous, and attack the nervous genre & the heart, & produce all the symptoms described" (163). Michael Ettmüller lived between 1644 and 1683, and his works were published posthumously by his son in Latin and French beginning in 1685.

54. Thomas Willis was also famous at the time for his work on chemistry. Debus (2001), 57–102.

55. Willis (1684), 38.

56. The terminology of animal spirits was commonly accepted at the time, even though its definition is not very clear. See Hoffmann (1971), 24–25. A useful source is Georges Canguilhem's study on animal spirits in Descartes: "Animal spirits are the most subtle part of the blood, a fluid body quickly moving throughout the organism; they are what flow incessantly from the arteries to the nerves and the muscles inside the brain, and move the entire machine. The composition of the spirits is that of blood, their physical nature is that of wind or of a flame. They are blood which has lost its shape." Canguilhem (1955), 35.

57. George Rousseau (2007), 360–61. According to Rousseau, for Willis "the subtly immaterial fluid within the animal spirits—Willis' 'vital spirits,' which differed from the animal spirits—originated in the brain; [sic] brain alone could manufacture them from the moment of conception."

58. Cunningham (1989), 164–90. See also Sydenham (1682). Sydenham (1624–89) gave his name to Sydenham's Chorea (also called "Saint Vitus' Dance"). His writings, first published in Latin, were translated into English, French, and other languages and reprinted repeatedly throughout the eighteenth century.

59. Sydenham (1742), 377.

60. Micale (1995), 20.

61. Vieussens (1674), 1:44–56.

62. From antiquity to the eighteenth century, the four temperaments—bilious, phlegmatic, melancholic, and sanguine—constituted the primary model for understanding the body. A single dominant temperament was thought to govern each person's body. Any imbalance between the temperaments would eventually provoke an illness. Up until the eighteenth century, most physicians explained all illnesses as a result of such imbalance. This framework was eventually abandoned, with the nerves becoming the new model for understanding the body. On the theory of temperaments, see especially Brockliss and Jones (1997); Laurinda Dixon (1995); Harkness (2006), 171–92; Pigeaud (1981), (2006); Riva (1992); Starobinski (1963), (1960); and Fernando Vidal (2006).

63. Lange (1689), 1–2.

64. See Debus (2001), (2006).

65. Lange (1689), 24–25.

66. Ibid., 174–76.

67. La Chambre (1662), 60.

68. Fibers, along with nerves, would play a great role in the understanding of the body in the eighteenth century. They included all the tissues of the body and were thought to be stronger and more resistant in men than in women. See Wenger (2007).

69. Anne C. Vila skillfully analyzes the concept of sensibility, which she situates between the mid-eighteenth century and the French Revolution, as a conception that would disappear with the end of the ancien régime: "Sensibility was endowed with an enormous deterministic power over both the moral and the physical natures of humankind and thus provided the philosophes with a key instrument in their efforts to improve contemporary society, along with the individuals within it. . . . At the height of its conceptual popularity, therefore, sensibility was situated somewhere between enlightenment and pathology: it was seen as instrumental in the quest for reason and virtue, but was also implicated in the epidemic of nervous maladies that seemed to be overtaking the population in France and of Europe in general." Vila (1998), 1. See also Barker-Benfield (1992); Riskin (2002).

70. Although three scholars have studied the role of gender in the conception of hysteria in eighteenth-century medical texts, the inquiry developed here differs radically in methodology. In her excellent article, Elizabeth Williams examines the construction of gender in treatises on hysteria. She looks at the main works on vapors between 1756 and the end of the century, finding a decisive move toward affirming the distinction between male and female gender throughout the period, when the illness became a female illness. She concludes with the works of Louyer-Villermay as the ultimate step toward a "gendering of the malady." Elizabeth Williams (2002). Elsa Dorlin (2006) articulates a series of associations developed about women in order to construe sexual difference, and considers the role played by social class and race in this regard in the eighteenth and nineteenth centuries, notably in the corpus on hysteria. She studies textual strategies in the move from a Galenic vision of the "cold complexion" of the woman's body to a later one considering it from the viewpoint of depraved sexuality. In her interpretation, two kinds of women were imagined: women defined by their temperament and women possessing a "mutant body" (i.e., with the masculine traits of a warmer and stronger body). Dorlin sees theorization about the pathology as an instrument of sexual difference. Finally, Mark S. Micale devotes a chapter to eighteenth-century conceptions of hysteria in *Hysterical Men: The Hidden History of Male Nervous Illness (2008)*. He demonstrates the resemblance between hysteric and hypochondriac affections in order to deconstruct conceptions of masculinity.

71. Joseph Raulin was Montesquieu's physician. This allowed him to settle in Paris, where he published works that brought him immediate celebrity. From the moment of the *Encyclopédie Méthodique's* publication, his reputation spread from the capital to all of France. On Raulin's treatise, see Vila (1998), 229–35; Dorlin (2006), 99–103, 127–29.

72. Raulin (1758), ix, x–xi.

73. Ibid., 43.

74. Ibid., xxi–xxiv.

75. Ibid., 21, 97.

76. Aristotle (2000). The Roman physician Galen used and developed the Aristotelian division between female and male in his works and was an inspiration for physicians in that matter from antiquity until the eighteenth century. Galen (2006).

77. Raulin (1758), xxxiv–xxxv.

78. Ibid., 42.

79. Ibid., 98.

80. On the development of conceptions of the role of fibers in the body's disorders, see Wenger (2007).

81. See Elias (2008); Barker-Benfield (1992).

82. See Fernando Vidal (2006), 305–6.

83. *Partes blanches*, or vaginal discharge.

84. Raulin (1758), xviii.

85. See Vila (1998), 13–28; Steinke (2005), 190–93; Duchesneau (1982), 171–234, Bound Alberti (2006), 1–21.

86. See Lawrence (1979), 25.

87. Rocca (2007), 88.

88. Whytt (1767), 3.

89. Ibid., 1.

90. Dumoulin (1703), 153.

91. This statement is based on extensive research in the collections of the Facultés and Académies de Médecine in Paris and Montpellier, of libraries in New York, Boston, and Galveston, of the Bibliothèque Nationale Française and the New York Public Library, and in the manuscript archives of municipal libraries in Rouen, Arles, Lyon, Avignon, and Montpellier.

92. Dufau (1768), 120–29. Pomme used the same noun in his response, published in the same journal, 273–75.

93. Hélian (1771).

94. Sauvages (1763). See Elizabeth Williams (2003), 80–111.

95. Sauvages (1771), 792.

96. Ibid., 791. In doing so, Goubert was actually following Sauvages's word choice in the first classification he had published much earlier, Sauvages (1731), where Sauvages used the word "hysteria" instead of the French *hystérie*.

97. The anonymous writer in Dumoulin's volume wrote about the presence of the same symptoms in hypochondriac men and described the vapors of a male patient. Dumoulin (1703), 157.

98. "constitution molle et efféminée"; Sauvages (1771), 4:133.

99. Sennert (1611).

100. Jonston (1644).

101. The literature on classification schemes in medicine is surprisingly sparse. See Jewson (1976); Martin (1990); Lester King (1958), in particular 193–214; (1966); Yeo (2003), 214–66.

102. Sauvages (1771), 1:95.

103. Ibid., 96.

104. Ibid.

105. Steinke (2005), 237.

106. Sauvages's interest in botany was such that he was awarded the Chair of Botanics at the Faculté de Médecine of Montpellier upon the death of Aymé-François Chicoyneau (1740).

107. Elizabeth Williams (2002) also suggests that Sauvages's choice of the word "hysteria" might have been in homage to Linnaeus.

108. Linnaeus (2003), 172, 175, 180.

109. "Partial tonics" (*toniques partiaux*) included all cases of rigidity or immobility of a limb or organ (such as a stiff neck). "General spasms" applied to the entire body and included tetanus and catalepsy. "Spasms marked by involuntary and forced agitation of an organ" included convulsions, palpitations, and limping. "General tonic spasms" included hysteria, shuddering, child convulsions, epilepsy, Saint Vitus' dance, and beriberi.

110. "White flowers" has been renamed "leucorrhea" in our contemporary medical discourse.

111. Sauvages (1771), 687.

112. Ibid., 585.

113. Ibid., 1:763, 737; 2:305, 306, 316, 326, 359, 615, 687.

114. As named by Tort (1989).

115. Cullen (1777–84), (1785), (1792).

116. Alibert (1817).

117. Hosack (1821).

118. Broussais (1821).

119. The English words "hystericism" and "hysteria" first appeared in 1710 and 1764, respectively. In 1615, the adjective "hysterical" saw its first use, soon followed by the term "hystericks," a noun used from 1649 to name therapeutics against hysterical symptoms, and "hysteric," which was used more widely from 1657. The first mentions of the noun "hysteria" are in Marryat (1764), 220. Andrew Duncan uses it as a chapter heading in (1781), 217, 234, 237–41. The term appears regularly from that point on. See, for example, Aitken (1784), 54, 61, 63; Cullen (1797), 83.

120. William Cullen (1710–90), holder of a diploma from the University of Edinburgh's Faculty of Medicine, was considered, along with Sydenham, the greatest English-speaking physician of the "prescientific" era. See Dechambre (1880). One of Cullen's works was translated into French by Philippe Pinel: Cullen (1785).

121. Robert James chose the noun "hysterica" for the first dictionary of medicine, published in 1743–45.

122. Vicq d'Azyr, *Traité* (1786).

123. Nysten (1814). Motherby included the noun "hysteria" in his dictionary, published in 1795.

124. See Elizabeth Williams (2002); Foucault (1972), 296–99.

125. On this topic, see Edelman (2003), 16–22; Riva (1992).

126. Louyer-Villermay, "Hypocondrie" (1818), 108.

127. Ibid.

128. Ibid., 248.

129. After returning from his work with Jean-Martin Charcot in Paris, Freud presented his paper on male hysteria to his home audience in Vienna. His recollection was that his audience found his remarks incredible. One of them, an old surgeon, exclaimed: "But, my dear sir. How can you talk such nonsense? *Hysteron* means the uterus. So how can a man be hysterical?" Gilman (1985), 402.

130. Georget (1820), 263. A student of Pinel and Esquirol, Georget (1793–1828) had a very short career.

131. Georget (1820), 263.

132. Le Pois (1563–1633) latinized his name as Caroli Pisonis. His *Selectiorum observationum et consiliorum: De Cognoscendis et curandis, præcipue internis humani corporis morbis* (1618) seems to be the first to localize the origins of "suffocation of the womb" in the brain.

133. Georget (1820), 192.

134. Ibid., 292–94.

135. Ibid., 262.

136. La Framboisiere (1613), 666.

137. Joly (1609), 1–2.

138. Rosen (1980), 166.

139. This point will be further illustrated in chapter 4 for Paumerelle's work *La philosophie des vapeurs.*

140. Vallot, Daquin, Fagon (2004).

141. Ibid., 194–95.

142. Marin (1997), 218–41.

143. Sévigné (1978), 583.

144. Ibid., 634.

145. Diderot (1821).

146. 5 F 196 Fonds Requin, Archives Départementales de Vaucluse (Avignon).

147. See George Sebastian Rousseau (1969); Porter and Porter (1988).

148. See Noyes (2011); George Sebastian Rousseau (2004), 246–99.

149. See Emch-Deriaz (1992). On Tissot, see chapter 5.

150. The first edition was Tissot (1761). A translation in English was published in 1765.

151. Tissot (1763), 414–21.

152. Tissot's text only uses the word "mother" to refer to "fits of the mother," as was common in medical texts of the time.

153. Vandermonde (1770).

154. Vandermonde (1758).

155. George Sebastian Rousseau (1991), 37.

156. Raulin (1758), viii. See Dorlin (2006), 102.

157. Mercier, *Le nouveau Paris* (1994); *Tableau de Paris* (1994).

158. Risse (1988).

159. E.g., Pomme (1777).

160. Collet, 1777, SRM 117, dr. 20; Académie de Médecine de Paris: Archives de la Société Royale de Médecine.

161. Ibid.

162. Pressavin, *Nouveau Traité* (1770), 245.

163. The contest rules were phrased as follows: "Describe the characters of Nervous Illnesses themselves such as hystericism, hypocondriacism, on which points they differ from analogous illnesses such as melancholy, what are their principal causes & which method is generally to be employed in their treatment." No prize was awarded that year. Manuscript Société Royale de Médecine SRM 197, dr. 8, 1786.

164. Urban (1787).

165. On this therapeutic practice, see Bertucci (2006); Heilbron (1979); Schaffer (1983).

166. Even though Mesmer did not aim to cure specifically nervous illnesses and advocated the use of animal magnetism to prevent all illnesses, caricatures always portrayed him with vaporous patients. See Trillat (1986); Ellenberger (1970); Veith (1973); Brockliss and Jones (1997), 791–94; Darnton (1968); Rausky (1977).

167. *Chansons choisies, avec les airs notés* (1782), 3:132. The poem is entitled "Air N° 32 Les Vapeurs."

168. In French, the song adds a lisp, *Z'ai* instead of *J'ai*, as a further marker of affectation. The author also plays on the meaning of *fadeurs,* "insipidness," which also connotes a hackneyed compliment.

169. De Baecque (1997).

170. See also Seth (2006); Thomas (1989).

171. See De Baecque (1993), (2000), (1997), (1989); De Baecque and Mélonio (2004).

172. The Princess de Lamballe was famous for frequent fainting spells and was the target of mockery for her swoons at the mere sight or smell of a lobster. She took numerous trips to the waters of Bourbonne, Vichy, and Plombières to attempt to cure her vapors. Andreas Saiffert, a

German physician, had more success in healing her than did his French colleagues, who pronounced the princess incurable. He wrote a detailed account of the princess's vapors in a work published long after her death: Saiffert (1804), 231–28.

173. See, for example, the memoirs attributed to Pidansat de Mairobert (from 1777 to 1779) and Mouffle d'Angerville (from 1779 onward), both secretaries of Bachaumont, and published under the name Louis Petit de Bachaumont (1784), 48.

174. Angulo (1979), 241.

175. Anonyme, "De l'Hystéricie," manuscript 5069–5070, 218, Bibliothèque Interuniversitaire de Médecine (Paris).

176. Such comments reappear in Van der Meulen (1793), 27; and a similar anecdote told by Victor du Bled in his volume devoted to the eighteenth century, published in 1908, attests that this interpretation circulated for a long time. Du Bled (1908), 135.

177. Duchesne (1792), 45.

178. Guibert (1806), 17.

179. Artaud (1790), 1.

180. Dartaize (n.d.), 28.

181. Ibid, 64.

182. Meilham (1965).

183. *Testament de la République et ses derniers adieux, liste et noms de ceux qui doivent en hériter et de ceux qui n'auront aucune part à la succession* (1799), 2–3. These metaphors reappear during the Paris Commune in 1870. See Beizer (1994), 205–26.

184. Burke (1999), 125–26.

185. Ibid., 278–79.

186. Ibid., 241.

187. See Holt (1820); Huish (1821); Macalpine and Hunter (1969).

188. As Jonathan Andrews and Andrew Scull argue, William Pitt the Younger's choice of Willis instead of John Monro, then director of the famous Bethlem Hospital, was probably the result of his search for a "cooperative medical man" who would provide "a diagnosis more to his liking." Andrews and Scull (2001), 258; also 256–61. See also Scull (2005); Thomas Dixon (2006).

189. See Kantorowicz (1989).

190. Macalpine and Hunter (1969), 54.

191. Thomas Dixon (2006).

192. The interrogations were published in complete versions and cheaply printed expurgated versions, which sold for a shilling. *Important Facts and Opinions Relative to the King* (1789). The interrogations were also regularly published in journals such as *Philological Society, The European Magazine,* and *London Review.*

193. *Important Facts and Opinions Relative to the King* (1789), 10.

194. See, for example, Pinel's affirmation in his 1791 *Nosographie* regarding "true medicine," which "is grounded on principles, which consists much less in the administration of medicines than in the deep knowledge of illnesses, and which has been practiced by physician-observers in all ages." Philippe Pinel, B MS b 13, Francis A. Countway Library, Harvard University (Cambridge, USA).

195. *Important Facts and Opinions Relative to the King* (1789), 42.

196. Charcot (1877); Charcot and Richer (1887). Charcot was drawing on Diderot's works.

197. See notably de Certeau (1975), (1982), (2005); Fissell (2004), 53–89; Foucault (1995); Mazzoni (1996); Gilman et al. (1993), 91–221.

198. Kramer and Sprenger (1487). The theological work *Malleus maleficarum* was not the first to describe the existence of witches, or to advocate torture, but it is the one that became most famous. It served as a key reference on the subject of possession for centuries to come, despite arousing some opposition from the Catholic clergy from its first publication. See Mackay (2006); Trevor-Roper (1967).

199. Weyer (1563), (1567), (1577), (1579).

200. Le Loyer (1608), 345–46; my translation.

201. Marescot quotes the synod's edict as follows: "Before undertaking exorcism, the priest must inquire diligently about the life of possessed persons, their social condition, their fame, their health, & other circumstances, and discuss it with wise, cautious, and well-advised people. For often the credulous are mistaken, & often melancholics, lunatics & the bewitched deceive the exorcist, saying that they are possessed and tormented by the Devil. Yet these need the physician's therapies more than the exorcists' ministry." Marescot (1599), 48. See Foucault (1994), 753–66.

202. See Scull (2009), 6–23.

203. Jorden (1991). See also Scull (2009), 1–23.

204. See Digby (1658), 125–26.

205. See Mandrou (1980); Chaunu (1969); Muchembled (2000), 199–248.

206. Willis (1681, 1684), preface (n.p.).

207. Bayle and Grangeron (1682), 59.

208. Réverend Père Pierre Le Brun (1733).

209. Ibid., 150.

210. "vapeurs peurs noires"; Le Brun (1733), 157.

211. Le Brun (1733), 160–61.

212. *Recueil général des pièces concernant le procez entre la Demoiselle Cadière, de la ville de Toulon, et le Père Girard, Jésuite* (1732).

213. See the statistics published by Daniel Vidal (1987), 174–75, 179–81; and Kreiser (1978).

214. Maire (1985).

215. Chapter 2 will show how a Jansenist physician, Philippe Hecquet, dissociated himself from the Convulsionaries. See Maire (1998); Lyon-Caen (2008).

216. Du Sault (1735), 15. See also "Copie de la Lettre de Mlle Agne's, Pensionnaire au Couvent d'Ollioules, adressée à Mr. L'Avocat Chaudon, de premier juillet 1731," in *Recueil général des pièces concernant le procez entre la Demoiselle Cadière, de la ville de Toulon, et le Père Girard, Jésuite, Recteur du Séminaire Royale de la Marine de ladite ville* (1732), 3; De Bonnaire (1733); Delan (1735); La Taste (1734); Fouillou (1733); La Boissiere (1787); Hecquet, *La Suceuse convulsionnaire* (1736), (1733), *Lettre sur la convulsionnaire* (1736).

217. Fouillou (1733), 33, 46–47.

218. "Rapport fait par les Sieurs Andry & Winslow Docteurs en Médecine de la Faculté de Paris, au sujet de la maladie & guérison d'Anne le Franc," in *Dissertation sur les miracles* (1731), 34.

219. See Lindsay Wilson (1993).

220. "Rapport fait par les Sieurs Andry & Winslow," in *Dissertation sur les miracles* (1731), 34.

221. Desessarts (1733), 8.

222. Ibid., 11.

223. See the interrogation of Suzanne Françoise Roubeau. Bernadac and Fourcassié (1983), 201.

224. See Lindsay Wilson (1993), 17–33.

225. Quoted in De Baecque (2000), 185-86.

226. Bressy (1789).

Chapter Two

1. "Voilà l'effet de la dénivelée du langage: substituer au sens d'un mot, le sens de la répétition de ce mot." Antoine Compagnon, *La seconde Main; ou, Le Travail de la citation* (Paris: Seuil, 1979), 86.

2. Brockliss and Jones (1997), 114.

3. Sydenham (1742, 1809), 377.

4. Pressavin, *Nouveau Traité* (1770), 201.

5. Lieutaud (1781), 394. Likewise, Whytt wrote: "Some doctors called any malady they did not understand a nervous disorder." Whytt (1765), iii.

6. Blackmore (1725), 111-12.

7. Raulin (1758, 1759), liv-lv.

8. On the question of observation, see Daston and Lunbeck (2011).

9. Manuscript by Broussonet, *Cours de Pathologie*, H 567, 281, Bibliothèque de *l'École* de Médecine de Montpellier.

10. Victor Broussonet (1761-1807) was the son of François Broussonet, himself a physician holding a diploma from Montpellier. Broussonet, *Cours de Pathologie*, H 567, 284.

11. The use of metaphors in medical and scientific discourse has already been a subject of inquiry. See especially Colin Jones (1996); Otis (1999); Seth (2008); Brandt (2005); Sontag (1978).

12. Foucault (1989), 142.

13. "fourmillements importuns"; Hunauld (2009), 105.

14. "mouvements reptiles." Pomme is among the authors choosing the image of reptiles: "Many say they feel the move of a ball from bottom to top; this ondulation many times (as I observed it myself) imitated that of a serpent, & is felt in the throat, which undergoes a more or less violent strangulation." Pomme, *Traité des affections vaporeuses des deux sexes: où l'on a tâché de joindre . . .* (1763), 4-5.

15. "la marche des fourmis." "For example, swarmings in the flesh . . . One feels small reptilian movements in different parts of the body, the sensation of which seems to resemble that provoked by ants walking on the skin." Révillon (1779, 1786), 6.

16. Fissell (2002).

17. At the end of the nineteenth century, the images of animals reappear, in the works of Charcot and Gilles de la Tourette. Charcot (1881); La Tourette (1891).

18. See, in particular, Mora (1991), xcii, 790, illus.

19. Planque (1750), 506-8. Planque's French translation of Freind's observation is reproduced in Arnaud, "Une maladie indéfinissable?" (2010).

20. Freind's "was ordinarily overcome by convulsions" is translated as "fut prise à l'ordinaire de ses convulsions." Planque (1750), 507. See also Freind's "the ordinary signs by which one knew the fit was about to begin."

21. In the French translation of Freind: "J'ai peine à exprimer ce que je vis, & ce que j'entendis, & il faut l'avoir vû pour en avoir une idée juste." Planque (1750), 507.

22. See, for example, Nynauld (1990).

23. For the image of the dog, see the manuscript SRM 163, dr. 10, no. 25, Académie de Médecine de Paris: Archives de la Société Royale de Médecine; the image of the wolf is found in the manuscript SRM dr. 1, no. 2.

24. Ettmüller (1698), 138–39.

25. Varandée (1666), 166.

26. La Chambre (1662), 51.

27. Digby (1658), 150.

28. La Chambre (1662), 61.

29. Mauriceau (1694), 447.

30. Finch (1723), preface (n.p.).

31. This metaphor is also used in Burton's work on melancholia: Burton (1621).

32. Purcell (1702), 1–2.

33. Hunauld (2009), 96.

34. Pomme, *Traité des affections vaporeuses des deux sexes: où l'on a tâché de joindre . . .* (1763), 2–3.

35. Perry (1755), preface.

36. Chambaud (1786).

37. Perfect (1780), 75.

38. Capuron (1817), 84.

39. Cheyne (1733), 135.

40. Retz (1790), 177.

41. Pomme, 1769, xiii.

42. Ibid., 209.

43. Sauvages (1772), 793.

44. The chameleon is one of the monsters listed and illustrated in Paré (1585).

45. Sydenham and Latham, trans. Greenhill (1848), 89. This work was first published in Latin by Sydenham (1682). Curiously, the image of the chameleon and of Proteus present in the Latin text and its translation by John Pechey disappear in John Swan's translation, the only one to be republished from 1742 on. These images are nevertheless found with great regularity in physicians' writings on hysteric illness. Jault's French translation is based on Swan and therefore does not include the metaphor: Sydenham (1774), 307.

46. Ficinus (1962), 1814–35. Theophrastus's text is lost. Ficinus's commentary has recently been translated into English and added to the text of Priscian: Ficinus (1983); Priscian, trans. Huby (1997).

47. On the conception of the imagination in the Renaissance, see Robert Klein (1960), 47–84; Garin (1989), 303–17.

48. "It [imagination] can produce phantasm far beyond the activity of the senses. It surpasses sensation, because in producing images without external stimulation, it is like Proteus or like the chameleon." Weyer (1991), 186–87.

49. Du Laurens (1600), 99.

50. Ibid., 104; see especially 99–122.

51. Chambaud (1786), 128.

52. Broussonet, *Cours de Pathologie,* H 567, 284, Bibliothèque de l'École de Médecine de Montpellier.

53. See, for example, this 1780 description by William Perfect, a surgeon at West Malling, in Kent: "A disorder peculiar to the Fair Sex, differing in most cases very essentially from the Hypochondriac Affection, both in Cause and Situation: The fits accompanying this complaint are, in general, very uncertain; in some they will return weekly, or monthly; in others, four, five, or six times in a year, or oftener upon any sudden commotion of the mind, or disturbance of

the spirits; by fear, grief, anger, or disappointment; wind, and acrid humours, vellicating the nerves of the stomach and intestines, will frequently produce the fits in women of a delicate habits, whose nervous system is naturally weak and irritable. The symptoms preceding the fits are different in different persons; those attending them, are well known, to be a difficult respiration, and sometimes weak and easy, as if the patient was asleep, convulsed agonies of the whole body, involuntary laughter and crying, paleness of the face, coldness of the extremities, oppression, anxiety, reaching to vomit, a violent rising in the throat, and often a strong intermitting pulse. After the fit is gone off the patient frequently complains of universal soreness, pain in the head, noise in the ears, dimness of sight, the pulse becomes quicker, and more regular, and the patient either relapses into another fit, or falls asleep, and, for that time, recovers. Much more be added, were it not almost impossible to describe and enumerate the variety of symptoms attendant on either this or the Hypochondriac disorder; for as the sagacious Sydenham has very wisely observed: 'The shapes of Proteus, or the colours of the Camelion, are not more numerous and inconsistent, than the variations of the Hypochondriac or Hysteric disease.'" Perfect (1780), 73–75.

54. On the relationship between, seeing, knowledge, symptom, and sign, see Foucault (1973), 107–51.

55. Pomme (1769), 60.

56. Tronchin to Mademoiselle Thellusson, 1758–1760, Bibliothèque Publique et Universitaire de Genève, fond manuscrit, 305.

57. Sue (1789), 102.

58. Daignan (1786), 182.

59. Boulogne and Delattre (2003), 213. "For the physician always pays attention to the disease and to nature, which are, so to speak, adversaries of each other. He makes a prognosis first of the patient's recovery or death from diagnosis of which is stronger, and then he determines the duration by how much one is stronger than the other." Dean-Jones (1993), 124.

60. See Cunningham and French (1990); Porter and Rousseau (1980); Barroux (2008); Chamayou (2008); Licoppe (1996); Keel (2002); Louis-Courvoisier (2000); Goldstein (2001); and Weiner (2002).

61. Pinel, B MS b 13, Francis A. Countway Library, Harvard University (Cambridge, USA).

62. Paré (1573), 189.

63. Given this perspective, it is clear why so many doctors followed Hippocrates in advising marriage as a therapy, so as to align the order of morals with natural order. It should be noted that Hippocrates does not use the image of the animal. A passage on uterine suffocation contains an explanation of the womb's movement as resulting from the emptiness of the organ. Hippocrates (1853), 33.

64. Plato, Timaeus, 91b-92b, quote at 91c.

65. Ibid., 91a-91c.

66. Compagnon (1979), 56: "un énoncé répété et une énonciation répétante"; see also 49–92.

67. "Historians need . . . to examine the ways in which gendered identities are substantively constructed and relate their findings to a range of activities, social organizations, and historically specific cultural representations." Scott (1988), 44.

68. Montaigne (1999), 859 (chap. 5). Rabelais, in contrast, ironically introduces a division between the sexes through Doctor Rondibilis's citation of Plato's animal during a description of symptoms of feminine desire in Pantagruel. Rabelais (1995), 309–11.

69. See the conceptions of suffocation of the womb or *mal de mère* in La Framboisière (1613);

Guyon (1673), 330 ff. On the image of the womb in sixteenth-century England and its role in the perception of women, see Fissell (2004), 35–89.

70. Liébault (1651), 433: "matrice vagabonde."

71. Liébault (1651), 411: "The cause of this movement, which is strange and counter to nature, varies, according to Hippocrates in *Maladies of Women*. First, when the womb of young girls being without the enjoyment of desire of conception that she naturally inherits, becomes indignant like an animal, thus flutters to and fro demanding the means by which it can satisfy concupiscence."

72. Varandée (1666), 172–73.

73. Many Jansenists distanced themselves from the Convulsionaries, observing that the movement did not reflect their faith. Some wrote as much publicly, in order to avoid being associated with their discredited fellows. One of these was Philippe Hecquet, who devoted a series of works in the early 1730s to proving the physiological origin of the Convulsionaries' spasms. Hecquet's religious convictions are demonstrated by his biography. Upon receiving his diploma in medicine in 1689, he retired to Port Royal to share in the communal life of the Jansenists there. After nine years, he left the abbey and obtained his certification in medicine. He would work for religious congregations his entire life. His most substantial work is *Le Naturalisme des convulsions* (1733).

74. He references the suffocation of the womb rather than the "sacred malady." Hecquet, *Le Naturalisme des convulsions* (1733), 149.

75. The text continues: "These are therefore hysteric passions in all girls, but in the Convulsionaries they are fits of the mother (as these names no longer cause offense to girls). We should also add uterine furors, as they give themselves too familiarly to the eyes, hands, & actions of men, gladly knowing the aid that men can give them." Hecquet, *Le Naturalisme des convulsions* (1733), 149.

76. Hecquet, *Le Naturalisme des convulsions* (1733), 164.

77. See also Hecquet, *Le Naturalisme des convulsions* (1733), 149.

78. De Bonnaire (1733), 43.

79. Ibid., 63–64.

80. Ibid., 40.

81. Diderot, "Sur les Femmes" (1994), 952.

82. Ibid., 952–53.

83. Chambon de Montaux (1798), 450.

84. "In fact, the womb does not enter into movement, like a wild animal leaving its lair, because it likes good odours and flees from bad ones; on the contrary, it curls up on itself because of constriction due to inflammation." Soranus of Ephesus (1988), 1:30. On this, see Cadden (2003), 28–30.

85. Helen King (1998), 223.

86. Chambon de Montaux (1798), 451.

87. Ibid.

88. Jouard (1804), 55.

89. Virey (1823), 262.

90. Ibid., 414–15.

91. On this point, the present study shares the approach of researchers such as Helen King, Joan Cadden, Monica Green, and Elsa Dorlin, among others. Green (2005), 1–46; Cadden (2003); Green (2001); Helen King (2005), 47–58; (1998); Laqueur (1990); see also Packard and Fissell (2008).

Chapter Three

1. Athena Vrettos, *Somatic Fictions: Imagining Illness in Victorian Culture* (Stanford, Calif.: Stanford University Press, 1995), 2.

2. See, for example, Foucault (1963); Roger (1963).

3. On objectivity, see Lorraine Daston's work: for example, Daston (2001); Daston and Galison (2007).

4. See Pomata (2011); (2010); Lockwood (1951); Nance (2001); Harkness (2006), 171–92.

5. Barras, Hächler, and Pilloud (2004); Barras and Rieder (2001), 201–22.

6. Wenger (2007).

7. Révillon was translated into German, the works by Dr. Rostaing and Noël Retz were reprinted twice, and Pierre Pomme's work eight times, not to mention translations into Spanish, Italian, and English.

8. Révillon claims to have read the works of many physicians on vapors: "Willis, Sydenham, Boerhaave, Viridet, Montanus, Mead, Hunauld, Robinson, Sauvage, Lorry, Pomme, Witt, Raulin, & and other special treatises." Eight of these thirteen works are foreign treatises, and only those of Sydenham, Boerhaave, and Whyt (here spelled Witt) are translated into French. A little later, he makes a similar assertion with regard to further untranslated authors: "Sanctorius in Italy, Keil in England, Gorter in Holland, Dodart in France . . ."; Révillon (1779), 14, 28. In his study of the library of Guillaume Descamps, an unknown physician from the countryside who died in 1758, Léo Barbé lists thirty-three titles in fifty-three volumes. He mentions editions of the Hippocratic works, Bayle and Baglivi published in Latin, and English works by Richard Borthon, the Scot Thomas Burnet, Sydenham, and Willis. Barbé (1991), 200–213.

9. Stolberg (2004), 89–107.

10. Bullein (1579).

11. Bovier de Fontenelle (1990).

12. See Mazauric (2007), 137.

13. See Goodman (1994).

14. In Molière and Goldoni, and Lesage's *Gil Blas*, to name just a few relevant texts, the physician is presented as a man of knowledge who is unable to make use of the experience of his patients and uses Latin to indulge in references inaccessible to his patients instead of listening to their sufferings.

15. Mandeville (1711), 233.

16. Hunauld (2009).

17. Ibid., 85.

18. At this point, he echoes advice given by Sydenham.

19. Hunauld (2009), 85.

20. Ibid., 89.

21. Ibid., 90.

22. The name Asclepiades is a reference to Asclepius, the son of Apollo and Coronis in Greek mythology. Asclepius learned the art of healing from the centaur Chiron, became a hero through his exploits, and was venerated as a god. The cult of Asclepius extended from Thessaly to Rome until Christian times. Many physicians took the name Asclepiades during antiquity.

23. Hunauld (2009), 102.

24. Ibid., 101.

25. Ibid., 102.

26. Ibid., 94.

27. Ibid., 104.

28. Ibid., 117.

29. Ibid., 89, 106, 141.

30. Ibid., 102–3.

31. Ibid., 109.

32. Ibid., 115.

33. Ibid., 117.

34. Ibid., 115.

35. See Schiebinger (1989).

36. Hunauld (2009), 97.

37. Ibid., 104.

38. Ibid., 126.

39. "Pituite" is an older term for gastrorrhea.

40. Hunauld (2009), 126.

41. Ibid., 106.

42. Ibid.

43. Ibid.

44. Ibid., 107.

45. Hunauld (1756), 43.

46. Ibid., 44.

47. Hunauld (2009), 118.

48. Ibid., 119.

49. Hunauld (1756), 170–71.

50. See Deleuze (1990).

51. Puységur's volume is not on hysteric illness or magnetism, but the case he recounts presents similar symptoms to the ones given in treatises on hysteric and hypochondriac illness—he provides no diagnosis. See Puységur (1986), 338–39; (1811), 339.

52. Cardano (2002).

53. See Lawrence and Shapin (1998), 156–201; Jordanova (2000).

54. Fuller (1705, 1750).

55. Ibid., 253.

56. Ibid., 253–54.

57. Ibid., 269–70.

58. Sydenham (1809), §116, 292–93.

59. See Porter (1991), 276–300.

60. Scottish physician George Cheyne (1671–1743) enjoyed ties with the most renowned mathematicians and physicians of his generation. He recounted spending a great deal of time in taverns to develop his clientele. Cheyne settled in Bath, where he became a famed doctor. He attacked understandings of the body in terms of temperament, instead using an iatrochemical approach that analyzed physiology in terms of solids and fluids. He thus viewed maladies as the result of a disturbance in the circulation of fluids. Cheyne (1733), 250. For scholarship on George Cheyne, see Guerrini (2000); Porter (1991); Shapin (2010); Shuttleton (1995), 316–35; Shuttleton (1999); Rousseau (1988); Dacome (2005), 185–204.

61. Cheyne (1733).

62. Graham John Barker-Benfield devotes several pages of his book *The Culture of Sensibility: Sex and Society in Eighteenth-Century Britain* to the work of George Cheyne. He presents Cheyne as "essentially torn between reform and consumerism" and considers that his "vision

was shaped by a pseudo-historical contrast between the putative health of a pastoral age and the sickness engendered by subsequent economic success." He examines how Cheyne's portrait of England belongs to the affirmation of a new economic abundance at the beginning of the eighteenth century. According to Barker-Benfield, Cheyne sees this richness as the source of a new refinement while also confronting the physiological troubles and moral wanderings to which it can lead. Barker-Benfield (1992), 11–12. Shapin underlines that "his patients' return to health was also enmeshed in a network of reciprocal moral obligations with their physician." Shapin (2010), 307. See also Porter (1991); Guerrini (2000); Shapin (2010); Shuttleton (1995), (1999); and Wild (2006).

63. Shapin (2010), 312.

64. Wild (2006), 147.

65. Similar claims were regularly the object of treatises—for example, that published by Sir Richard Blackmore, fellow of the Royal College of Physicians in London, eight years earlier. Blackmore (1725), A2.

66. Cheyne (1733), 34–35.

67. For a history of food symbolism in relation to class in the early modern period, see Albala (2002), 184–216.

68. See Blackmore (1725), 97.

69. Cheyne (1733), 110.

70. Ibid., 36.

71. See Benveniste (1974), 82.

72. Cheyne (1733), 1–2.

73. Sydenham (1742), 374.

74. He adds: "Of this every one who has liv'd any time in the World may have seen Instances, from the *Hero* to the City Girl." Cheyne (1733), 1–2.

75. Shapin (2010), 307.

76. See Klein and LaVopa (1998); Mee (2003); Heyd (1995).

77. Cheyne (1733), 250.

78. See Shapin (1998), 21–50.

79. Cheyne (1733), 3–4.

80. Ibid., 342.

81. See Wild (2006), 118 ff.; Guerrini (1999). I differ here from Anita Guerrini, who insists on the role of faith in Cheyne's transformation: "A pervasive sense of sin underlies Cheyne's discussion, culminating in his autobiography." Guerrini (2000), 149.

82. Cheyne (1734), 331.

83. Ibid., 327.

84. Ibid., 368.

85. Moody Library, University of Texas Medical Branch. Cheyne (1734) edition with marginalia owned by the Blocker Collection of the Moody Library, University of Texas Medical Branch, Galveston.

86. See Chartier (1989), 154–75; Parent (1984), 801–21.

87. For example, on p. 337 a note indicates "p. 361," and p. 337 indicates p. 361; p. 340 refers to pp. 306–307, and, conversely, p. 307 to pp. 340–357.

88. Cheyne edition owned by the Moody Library, University of Texas Medical Branch, 370.

89. For an analysis of the limits of the opposition between physicians and charlatans, see Ramsay (1988).

90. "mélancolie profonde et de vapeurs noires."

91. Gamet (1772), 272.

92. Ibid., 275.

93. Ibid., 306.

94. Révillon (1786), 23.

95. *L'Esprit des journaux, françois et étrangers* (1786): 106–16.

96. It is the *Journal des savants* that gives the highest praise (1786): 1471–72.

97. *Mercure de France* (1786): 190–91.

98. Smollett (1781).

99. Révillon (1786), 85.

100. Révillon (1779), 4.

101. Révillon (1786), 9; also 2.

102. Révillon (1779), 63.

103. Benvéniste (1971), 260.

104. Révillon (1786), Lettre VI, 45–46.

105. Ibid., 23.

106. Révillon (1779), 41–42.

107. Ibid., 25.

108. Ibid., 35.

109. Ibid., 62.

110. Ibid., 38, 41.

111. Révillon (1786), Lettre XVII, 97.

112. Ibid., 29–30.

113. Ibid., viii–ix.

114. Ibid., xiij.

115. Ibid., 1.

116. Révillon (1779), 11.

117. Révillon (1786), 30.

118. "souffrant," "faible," "assez bien," "mieux."

119. Révillon (1786), 19–20, 63.

120. In the preface, Révillon wrote of vaporous people: "When they are attentive, they can, on waking up, recognize and name the wind that reigns on their horizon by the moral and physical dispositions in which they find themselves." Révillon (1786), xiij.

121. Santorio (1670). On this, see Dacome (2012). At the time a shift from observation toward experimentation was under way. See Steinke (2005), especially 127–64. See also Gooding, Pinch, and Schaffe (1989); Licoppe (1998).

122. Golinski (1999).

123. Révillon (1786), 127.

124. Brockliss (1994), 80.

125. Wayne Wild has looked in detail at Cheyne, Whytt, and Cullen; this chapter focuses on Tronchin, Pomme, and Tissot. See especially Wild (2006); Barras, Hächler, Pilloud (2004); Barras and Rieder (2001); and Rieder (2010).

126. On Tronchin, see Tronchin (1906); and Voltaire (1950).

127. Brockliss (1994), 81–84.

128. Tronchin, manuscript, Registre de Consultations (1758), 119, Bibliothèque de Genève.

129. Ibid., 125.

130. On earlier writings about regimen, and especially Renaissance writing on dietary regimes, see Albala (2002).

131. Pomme, manuscript, Arles.

132. Barthez, manuscript, Montpellier.

133. Pomme, manuscript, Arles.

134. Letter, April 16, 1790, Fonds Tissot, Bibliothèque Universitaire de Lausanne.

135. Masturbation has been widely studied in the last twenty-five years. See notably Jordanova (1987); Wagner (1988); Stolberg (2004), 89–107.

136. On the use of medical terminology and models from the treatises of Tissot in writing biographical memoirs, see Barras, Hächler, and Pilloud (2004); Barras and Rieder (2001).

137. Pomme, manuscript, Arles.

138. Until the end of the eighteenth century, physicians often used the word "recipe" when talking about prescriptions that had to be prepared, either by the apothecary or by patients themselves just before taking them. In the sixteenth and seventeenth centuries, cooking recipes and medical potions were often published in the same volumes.

139. Tronchin, manuscript, Registre de Consultations (1756–1759), Bibliothèque de Genève.

140. Pierre Chirac (1650–1732) was Philippe d'Orléans's physician, ennobled in 1728 before becoming King Louis XV's First Physician. See Michaud (1813).

141. Tronchin, manuscript, Registre de Consultations (1756–1759), Bibliothèque de Genève.

142. See Derrida (1972).

143. Doppet, Le Médecin de l'amour (1787), 224–25.

144. Ibid., 15.

145. On Julien Offray de la Mettrie (1709–51), see Wellman (1992).

146. Jean-Baptiste Silva was a doctor famous for treating the vapors among aristocratic women. See La Mettrie (1748, 1768), 45–46.

147. La Mettrie (1748, 1768).

148. Bourdelin (1776), 75–76. The attribution to Louis Henri Bourdelin, or Bourdelin le jeune, remains uncertain, and no biographical information is available about him.

149. Sue (1789), 94.

150. The Hôpital des Petites-Maisons, established in 1557, was the oldest hospital for those considered mad. First created for elderly people, it was later extended to receive the "mad" in return for a pension.

151. Sue (1789), 96.

152. "vapeurs"; Trévoux (1752).

153. Raulin (1766), vii–viii.

154. Ibid, 97.

155. Goldstein (2001), 324.

Chapter Four

1. Jean Louis Schefer, L'espèce de chose mélancholie (Paris: Flammarion, 1978), 153.

2. Goodman (1994), 136.

3. For an important discussion of the role of salons, see Goodman (1994).

4. The adjective "vaporous" is widely used in French works from the eighteenth century on, most notably in Prévost, Madame Riccoboni, Diderot, Mirabeau, Mercier, and Madame de Charrière.

5. See Imbroscio (2002).

6. Montenoy (1755).

7. Dramatist Charles Dufresny's complete works, first published in 1699, were reprinted regularly in the eighteenth century. One of his plays presents patients with imaginary illnesses who are reluctant to use the word "vapors" to describe their distress, fearing they will not be taken seriously. The female patient says: "Pooh, Sir, I have an aversion to women with vapors; as far as I'm concerned it is some kind of quivering . . . of horror . . . there . . . a dejection." *Œuvres* (1779), 2:71. Many of Goldoni's plays, and those of his emulators, offer a rapid characterization of a vaporous woman in the direct heritage of Molière. Using the vapors, playwrights delineate the distress of a young woman in love as a victim of thwarted desires, or a pathology simulated to move fathers, tutors, or husbands. Here the vapors are the playwright's faithful servant and, in the guise of a doctor, pave the way for the dénouement. Scene after scene portrays refined or smooth-tongued doctors applied to by patients who seek attention or indulgence. See also, for example, La Molinière (1753); Saurin (1788). Charles Palissot de Montenoy pursues this tradition with unequaled derision in a play first staged in Nancy in 1755, for the unveiling of a sculpture of Louis XV. Montenoy (1755), 225–28, scene 11.

8. See Du Freny (1781), 71; La Molinière (1753); Saurin (1788); Beaumarchais (1988), 113–14.

9. Kéralio (1792), 51.

10. The Duc de la Morlière presented an approach to vapors in terms of simulation and instrumentalization. His most famous publication, the tale *Angola*, was first published in 1746 and reprinted fifteen times before 1789. Set in an Indian court, the tale recounts the tumultuous sentimental life of princes and fairies.

11. Dorat (1993), 955.

12. Diderot (1965), 76.

13. Caraccioli (2005).

14. Caraccioli (1759), A2.

15. Ibid.

16. Ibid., xv–xvj.

17. Caraccioli (1757), 32–33.

18. Caraccioli (1759), xvij–xix.

19. Ibid., xx.

20. For example, in the games played at court. See De Baecque (1988), 175.

21. Bethmont de Paumerelle, in Arnaud, La philosophie des vapeurs (2009).

22. Ibid., 35.

23. "Ceci n'est pas une pipe"; see Foucault's analysis (1973).

24. Paumerelle, in Arnaud, La philosophie des vapeurs (2009), 38.

25. Rosen (1980), 166.

26. Ibid., 21.

27. Ibid., 28.

28. Ibid., 74.

29. Ibid., 61.

30. Ibid, 49.

31. Ibid., 37.

32. On the boudoir in the eighteenth century, see Delon (1999).

33. Laclos (1989).

34. Ibid., 242–48. After seducing Prévan in her boudoir, the marquise calls her domestics and tells them that he is hiding there, waiting to rape her. Found naked in the boudoir and unable to explain himself, Prévan is sentenced to prison. See Laclos (1812), Letter 85, 211–35.

35. Paumerelle, in Arnaud, La philosophie des vapeurs (2009), 50.

36. Ibid., 37.

37. On the game as a substitute for war in the royal court, see Apostolidès (1981).

38. See Jordanova (1989); Schiebinger (1993); Castle (1995).

39. See Beauvoir (1985); Butler (1990).

40. Gracián (1993), 34.

41. See Goodman (1994).

42. Paumerelle, in Arnaud, La philosophie des vapeurs (2009), 74.

43. On Pomme, see chapter 6.

44. See Chauvot de Beauchêne (1781); Boyveau-Laffecteur (1794).

45. The reading of personal letters was common practice in salon societies, and much correspondence was written to be circulated later (see that of Madame du Deffand with the Abbé Galiani and Voltaire, and the letters between Madame de Lespinasse and Condorcet, among others). But the disclosure of private letters also appears as an act of betrayal, in both novels (Laclos, *Les Liaisons dangereuses*; Marat, *Le Comte Potowski*) and autobiographical accounts (Goldoni, *Mémoires De Goldoni: Pour Servir a L'histoire De Sa Vie Et a Cellede Son Théatre*).

46. Roger (1997), 453.

47. Buffon (1797), 57–58.

48. Diderot (1971), 190.

49. Ibid., 186.

50. Ibid., 192.

51. Ibid.

52. Ibid., 190.

53. Ibid.

54. Ibid.

55. Ibid., 196.

56. This will be further discussed in chapter 6.

57. La Mettrie (1748). Other works by La Mettrie develop similar arguments on the force of imagination: La Mettrie (1747) (see especially 97–113), and he added *Lettre Critique de Mr. De La Mettrie sur l'Histoire Naturelle de l'Ame A Madame la Marquise du Chattelet* (1747) and *Traité du vertige avec la description d'une catalepsie hystérique, & une lettre à Monsieur Astruc, dans la quelle on répond à la critique qu'il faite d'une dissertation veneriennes* (1738), about therapeutic methods in a case of catalepsy.

58. La Mettrie (1912), 90, translation slightly modified. On the relationship of La Mettrie's works to Haller's, see Steinke (2005), 194–97.

59. La Mettrie (1912), 89.

60. Le Camus (1753), X.

61. Ibid., 2:35. See Wenger (2007), 90–95.

62. Le Camus (1753), 2:54.

63. See Carson (2007).

64. Le Camus (1753), 2:143.

65. See especially Casaubon (1970); More (1966); For scholarship, see Hawes (1996); Heyd (1995); Irlam (1999); Klein and LaVopa (1998); Mee (2003).

66. Plato (1925), 421.

67. Le Camus (1753), 2:144–45.

68. The author borrows the name of Epimenides, a poet who was part of the movement of Orphism in the sixth century AD and contested political and religious power. Chassaignon (1779).

69. Chassaignon (1779), 65–67.

70. Ibid., 24n.

71. Jean-Jacques Rousseau (2011), 92; (1997), 117.

72. Jean-Jacques Rousseau (2011), 92; (1997), 117.

73. See Starobinski (2004).

74. Jean Starobinski, among others, has analyzed Jean-Jacques Rousseau's relationship with creation and with sickness. See his *La Transparence et l'obstacle* (2004), translated by Arthur Goldhammer as *Transparency and Obstruction* (1988).

75. Chassaignon (1779), 125.

76. Ibid., 15.

77. Ibid., 15n.

78. Ibid., 68.

79. The "mole" is a quantity of fetal tissue lodged in the uterus, which may take up the volume of a fetus even though it is completely devoid of life, and which might lead doctors and women themselves to believe they are pregnant.

80. Chassaignon (1779), 68.

81. La Bretonne (1987), 36.

82. The word in French would be *désœuvré*, as discussed by Maurice Blanchot, that is *désœuvré*, un-worked, lacking work, waiting for inspiration, and after the experience of inspiration, always continuing in the impossibility of continuing, dying again and again, but never dead. Blanchot (1969), 289–99, translated by Hanson (1993), 194–95.

83. Bataille (1954), 7. Georges Bataille reports this reply by Maurice Blanchot to the question of the possibility of giving authority to experience. Bataille (1992), 19.

84. M. J. F. (1782), 99–102.

85. See chapter 6.

86. La Mettrie and Casanova both play ironically with an association of the brain with the womb. La Mettrie, *Histoire naturelle de l'ame, traduite de l'anglois* (1747), 11. See also Casanova (1998), esp. 29–33. See Schiebinger (1989).

Chapter Five

1. Louis Marin, *Le récit est un piège* (Paris: Éditions de Minuit, 1978), 8.

2. Vila (1998).

3. Ender (1995), 22.

4. See Barker-Benfield (1992); Mullan (2002); Porter and Roberts (1993).

5. Pope (1971); Shin (1996).

6. Haywood (1999); Bocchicchio (2000).

7. Smollett (1999), (1983), (1902); Labrude-Estenne (1995).

8. Grosvenor Myer (1984).

9. Cutting-Gray (1992).

10. Radcliffe, *The Italian* (1998), *The Mysteries of Udolpho* (1998), (1999), (1993); Lewis (1994); Bronfen (1994). See Miles (1994), (1995).

11. Fanny Burney's works are particularly rich in expressions for hysteric affections: "frightened into violent hysterics" and "to vapour him to death" in Burney, *Cecilia* (1999), 80, 935; and "talk to me into the vapours," "sigh me into the vapours," and "horribly vapoured" in Burney, *Camilla* (1999), 380, 399, 472; the verb "convulse" is used in the literal and figurative sense in expressions such as "Convulse my fairest hopes" in *Camilla* (1999), 575.

12. This applies, at least, to the work of John Cleland, Crébillon, Fougeret de Monbron, Nerciat, Réstif de la Bretonne, and Sade.

13. Fougeret de Monbron first uses the word *Mylord*, then plays with the adjective *lourd*, meaning "heavy."

14. Monbron (1992), 112.

15. Ibid., 98.

16. The theme of sensibility in the late eighteenth century has attracted much attention in recent decades: Barker-Benfield (1992); Crane (1934); Markman (1996); Goring (2005); Chris Jones (1993); McGann (1996); Nagle (2007); Pinch (1996); George Sebastian Rousseau (1976); van Sant (1993).

17. Mullan (2002), 201.

18. "In Richardson's last novel, *Sir Charles Grandison* (1753), the willowy heroine Clementina endures the three stages of 'vapours' Cheyne described in *The English Malady*, proceeding from fits, fainting, lethargy, or restlessness to hallucinations, loss of memory, and despondency (Cheyne recommended bleeding and blistering at this stage), with a final decline toward consumption. To cure her, Sir Charles follows Cheyne, prescribing diet and medicine, exercise, diversion, and rest, and the story is considerably affected when Clementina's parents adopt unquestionably Dr Robert James's further recommendation that 'in Virgins arrived at maturity, and rendered mad by Love, Marriage is the most efficacious remedy.'" George Sebastian Rousseau (1993), 152. See also Shuttleton (1999), 59–79.

19. See Schmid (1997).

20. See Lennox (1970), 300.

21. Jean Baudrillard's analysis of the simulacrum can be applied to this eighteenth-century novel. Baudrillard (1994).

22. Lennox (1970), 23.

23. Ibid., 133.

24. Ibid., 136.

25. Godwin (1985), 5.

26. In Peter Logan's *Nerves and Narrative*, devoted to the role of nervous maladies and hysteria in nineteenth-century English literature, the pathology in *Caleb Williams* is analyzed as a form of social and political resistance. Logan is particularly interested in the pathology's function of affirming social rank and sexual difference. Quite unlike the interpretation pursued here, Logan sees Caleb as a hysteric character, and interprets his hysteria as the result of what he calls a "feminization" of his character—itself the consequence of his being denied the opportunity to speak. Logan (1997). See also Forge (2003).

27. Godwin (1985), 12.

28. Ibid., 7.

29. Ibid., 8–9.

30. Ibid., 122.

31. Ibid., 118.

32. See Godwin (1985), 114.

33. Godwin (1985), 293.

34. Ibid., 294.

35. Ibid., 142.

36. Ibid., 331.

37. Ibid., 334.

38. Ibid., 5.

39. Diderot published only a part of *La Religieuse* before his death, and many versions were found. The final revisions were made for the first publication of excerpts between 1780 and 1783 in *Correspondance littéraire* (Literary Correspondence). This was followed by a posthumous publication of the work in 1796, an English translation printed in Dublin in 1797. Dieckmann (1959); Tunney (1992); Diderot (1797), (1974). English references are to the 1974 translation by Tancock unless otherwise noted.

40. Vila (1998), 168.

41. "The hand she had rested on my knee wandered all over my clothing from my feet to my girdle, pressing here and there, and she gasped as she urged me in a strange, low voice to redouble my caresses, which I did. Eventually a moment came, whether of pleasure or of pain I cannot say, when she went as pale as death, closed her eyes, and her whole body tautened violently, her lips were first pressed together and moistened with a sort of foam, then they parted and she seemed to expire with a deep sigh. I jumped up, thinking she had fainted, and was about to go and call for help." Diderot (1974), 137–38. See Deneys-Tunney (1992), 160; Tallent (2002).

42. Diderot (1994), 365. According to Anne Vila, Suzanne's innocence and momentary insensibility force the reader to feel what she does not during her encounters with the mother superior. Vila (1998), 175.

43. Diderot (1994), 362, "folle"; (1974), 134.

44. Diderot (1994), 363, "extravagante"; (1974), 135.

45. Diderot (1974), 71.

46. On the internal and external gaze, see Deneys-Tunney (1992), 149.

47. Diderot (1974), 92.

48. Diderot (1994), 332; (1974), 94.

49. Diderot (1974), 94.

50. Ibid., 165.

51. While Raulin dedicated a whole volume to his observations, *Observations de médecine* (1754), most physicians included them in their treatises. Some of those who made particularly heavy use of consultations in their treatises are Pomme (see chapter 6); Pressavin, with personal observations filling more than ten octavo pages (1770), 268–78; and most notably Tissot, with the four-volume *Traité des nerfs et de leurs maladies. De la catalepsie, de l'Extase, de l'Anæsthesie, de la Migraine et des Maladies du cerveau* (1778–80), *Traité de l'épilepsie* (1770), where he often relies on cases recounted by other doctors and not only his own.

52. Sauvages illustrates all the diagnoses classified in his nosologies with specific cases, and often gives two or three to show different manifestations of the disease. Pinel also used observations in all his works, often borrowing them.

53. Bachelard (2002); Daston (2011); Pomata (2010); Pomata and Siraisi (2005); Harkness (2006), 171–92; Lockwood (1951); Nance (2001).

54. See especially Barbara Duden, who analyzes the consultations of an eighteenth-century German physician with special attention to his approach to patients (1987, 1991).

55. On the use of narrative practice, see especially Davis (1987).

56. See Peter (1976), (1967); Foucault (1973).

57. Roy Porter was one of the first to advocate a history of medicine centered on the patient, leading to a whole new set of inquiries in history. Porter (1985), 1–22, 129–44. See also Guthrie (1945).

58. On the medicine of charlatans and empirics in eighteenth-century France, see Ramsay (1988); Brockliss and Jones (1997), 480–669.

59. Léonard (1988), 73–94.

60. Guindant (1768), 56–62.

61. Ibid., 115.

62. He adds: "Unfortunate towns, where such dangerous prejudices reign, who host Charlatans and Empirics; who retain Dogmatics only to crush them under the ruins of calumny. You prefer the surface of things to an enlightened science, and close your eyes to candor, open them to hypocrisy. You, in a word, protect and strengthen the throne where the Empiric sits, to establish and announce his laws, his precepts, and his maxims! . . . You lock so many victims in your surrounding walls, and are pleased to nourish the hangmen! In one neighborhood cachexia can be seen to dominate; in another vapors; in one family icterus, hydropsia, phthisis; in another hysteria & hypochondria." Guindant (1768), 74–75. See also Raulin (1758), 232. On the development of these polemics, see Brockliss and Jones (1997).

63. Whytt (1777), 2nd ed., Publisher's note, iii.

64. See Brockliss and Jones (1997), 760–79; Peter (1967).

65. On the academies, see the pioneering work of Roche (1978).

66. Thanks to Lassone's and Vicq d'Azyr's intervention, the Société Royale de Médecine was created on August 20, 1778, after a series of difficulties arising from the opposition of the Faculty of Medicine in Paris. Pierre Chirac, First Physician to King Louis XV, initiated the first project in 1730. Louis XVI's decision to establish a doctors' commission on April 29, 1776, and to hold a correspondence with the physicians from the provinces on epidemics and epizooties were the main steps. See Hannaway (1974); Pauthier (2002); Mandressi, Vicq d'Azyr, www.bium.univ -paris5.fr/histmed/medica/vicq.htm.

67. This abbreviation of the *Journal de Médecine, Chirurgie, Pharmacie* was commonly used in the eighteenth century.

68. Not long afterward, in 1744, the Académie des Sciences, Belles Lettres et Arts was created with Louis XV's patent. It was said that Claude-Nicolas Le Cat refused La Peyronie's offer of an establishment in Paris in order to pursue his work in Rouen. On public practices of anatomy, see Guerrini (2004), (2006), (2003), (2009). On electricity, see Pancaldi (2003); Fara (2002); Schaffer (1983).

69. See Ramsay (1988).

70. Robertson (2000).

71. Marrigues, Manuscript Rouen, 1er X, 1762, C. 17, Bibliothèque Jacques Villon (Rouen): Archives de l' Académie des Sciences, Belles Lettres et Arts de Rouen.

72. Pomata (2005), 136.

73. Brockliss (2002).

74. Observation by Dr. Calvet, July 1, 1778, SRM 122, dr. 5, no. 2, Société Royale de Médecine.

75. Ibid.

76. Denis, 1775–1779, SRM 133, dr. 25, no. 1, Académie de Médecine.

77. Kafka (1948).

78. De la Porte, 1775–1779, SRM 184, dr. 9, no. 4.

79. Collet, 1777, SRM 117, dr. 20.

80. Ibid.

81. Ibid.

82. Ibid.

83. Ibid.

84. In fevers, crises were seen as the moment when the body started to recover. See Bordeu's theory of crises in his definition of this word in the *Encyclopedie des arts et des lettres*. See also Bordeu (1768).

85. Anonyme (1786), SRM dr. 1, no. 2.

86. Retz (1786), iv–v.

87. Lavallée (1786).

88. Ibid.

89. Ibid.

90. Chirac (1744–55), 399.

91. Retz (1786), 313–17.

92. The original work is Bowen (1783).

93. Robin (1787), 55–56.

94. Anonyme (1786), SRM dr. 1, no. 2.

95. Yellow bedstraw is an herb in the family of Rubiaceae (Galium verum). The "flowers of nercau" could not be identified.

96. Anonyme (1786), SRM dr. 1, no. 2.

97. Ibid.

98. Goulemot (1994); Groneman (1994); Jaquier (1998); Jordanova (1987); Sheriff (1998); Stengers and Van Neck (1984); Stolberg (2004), 89–107; Wagner (1988), 46–68; Wenger (2007); Laqueur (2005).

99. Lafage (An VIII), 19.

100. William Cullen, with many contemporaries, describes the feeling of a ball in hysteric fits: "LXII. Hysteria: rumbling of the bowels; sense of a ball rolling in the abdomen, rising toward the stomach and gullet, and there giving a sense of strangulation." Cullen (1800), 128.

101. Louyer-Villermay (1802), 41–44.

102. On practices of avowal, see Foucault (2003), 23–40, 267–98; Benrekassa (2007).

103. The term "refiguration" is based on Ricoeur (1983), 110.

104. Of course, the interpretation of the pathology as an effect of emotions or movements of the soul did not appear only in the eighteenth century, but the narrative format made it possible to rethink their relationship. On the relationship between pathology and soul, see Pigeaud (1981); Berriot-Salvadore, (1995), 257–69; Imbroscio (2002); Fernando Vidal (2011); Vila (1998).

105. This point is posited on the basis of Foucault's conception of the gaze: "The observing gaze refrains from intervening: it is silent and gestureless. Observation leaves things as they are; there is nothing hidden to it in what is given." Foucault (1989), 131.

106. *Traité de la Maladie d'Amour ou Mélancholie Erotique* (1623). Jacques Ferrand practiced medicine in Toulouse, without becoming particularly famous. His few other works have been completely forgotten. This one was mentioned a few times in subsequent centuries for its collection of anecdotes and its style, which some described as full of verve and close to that of Montaigne.

107. Trillat (1986); Ellenberger (1970); Micale (1995); Veith (1973): Brockliss and Jones (1997), 791–94; Darnton (1968); Peter (1999), 9–88; Rausky (1977).

Chapter Six

1. Elaine Showalter "Hysteria, Feminism, and Gender," in *Hysteria beyond Freud*, ed. Sander L. Gilman et al. (Berkeley: University of California Press, 1998), 335.

2. Roger (1803).

3. Trotter (1807).

4. Léonard (1992), 210.

5. Vieussens (1698), 5.

6. Nancy Siraisi has studied a sixteenth-century physician in terms of a similar imagining. Siraisi (1991).

7. As was argued by, for example, Livi (1984); Kniebiehler and Fouquet (1983).

8. For further biographical information on Pomme, see Aubert (1998). The biographical dates provided are subject to variation; those given here are indicated by the *Departmental Encyclopedia* of the Bouches-du-Rhône, from the volume *Biographies*, which provides the dates June 5, 1728–April 7, 1814. The article is signed by Emile Fassin.

9. Pomme (1765).

10. Pomme (1775).

11. Pomme (1760, 1765, 1767, 1769, 1771, 1782, 1798–99, 1804), (1776, 1786, 1794), (1777).

12. The word "quinine" is used in the French text, though it was not in common use in English at the time.

13. Roche (2000). See also Roche's chapter on the increased attention to and consumption of water, and the fashion for spas in the eighteenth century.

14. Wenger (2009).

15. Voltaire (1878), 235–36.

16. Ibid., 336–37.

17. Pomme (1782), 17n. Following this reasoning, Pomme explains the case of many patients: "If the patient has floated for two months, it is because it has taken such time for the water to penetrate and restore to the body its first weight" (101–4).

18. Pomme's response is reprinted in the *Journal Encyclopédique*, 463, letter from Pomme to Voltaire. See also the manuscripts of Abbé Bonnemant, Archives Municipales d'Arles.

19. Voltaire (1986), 423. On Madame Racle, see also the letter from Pomme to the Marquis de Grimaldi (Publisher's note).

20. Luke 8:43–48.

21. Pomme (1771), 2:75; reprinted in (1804), 75–76.

22. Pomme (1782), Avant-Propos, v.

23. Pomme (1771), vol. 2.

24. Voltaire (1986), 704.

25. Later in the same letter, he writes: "The ferment is as strong in the provinces as in Paris, and will likely produce only decrees that will not survive, and very useless protests without which France would be the fable of Europe."

26. Pomme (1771), 309.

27. "I could support this assertion with striking examples; but the tableau I would present would be too frightening, it would be a true martyrology." Pomme (1804), 14.

28. See, for example, Pomme (1806), 20.

29. See Bret (1806), 81. In the third edition of *Réfutation de la doctrine médicale du docteur Brown, médecins écossais, suivie d'une notice sur l'électricité, le galvanisme et le magnétisme, sous le rapport des maladies nerveuses* in 1808, he blames his youthful inexperience and claims: "I prepare myself to make amends for my faults; I will do so by preaching my example for the future." Pomme (1808), 67.

30. "On my return to Paris last year in the month of October, I was stopped several times in my journey by the weak; if not by the master of the home, or the mistress, then by postilions and other domestics who were lying on their pallet, again through the effect of quinine employed

at the wrong time and with the greatest profusion; which indicated to me that the epidemic was spreading from Paris to Arles, and even to Nice, and that febrifuge expanded all around." Pomme (1804), 99.

31. "indigens"; Pomme (1806), 15.

32. "saintes vérités"; Pomme (1804), 4.

33. Pomme (1804), 101.

34. This note is repositioned in subsequent editions. It first appears in the second edition of the treatise, Pomme (1765), 443–44n. It reappears in the Post-Scriptum to "Régime du Tempérament Vaporeux," 470–79, Pomme (1767), 475; and then in the 1769 edition in "Réponse à l'Auteur du Journal des Savants," 443–44.

35. Pomme (1771), 309.

36. Pomme (1804), 3:76–77.

37. He also speaks of the recovery of the Marquise de Bezons after fifteen months of treatment in terms that again insist on the miraculous character of the event. He later describes her as "resuscitated" (ressuscitée). Pomme (1771), 186, 188.

38. Astruc (1761); Daignan (1786); Gamet (1772); Duvernoy (1801); Falconer (1796); Marquet (1770); Chambaud (1784); Perry (1755); La Roche (1778); Tissot (1795); Trotter (1807); Whytt (1765).

39. This is the line proposed by Veith (1965); Trillat (1986); Ellenberger (1970). See also below.

40. Daston and Park (1998), 137–45; Castle (1995).

41. Mesmer (1779), 74, Proposition 1. See also Mesmer (1781), (1785).

42. Mesmer (1779), 74, Propositions 1 and 2. Some physicians gathered around Mesmer and published in support of his claims. See Bergasse (1784); Delandine (1785); Fournel (1785); Thouret (1784).

43. Rausky located its source in Mead's dissertation, "Of the Power and Influence of the Sun and Moon on Humane Bodies." Rausky (1977).

44. See Rommevaux (2010).

45. van Helmont (1621).

46. Planque (1750), 503.

47. Riskin (2002),198.

48. Brockliss and Jones (1997), 792.

49. Veith (1973). Trillat (1986), follows a similar trajectory.

50. In 1976 François Azouvi queried this kind of reading, showing that by making these men the inventors of a psychological medicine, historians were in fact only echoing the thesis of Mesmer's adversaries, who accused him of healing patients by psychological suggestion. Mesmer himself strove to demonstrate the opposite, locating the validity of his therapy in universal magnetism.

51. Mesmer (1779), 81.

52. Ibid., 36–37.

53. D'Eslon (1780), 94.

54. Numerous pamphlets were published: Doppet (1785); Cailleau (1785); Deduit (1786); Vernet (1816); Radet (1784).

55. Membres de la Société du Magnétisme (1817), 154.

56. Ibid., 156–57.

57. "This convulsive state which accompanied the sibyls is not foreign to magnetism. Mesmer's tubs are the proof of it. The progress that wise magnetizers have made science achieve

has freed it from these frightening paroxysms." Membres de la Société du Magnétisme (1817), 162–63.

58. Membres de la Société du Magnétisme (1817), 85.

59. Particularly useful in this regard are the case of Nanette Leroux, edited and analyzed by Goldstein (2001), and the later case of Lady Lincoln presented by Edelman, Peter, and Montiel (2009).

60. Petetin (1808), x.

61. Because the electricity used by Petetin is what would today be called "magnetic electricity," there was no danger for his patients.

62. Petetin (1808), 252.

63. See Petetin (1787).

64. Petetin (1808), 162.

65. Ibid., 288.

66. Puységur (1811), 425–26.

67. See also Daston (1991).

68. Gabriel Bègue to Tissot (1792), Fonds Tissot, Bibliothèque Universitaire de Lausanne.

69. On the attempts by aristocrats to find protection in Jacques Belhomme's institution, see Sournia (1989).

70. An aristocrat to Tissot (1792), Fonds Tissot, Bibliothèque Universitaire de Lausanne.

71. For the medical profession, see Weiner (2002); Goldstein (2001), 1–40. For other sciences, see Salomon-Bayet (1978).

72. See especially Ackerknecht (1986); Coleman (1982); Staum and Larsen (1981); Goldstein (2001); Gelfand (1980); La Berge (1992); Weiner (2002); Quinlan (2007).

73. Keel (2002).

74. La Berge (1992).

75. See especially Weiner (2002), 79–101.

76. The term "actuality" was theorized by Kant (2009); and Foucault (1984), 32–50.

77. Venette (1721), preface.

78. Quinlan (2007), 20.

79. Raulin (1766), v.

80. Ibid., v, vi, viii, xii, xviii.

81. Raulin (1758), 42. See also Lignac (1773), 17.

82. Lignac (1773), 10–11.

83. "The hysteric and hypochondriac disorders, once peculiar to the chambers of the great, are now to be found in our kitchens and workshops. All these diseases have been produced by our having deserted the simple diet, and manners of our ancestors." Rush (1779), 57. See also Rush (1774).

84. Pressavin, *Nouveau Traité des vapeurs (1770)*, avant-propos (n.p.).

85. Hugues Maret (1726–86), surgeon and doctor, taught chemistry and medicine in Dijon and was famous for being extremely active. He published memoirs on a wide variety of medical topics such as smallpox, epidemy, and air, and was a member of a great number of French medical academies. For a detailed biography and list of publications, see *Encyclopédie Méthodique* (1808) and *Dictionnaire des sciences médicales. Biographie médicale Panckoucke* (1824).

86. The dissertation was published in 1772.

87. Maret (1772), 89–91.

88. Ibid., 96.

89. Ibid., 95–99.

90. Chauvot de Beauchêne (1781), 29–30.

91. See Daston and Vidal (2003).

92. Elizabeth Williams (1992).

93. See the work of Weiner (2002).

94. Bergasse (1784), 65–67.

95. See Darnton's chapter "Mesmerism as a Radical Political Theory," in Mesmerism and the End of the Enlightenment in France (1968), 106–25.

96. Joseph-Michel-Antoine Servan (1784), 101–102, cited in Darnton (1968), 84.

97. Bergasse (1781). See also Darnton's discussion of Bergasse. Darnton (1968), 151–73.

98. Bergasse (1781), 65–67.

99. Ibid., 67n.

100. Ibid.

101. Ibid., 83.

102. Ibid., 63.

103. Ibid., 72.

104. Paul-Joseph Barthez (1734–1806) is known mainly for his *Nouveaux Éléments de la science de l'homme* (1778). On his theories and adherence to vitalism, see Elizabeth Williams (1994), 168–71.

105. In the medical terminology of the day, diathesis is a series of affections happening simultaneously or successively, affecting several anatomical regions in an identical way.

106. Petit (1806), 133–34.

107. Ibid., 123–24.

108. Ibid., 119–20.

109. The ideologues, men of letters with differing backgrounds, met regularly to discuss everything from medicine and physiology, anthropology and agronomy, to politics. See Chappey (2002).

110. See Chappey (2002).

111. Guillon-Pastel (1800).

112. Goldstein (2001), 100.

113. In the United States in the same period, Benjamin Rush published his analysis of the encounters between the political, the economic, and the pathological. In an article entitled "An Account of the Influence of the Military and Political Events of the American Revolution upon the Human Body," he describes the effects of the financial devaluation on the minds of the people during the War of Independence. Agitated by the new increase in money, many fear neither cold nor fatigue in the instability of war. The conclusion of the war brings a return to health for hysterical women, who savor the calm of the return of peace: "Many persons of infirm and delicate habits were restored to perfect health, by the change of place or occupation, to which the war exposed them. This was the case in a more especial manner with hysterical women, which were much interested in the successful issue of the contest." Rush (1779), 223. In contrast, the foes of independence fall into hypochondria. Rush reads the mechanism of the passions through a physical law, and their effects as the result of a catalysis that only human reconciliation can stop. Rush (1779), 226–27. See also Rush (1797), vol. 1, 274–75, 277–78; vol. 2, 26.

114. Lanthénas (1792), 28; see also 26.

115. In an irony of history, when Lanthénas risked death on May 31 and June 2, 1793, with the arrest of the main Girondist leaders (notably Brissot, Condorcet, and Roland), Marat (Montagnard) is said to have saved him with a sally: "Everybody knows that the doctor Lanternas [*sic*] is

a half-wit." (Tout le monde sait, dit-il, que le docteur Lanternas [*sic*] est un pauvre d'esprit.) In French, the sobriquet *lantern* is given to people who are very slow to understand what is going on (a verb, *lanterner*, means "to dawdle"). Lanthénas then became secretary of the Convention and in the year VI resumed his activities as physician.

116. Elizabeth Williams (1992), 88. See also Williams's detailed analysis of medicine as anthropology, where she states: "In their effort to reformulate medicine as a broadly gauged human science, the medical revolutionaries of the 1790s frequently took their lead not from medicine proper but from the general intellectual movement of Ideology, whose importance in the larger history of the medical science of man can hardly be exaggerated." Elizabeth Williams (1994), 76–77; see Staum (1993); Chappey (2002).

117. This kind of positioning was not entirely new. Michel Foucault (1963) examined the presence of a political consciousness in medical activity before the Revolution and its redefinition after the Revolution.

118. Sue (1799), 3.

119. Cabanis (1989), 8.

120. Ibid., 10.

121. Cabanis's work has attracted much scholarship. Jacques Léonard noted that his inspiration came from Condillac, the sensualists, the encyclopedists, Condorcet, and the Ideologues, and that he "incarnates marvelously the inheritance of the Société Royale de Médecine: the concern for medical geography and meteorology, the lasting vogue for medico-hygienic topographies, the account taken of biological parameters such as age, gender, temperament, diet." Léonard (1981), 29. See also especially Elizabeth Williams (1994), 78–114; Quinlan (2007); La Berge (1992); Staum (1980); Starobinski (2003), 145–159.

122. Bacher (an XI); Rousseau (2007), 21.

123. Jacyna (1982), 233.

124. Pinel was famous for having removed the chains of the insane at Bicêtre with his assistant Jean-Baptiste Pussin, in some cases replacing the fetters with straitjackets and in others simply with moral authority. This narrative was circulated by the subsequent generation and immortalized in a painting by Robert Fleury. Recent scholarship has established that it was a myth and that Pussin was the one to "free" the insane. See Garrabé (1994); Weiner (2002), (1980).

125. Scull uses this expression to describe the transformation of the imagining of madness and of hospital practices that started in the nineteenth century, notably with Tuke and the circulation of an etching of insane people drinking tea in the asylum. Whereas Roy Porter sees an expansion of up-market madhouses for wealthy patients in the eighteenth century, Scull sees a switch from the drastic techniques used on maniacs to "Tuke's famous image of the inmate of the Retreat calming sipping tea and exchanging social pleasantries," which, he argues, "found its echo in the county asylum reports of the mid-century." Scull (1989), 77.

126. Quinlan (2007), 54.

127. Quinlan (2007), 57. On the growth of the population, see Coleman (1982).

128. Edelman (2003); Staum (1980); Schiebinger (1989); Laqueur (2003); Jordanova (1999), (1993).

129. One clear exception to this is *Medical Researches Being an Enquiry into the Nature and Origin of Hysterics in the Female Constitution, and into the Distinction between that Disease and Hypochondriac or Nervous Disorder*, a treatise published in 1777 by Andrew Wilson, which discusses the singularity of women's physiology and their proneness to hysteria, but in order to present this as a force.

130. Elizabeth Williams (2002), 1–16; Jordanova (1989), especially "Natural Facts, an Historical Perspective on Science and Vision," 19–42.

131. See Elizabeth Williams (1992). On hygiene more generally, see Jordlan (2010); Rey (2000); Brockliss and Jones (1997).

132. Vila (1995).

133. On this topic, see especially Jordanova (1999); Groneman (1994).

134. This point was made by Rosen (1974), 242, and elaborated by Weiner (2002). See also La Berge (1992).

135. La Berge (1992).

136. See Elizabeth Williams (1994), 168–71.

137. Le Jeune (1801).

138. Boyveau-Laffecteur (1794), 1–2.

139. Ibid.

140. Ibid.

141. Ibid., 2.

142. Ibid.

143. Ibid., 3.

144. Ibid., 20 ff.

145. La Mettrie (1737).

146. Le Camus (2008). On this, see Vila (1998).

147. Vila (1998), iii.

148. Boyveau-Laffecteur (1794), VI.

149. Ibid.

150. Ibid., vij.

151. Ibid., 133.

152. Ibid., 11.

153. Ibid., 63–64.

154. Ibid., 65.

155. Ibid., 128.

156. Ibid., 136.

157. Ibid., 80.

158. Virey (1825), 262.

159. On pregnancy and insanity, see Marland (2006), 53–78.

160. Louyer-Villermay (1816), 91.

161. Petit, *Traité des maladies des femmes enceintes* (An VII), 1:266; 2:17. Antoine Petit became inspector of the military hospitals and was granted the chair of anatomy and surgery for the king. One of his most impressive texts is reprinted in Peter (1993).

162. Géraud (1799), 6.

163. Ibid., 4.

164. Virey (1824, 1825), 13; see also 85, 109, 129.

165. Virey (1825), 129.

166. Ibid., 176.

167. Jouard (1804), 80.

168. Ibid.

169. Ibid., 80.

170. Ibid., 83.

171. Ibid., 65.

172. Astruc (1762).

173. The general description of the illness first indicates: "That the *uterus* receives many nervous filaments, from the extremities of the intercostal nerves; and the pairs of the *medulla spinalis*, which come out between the *vertebrae* of the loins." Astruc (1761), 3. Astruc elaborates: "There will happen, in this case, attacks of the hysterical passion, greater, longer, and more frequent, than in any other kind of suppression: because the sensations then excited in the nerves of the *uterus*, whether from the blood-vessels in which the blood, that is suddenly detained there, regurgitates; or from the muscular fibres which are in an *erebismus*, or violent state of contraction, will be stronger, longer, and more frequent." Astruc (1761), 149.

174. For a review of the other changes undertaken and their reception by Pinel's students and colleagues, see Weiner (1993), 259–310.

175. Pinel (1798), 288–89.

176. Pinel (1980).

177. Pinel (1804).

178. Pinel (1816), 267 (the sixth edition reproduces the text of the second edition).

179. Pinel (1816), 282.

180. Du Laurens (1609); Ferrand (1627); Varandée (1666); Pseudo-Aristotle (1668); Ettmüller (1698).

181. Capuron (1817), 83–84. Joseph Capuron (1767–1850) wrote several works on generation, venereal illnesses, and legal medicine. The biographer in the *Dictionnaire Dechambre* presents him as an orator with serious and dangerous rhetorical skills.

182. See Fissell, "Hairy Women" (2003), "Making a Masterpiece" (2003), (1995).

183. Peyre (2002), 111–27.

184. Bauduit (1822), 21.

185. For further analysis, see Jordanova (1999), 163–82.

186. Bauduit (1822), 23.

187. Laqueur (1987), 1–41; Matlock (1994).

188. Madame de Staël, *Corinne* (1985).

189. Ibid., 375.

190. In fact Virey uses Diderot's text (without citing him), but reverses his argumentation. He also caricatures Diderot's argument that piety could grant new life to women who have lost their youth ("La femme hystérique dans la jeunesse se fait dévote dans l'âge avancé,"): "Hysterical old maids are in fact an excellent instrument for any founder of a new religious sect; they bring to it an impetuous zeal that is unafraid to immolate itself for the propagation of new truths. La Bourignon and the Guyon woman, and many others of the devout, abandoning themselves to the pious works of the convulsionaires, made themselves famous by the intrepid fervor of their religious sentiments." Virey (1825), 124; Diderot (1994), 952.

191. Le Pois (1618).

192. In his nosology, Hosack distinguishes two kind of hysteria, *hysteria mentalis*, arising from the passions of the mind, and *hysteria corporea*, originating in the organs of the body. Hosack (1821).

193. See, for example, Thomas Trotter, who already favors this interpretation at the very beginning of the eighteenth century (1807). On Trotter, see Logan (1997).

194. For example, John Cooke focuses on apoplexy and gives a long list of pathologies arising from the nervous system: Cooke (1820), first vol.; (1823), second vol.; Rayer (1819); Hallaran (1818); Spurzheim (1817).

195. Doerner (1981).

196. Bichat (1955).

197. Rosen (1953); (2008), originally published in Centaurus 5, no. 2 (June 1957): 97–113; (1974); Foucault (1973), 24–43; Brockliss and Jones (1997), 734–59; Jordanova (1999), 143–59.

198. Goldstein (2001), 332.

199. See Carroy-Thirard (1982), (1981), (1980); Micale (1995).

200. Beizer (1994).

201. I have seen such a description only referring to Princess Lamballe's fits and in a doctor's manuscript about Madame de Vandermonde sent to Tissot.

202. George Beard still presented his seventy-five-page list of symptoms as incomplete in *A Practical Treatise on Nervous Exhaustion* (1880), 11–85.

Bibliography

Manuscripts

Académie de Médecine de Paris: Archives de la Société Royale de Médecine
4 pieces, 1775–1779, dr. 9, no. 22, Société Royale de Médecine.
Collet, 1777, SRM 117, dr. 20.
Concours de la Société Royale de Médecine, dr. 1, no. 2, mars 1786.
De la Porte, 1775–1779, SRM 184, dr. 9, no. 4.
Denis, SRM 133, dr. 25, no. 1, Académie de Médecine.
SRM 163, dr. 10. no. 25. Société Royale de Médecine.
SRM 197, dr. 8, 1786. Société Royale de Médecine.
Archives de la Bibliothèque du Patrimoine Montpellier, Médiathèque
    Barthez, Paul-Joseph. *Consultations Médicales.* Manuscrit.
    Notes de Médecine Physiologique recueillies au Cours de Moral Broussais par Emile Rouet
    D.M.M.
Archives Départementales de Vaucluse (Avignon)
    Fonds Requin (5 F 196).
Archives Municipales d'Arles
    Manuscripts of Abbé Bonnemant, *Extrait des singularités historiques, littéraire, politiques et sacrées et prophanes de la ville d'Arles,* vol. 225, 1st part (Library of the City of Arles).
Bibliothèque Historique de l'Assistance Publique (Paris)
    Registres de Bicêtre, QUE, 6 Q 6 6, 1 Q 2 84.
Bibliothèque Publique et Universitaire de Genève
    Tronchin, Théodore de, Registre de Consultations (1756–1759).
Bibliothèque de l'Arsenal, Bibliothèque Nationale Française (Paris)
    Archives de Police de la Bastille: 10 204, 5802 (R142564).
Bibliothèque de l'École de Médecine de Montpellier
    Berthe. *Pathologie et Thérapeutique, Choix de Consultations.*
    Broussonet, Victor. *Cours de Pathologie.* H 567.
    Damon, E. J. H 589, dr. 17, fol. 102.
    Varia, H 630.
Bibliothèque Interuniversitaire de Santé (Paris)
    Anonyme. "De l'Hystéricie." 5069–5070.

Bibliothèque Jacques Villon (Rouen)
  Archives de l' Académie des Sciences, Belles Lettres et Arts de Rouen: C. 17.
Bibliothèque Cantonale et Universitaire-Lausanne
  Fonds Tissot.
Francis A. Countway Library, Harvard University (Cambridge, USA)
  Philippe Pinel. B MS b 13.
Moody Medical Library, University of Texas Medical Branch
  Cheyne, George. *The English Malady; or, A Treatise of Nervous Diseases of all Kinds as Spleen, Vapours, Lowness of Spirits; Hypochondriacal and Hysterical Distempers, &c. printed for G. Strahan and J. Leake.* 1734. Edition with marginalia owned by the Blocker Collection of the Moody Library, University of Texas Medical Branch, Galveston.

Primary Sources

Académie Royale des Sciences. *Mémoires pour servir à l'histoire naturelle des animaux.* Paris, 1671.

Aitken, John. *Principles of Midwifery; or, Puerperal Medicine.* Edinburgh, 1784.

Algarotti, Francesco. *Il Newtonianismo per le dame: Ovvero Dialoghi sopra la luce e i colori.* Naples, 1737. Translated by Elizabeth Carter as *Sir Isaac Newton's Philosophy Explain'd for the Use of the Ladies: In Six Dialogues on Light and Colours.* London: E. Cave, 1739.

Alibert, Jean-Louis. "De l'Influence des causes politiques sur les maladies et la constitution physique de l'homme." *Magasin encyclopédique; ou, Journal des sciences, des lettres et des arts* 5 (1795): 298–305.

———. *Nosologie naturelle: Ou, Maladies du corps humain distribuées par familles.* Paris: Imprimerie Crapelet, 1817.

Allen, John. *Abrégé de toute la médecine pratique où l'on trouve les sentimens des plus habiles médecins sur les maladies, sur leurs causes, & sur leurs remèdes; avec plusieurs observations importantes.* Paris: Pierre-Michel Huart, 1737.

Andree, John. *Cases of the Hysteric Fits, and Système Vitus Dance, with the Process of Cure.* London: Printed for Meadows and J. Clarke, 1746.

Andry, Nicolas. *Recherches sur la mélancholie extrait des registres de la Société Royale de Médecine années 1782–1783.* Paris: de l'Imprimerie de Monsieur, 1785.

Angulo, Carlos de, ed. *Mémoires de Madame Campan, première femme de chambre de Marie-Antoinette.* Paris: Ramsay, 1979.

Aristotle. *On Generation and Corruption.* Translated by Harold Henry Joachim. Adelaide, Australia: The University of Adelaide Library, 2000.

Aristotle [Pseudo-]. *Les Problèmes d'Aristote, traitant de la nature de l'homme, & de la femme, des principes de la génération, de la formation des enfans au ventre de leur mere, et de l'usage de toutes les parties du corps humain.* Rouen: Pierre Caillove, 1668.

Arnauld. *Testament de la République et ses derniers adieux, liste et noms de ceux qui doivent en hériter et de ceux qui n'auront aucune part à la succession.* N.p., 1799.

Arnold, Thomas. *Observations on the Management of the Insane.* London, 1792.

Artaud, Jean-Baptiste. *Taconnet ressuscité, voyageant à Paris et en province.* Bordier, 1790.

Astruc, Jean. *Traité de la cause de la digestion.* Toulouse: Ant. Colomiez, 1714.

———. *Traité des maladies des femmes, où l'on a tâché de joindre à une théorie solide la Pratique la plus sûre et la mieux éprouvée.* Vol. 1. Paris: P. Guillaume Cavelier, Libraire, 1761.

———. *Treatise on the Diseases of Women in Which it Is Attempted to Join a Just Theory to the Most Safe and Approved Practice.* London: J. Nourse, 1762.

Audin Rouvière, Joseph Marie. *Essai sur la topographie physique et médicale de Paris.* Paris, 1794.

Bachaumont, Louis Petit de. *Mémoires secrets pour servir à l'histoire de la République des lettres en France depuis 1767 jusqu'à nos jours.* Vol. 9. London: John Adamson, 1784.

Bacher, Alexandre-André-Philippe-Frédéric. *De la Médecine considérée politiquement.* Paris: Mme Huzard, an XI.

Bailly, Jean Sylvain, ed. *Exposé des expériences qui ont été faites pour l'examen du magnétisme animal. Lu à l'Académie Française par M. Bailly en son nom et au nom de MM Kranklin, Le Roy, de Bory and Lavoisier, le 4 septembre 1784, imprimé par ordre du Roi.* Paris: Imprimerie Royale, 1784.

Banau, Jean-Baptiste, and François Turben. *Mémoire sur les epidémies du languedoc.* Paris, 1766.

Barbeyrac, Charles de. *Dissertations nouvelles sur les maladies de la poitrine, du cœur, des femmes, Vénériennes & Quelques maladies particulières.* Amsterdam: Janssons a Waesberge, 1731.

Barthez, M. *Médecin de sa Majesté l'Empereur et du Gouvernement; Et de MM. Bouvart, Fouquet. Lorry et Lamure.* Paris: Leopold Collin, 1807.

Barthez, Paul-Joseph. *Correspondance inédite de Albert de Haller, Barthez, Tronchin, Tissot avec le Dr. Rasr, de Lyon.* Édition Verbay. Lyon: Imprimerie d'Aimé Vingtrinier, 1856.

———. *Discours académique sur le principe vital de l'homme, prononcé le 31 octobre 1772 à la séance solennelle de rentrée du Ludovicée médical de Montpellier.* Montpellier: Boehm et Fils, 1703.

———. *Nouveaux Éléments de la science de l'homme.* 2nd ed. 2 vols. Paris: Gougon, 1806.

———. *Nouvelle Mécanique des mouvements de l'homme et des animaux.* Carcassonne: Pierre Polère, 1798.

Bartholin, Thomas. *Institutions anatomiques, augmentée et enrichie pour la seconde fois, tant par des opinions et observations nouvelles des modernes dont la plus grande partie n'a jamais été mise en lumière, que de plusieurs figures en taille douce.* Translated by A. du Prat. Paris: Mathurin Hénault and Jean Hénault, 1647.Battie, William. *A Treatise on Madness.* London: Dawsons, 1962.

Bartholin, Thomas fils. *Anatomia ex Caspari Bartolini parentis institutionibus.* Batavorum: Apud Franciscum Hackium, 1641.

Bauduit, P. *Considérations médicales sur le mariage, thèse présentée et soutenue à la faculté de médecine de Paris, le 12 août 1822.* Paris: Didot, 1822.

Bayle, François, and Henri Grangeron. *Relation de quelques personnes prétendues possédées faite d'autorité du Parlement de Toulouse.* Toulouse: Veuve Fouchac & Bely Marchands Libraires, 1682.

Bayle, Pierre. *Dictionnaire historique et critique.* Rotterdam: Reinier Leers, 1679.

Baynard, Edward, and John Floyer. *History of Cold Bathing, Both Ancient and Modern in Two Parts.* London: William and Johns Innys, 1722.

Beard, George. *A Practical Treatise on Nervous Exhaustion.* New York: William Wood & Company, 1880.

Beaumarchais. *Le Barbier de Séville.* Paris: Hachette, 1988. First published 1775.

Bellet, Isaac. *Lettres sur le pouvoir de l'imagination des femmes enceintes.* Paris: Les Frères Guérin, 1745.

Bénech, L. V. *Considérations sur les rapports du physique et du moral de la femme.* Paris: Gabon and Delaunay, 1819.

Bergasse, Nicolas. *Considérations sur le magnétisme animal ou sur la théorie du monde et des êtres organisés d'après les principes de M. Mesmer, avec des pensées sur le mouvement par M. le Marquis de Chastellux, de l'Académie Françoise.* The Hague, 1784.

————. *Lettre d'un médecin de la faculté de Paris, à un médecin du collège de Londres. Ouvrage dans lequel on prouve contre M. Mesmer que le Magnétisme animal n'existe pas.* The Hague, 1781.

Bernadac, Christian, and Sylvain Fourcassié. *Les Possédées de Chaillot.* Paris: J. C. Lattes, 1983.

Bertholon, Abbé. *De l'Electricité du corps humain dans l'état de santé et de maladie, ouvrage couronné par l'Académie de Lyon dans lequel on traite de l'électricité de l'atmosphère, de son influence & de ses effets sur l'économie animale.* Lyon: Bernuset, 1780.

Bertrand, Alexandre-Jacques-François. *Du Magnétisme animal en France et des jugements qu'en ont portés les sociétés savantes, avec le texte des divers rapport fait en 1784 par les commissionaires de l'Académie des Sciences, de la Faculté et de la Société Royale de Médecine.* Paris, 1826.

Bertrand, Antoine-Alexandre. *Essai médico-légal sur la stérilité des femmes, précédé de l'exposition physiologique de quelques systèmes qui ont eu lieu sur la génération.* Montpellier: G. Izar and A. Ricard, 1800.

Bichat, Xavier. *Recherches physiologiques sur la vie et la mort, 1800.* Paris: Gauthier-Villars, 1955.

Bienville, D. T. de. *De la Nymphomanie ou Fureur utérine.* Edited by Jean-Marie Goulemot. Paris: Le Sycomore, 1980.

Blackmore, Richard. *A Treatise on Spleen and Vapours; or, Hypochondriacal and Hysterical Affections.* London: J. Pemberton, 1725.

Blondel, James. *The Strength of Imagination in Pregnant Women Consider'd.* London, 1726.

Boerhaave, Hermann. *Commentaires des aphorismes de médecine d'Hermann Boerhaave sur la connaissance et la cure des maladies par M. Vans-Swieten.* Translated by M. Moublet. Avignon: Roberty and Gailermont, 1766.

Boissieu, Barthélemy-Camille. *Mémoire sur les méthodes rafraîchissante et echauffantes.* Dijon, 1772.

Bonnet, Charles. *Essai analytique sur les facultés de l'âme.* In *Œuvres d'histoire naturelle et de philosophie.* Neûchatel, 1779–83.

————. *Essai d'application des principes psychologiques de l'auteur.* In *Œuvres d'histoire naturelle et de philosophie.* Neûchatel, 1779–83.

Bordeu, Antoine. *Dissertation sur les eaux minerales du Bearn.* G. F. Qillau, 1750.

Bordeu, Théophile de. *Correspondance.* 4 vols. Edited by Martha W. Fletcher. Montpellier: CNRS, Université Paul Valéry, 1977–79.

————. *Lettres contenant des essais sur l'histoire des eaux minérales du Béarn, et de quelques-unes des provinces voisines ausquelles elles conviennent et sur la façon dont on doit s'en servir.* Amsterdam: Chez les Frères Poppé, 1746.

————. *Oeuvres complètes de Bordeu.* 2 vols. Edited by A. B. Richerand. Paris: Auguste Ghio, 1882.

————. *Recherches sur le pouls par rapport aux crises.* 2nd ed. Paris: Pierre-Fr Didot le Jeune, 1768.

Bouillon-Lagrange, Edme-Jean Baptiste. *Réflexions sur les affections nerveuses et sur quelques médicamens usités dans ces sortes de maladies. Présentées et soutenues à l'Ecole Spéciale de Médecine de Strasbourg, le 2 Frimaire, an XIV à midi.* Strasbourg, 1804.

Boulogne, Jacques, and Daniel Delattre, trans. and ed. *Galien: Systématisation de la médecine.* Lille: Presses Universitaires du Septentrion, 2003.

Bourdelin, Ludovicus Henricus. *L'Art iatrique.* Amiens, 1776.

Bovier de Fontenelle, Bernard le. *Entretiens sur la pluralité des mondes.* Amsterdam: P. Mortier, 1686. Translated by H. A. Hargreaves as *Conversations on the Plurality of Words* (Berkeley: University of California Press, 1990).

Bowen, Thomas. *An Historical Account of the Origin, Progress, and Present State of Bethlem Hospital: Founded by Henry the Eighth, for the Cure of Lunatics, and Enlarged by Subsequent Benefactors, for the Reception and Maintenance of Incurables.* London: S.n., 1783.

Boyer, attribué. *Les Abus de la saignée démontrés.* Paris: Vincent, 1759.

Boyveau-Laffecteur, Pierre. *Essai sur les maladies physiques et morales des femmes.* Paris, 1794.

Bressy, Joseph. *Recherches sur les vapeurs.* London, 1789.

Bret, Louis. *Analyse comparée de quelques observations citées par M. Pomme, Dr. en Médecine.* Arles: Gespard Mesnier, 1806.

Bright, Timothy. *A Treatise of Melancholie.* London: Thomas Vautrolier, 1586.

Brisseau, Jacques Pierre. *Traité des mouvements sympathiques, avec une explication de ceux qui arrivent dans le vertige, l'epilepsie, l'affection hypocondriaque, et la passion hystérique.* Paris: E. de la Roche, 1692.

Brissot, Jacques Pierre. *Nouveaux Voyages dans les Etats-Unis de l'Amérique septentrionale, faits en 1788.* Paris, 1791.

Broussais, François-Joseph-Victor. *De l'Irritation et de la folie, ouvrage dans lequel les rapports du physique et du moral sont etablis sur les bases de la médecine physiologique.* Paris and Brussels, 1828.

———. *Examen des doctrines médicales et des systèmes de nosologie.* Paris: Méquignon-Marvis, 1821.

Browne, Richard. *Medicina Musica; or, A Mechanical Essay on the Effects of Singing, Musick, and Dancing, on Human Bodies, Revis'd and Corrected.* London: John Cooke, 1729.

Bruguière, Jean-Guillaume. *Tableau encyclopédique et méthodique des trois règnes de la nature. Contenant l'helminthologie, ou les vers infusoires, les vers intestins, les vers mollusques, &c. septième livraison.* Paris: Panckoucke, 1791.

Brulley, Claude Antoine. *Essai sur l'art de conjecturer en médecine.* Paris, 1801.

Buc'Hoz, Pierre-Joseph. *Histoire naturelle physique et medicinale de l'homme.* 2nd ed. Paris, 1785.

Bucquet, Jean Baptiste Michel. "Observations sur l'analyse de l'opium." In *Histoire de la Société Royale de Médecine, année 1776.* Paris: Didot le jeune, 1789.

Buffon, George-Louis Leclerc de. *Buffon's Natural History: Containing the Theory of the Earth, a General History of Man, of the Brute Creation, and of Vegetables, Minerals, &C. &C. &C.* Vol. 5. London: Printed for the proprietor, 1797.

———. *De l'Homme.* Edited by Michèle Duchet. Paris: François Maspero, 1971.

———. *Discours sur la nature des animaux.* Paris: H. Lecène et H. Oudin, 1888.

———. *Histoire naturelle, générale et particulière, avec la description du Cabinet du Roy.* Paris: Imprimerie Royale, 1749.

Bullein, William. *Bulwarke of Defence Against All Sicknesse, Soarenesse, and Woundes that Doe Dayly Assaulte Mankinde: Which Bulwarke Is Kept with hilarius the Gardener, Health the Physician, with the Chirurgian, to Helpe the Wounded Souldiours. Gathered and Practiced from the Most Worthy Learned, Both Olde and New: To the Great Comfort of Mankinde.* London: Thomas Marthe, 1579. First published 1562.

Burke, Edmund. *Reflections on the Revolution in France.* New York: Penguin, 1999. First published 1790.

Burney, Fanny. *Camilla.* New York: Oxford, 1999. First published 1796.

———. *Cecilia.* New York: Oxford, 1999. First published 1782.Cabanis, Pierre-Jean-Georges. *Coup d'oeil sur les révolutions et sur la réforme de la médecine.* Paris, 1804.

———. *Du Degré de certitude de la médecine.* Edited by Jean-Marc Drouin Champion-Slatkine. Paris and Geneva: Éd. de la Cité des Sciences et de l'Industrie, Paris, 1989.

————. "Rapports du physique et du moral de l'homme." In *Œuvres philosophiques*, edited by C. Lehec-J. Careneuve. Paris: PUF, 1956. First published 1802.

Campan, Madame, Jean Chalon, and Carlos de Angulo. *Mémoires de Madame Campan, première femme de chambre de Marie-Antoinette.* Paris: Ramsay "image," 1979.

Capuron, Joseph. *Traité des maladies des femmes, depuis la puberté jusqu'à l'âge critique inclusivement.* Paris: l'auteur, 1812; Paris: Croullebois, 1817.

Caraccioli, Louis-Antoine. *Le livre à la mode.* Paris, 1757, 1759.

————. *Le Livre à la mode: Suivi du livre des quatre couleurs.* Edited by Anne Richardot. Saint-Étienne: PUSt Étienne, 2005.

————. *Le Livre à la mode verte-feuille.* Paris, 1760.

————. *Livre de quatre couleurs: Aux quatre éléments, de l'imprimerie des quatre saisons, en 4444.* Paris: Duchesne, 1760.

Caratery, Charles-Antoine. *De l'Influence des passions sur l'economie animale.* Avignon: François Guibert, 1785.

Cardano, Girolamo. *De Vita propria.* First published 1643. Translated by Jean Stoner as *The Book of My Life* (New York: New York Review Books, 2002).

Carrère, Thomas. *Traité des eaux minérales du roussillon.* Perpigan: Impr. J.-B Reynier, 1756.

Casanova, Giacomo. *Lana Caprina: Lettre d'un lycanthrope.* Bibliotheca Casanoviana. Paris: Éditions Allia, 1998. First published 1771.

Casaubon, Meric. *A Treatise Concerning Enthusiasm.* Gainesville: Scholars' Facsimiles & Reprints, 1970. First published 1655.

Cazeles, Masard de. *Second Mémoire sur l'electricité médicale, et sur l'histoire du traitement de quarante-deux malades, entièrement guéris ou notablement soulagés.* Paris: Méquignon, 1782.

Celsus, Aurelius-Cornelius. *De la Médecine.* Vol. 1. Translated by Guy Serbat. Paris: Les belles Lettres, 1995. First published 1478.

Chambaud, Jean Jacques Menuret de. *Essai sur l'histoire médico-topographique de Paris; ou, Lettres à M. d'Aumont, professeur en médecine à Valence, sur le climat de Paris, sur l'état de la médecine, sur le caractère et le traitement des maladies, & particulièrement sur le petite vérole et l'inoculation.* Paris: Rue et Hôtel Serpente, 1784.

————. *Essais sur l'histoire médico-topographique de Paris; ou, Lettres à M. d'Aumont, . . . sur le climat de Paris, sur l'état de la médecine . . . et particulièrement sur la petite vérole et l'inoculation.* Paris: Rue et Hôtel Serpente, 1786.

Chambon de Montaux, Nicolas. *Des Maladies de femmes.* 2 vols. Paris: rue et Hôtel Serpente, 1784.

————. *Encyclopédie méthodique, médecine, par une société de médecins.* Vol. HA–JUS. Paris, 1798.

Chansons choisies, avec les airs notés. Vol. 3. Geneva, 1782.

Charcot, Jean-Martin. *Leçons sur les maladies du système nerveux faites à la Salpêtrière.* Paris, 1874.

————. *Lectures on the Diseases of the Nervous System, Delivered at La Salpêtrière.* Translated by George Sigerson. Vol. 1. London: New Sydenham Society, 1877. First published 1876.

————. *Lectures on the Diseases of the Nervous System, Delivered at La Salpêtrière.* Vol. 2. London: New Sydenham Society, 1881.

Charcot, Jean Martin, and Paul Marie Louis Pierre Richer. *Les démoniaques dans l'art.* Paris: Delahaye et Lecrosnier, 1887.

Chassaignon, Jean-Marie. *Cataractes de l'imagination, déluge de la scribomanie, vomissement lit-*

téraire, hémorrhagie encyclopédique, par Epiménide l'Inspiré. Dans l'antre de Trophonius, au pays des visions. Lyon, 1779.

Chastelain, Matthieu. *Traité des convulsions et des mouvemens convulsifs, qu'on appelle à présent vapeurs.* Paris: Janisson, 1691.

Chauvot de Beauchêne, Edme-Pierre. *De l'Influence des affections de l'âme dans les maladies nerveuses des femmes, avec le traitement qui convient à ces maladies.* Paris: Méquignon, 1781.

Chevalier, M. *Mémoires et observations sur les effets des eaux de Bourbonne: Les Bains, en champagne dans les maladies hystériques & chroniques.* Paris: Vincent, 1772.

Cheyne, George. *The English Malady; or, A Treatise of Nervous Diseases of all Kinds as Spleen, Vapours, Lowness of Spirits, Hypochondriacal and Hysterical Distempers, &c.* London and Dublin: S. Powell, 1733.

———. *An Essay on Health and Long Life.* Bath: J. Leake, 1724.

———. *The Natural Method of Curing the Diseases of the Body and the Disorders of the Mind, Depending on the Body.* London: Strahan & Leake, 1773.

———. *Remarks on Two Late Pamphlets Written by Dr. Olliphant against Pitcairn's Dissertations.* Edinburgh, 1702.Chirac, Pierre. *Dissertations et consultations médicinales.* 3 vols. Paris: Durand, 1744–55.

Clerc, Nicolas Gabriel. *Histoire naturelle de l'homme dans l'état de maladie.* 2 vols. Paris, 1767.

Colman, George. *The Spleen; or, Islington Spa; A Comick Piece of Two Acts. As it Is Performed at the Theatre Royal, in Drury-Lane.* London: T. Becket, 1776.

Colombier, Jean, and François Doublet. "Instruction sur la manière de gouverner et de traiter les insensés." *Journal de médecine* (August 1785).

Consultation sur les Convulsions. Paris, 1735.

Cooke, John. *A Treatise on Nervous Diseases.* 2 vols. London: Printed for Longman Hurst Rees Orme and Brown, 1820.

Crêpe, le P., and Henry Charles Lea Library. *Notion de l'oeuvre des convulsions et des secours.* Paris, 1788.

Crichton, Alexander. *An Inquiry into the Nature and Origins of Mental Derangement, Comprehending a Concise System of Physiology and Pathology of the Human Mind, and the History of the Passions and their Effects.* London: Cadell and Davies, 1798.

Crooke, Helkiah. *Mikpokosmotpapia: A Description of the Body of Man Together with the Controversies Thereto Belonging.* London: William Iagard, 1615.

Cullen, William. *Clinical Lectures, Delivered in the Years 1765 and 1766 by William Cullen, Taken in Short-Hand by a Gentlemen Who Attended.* London: Lee and Hurst, 1797.

———. *Elements de médecine pratique, de M.D. Cullen . . . avec des notes, dans lesquelles on a refondu la nosologie du meme auteur, décrit les différentes espèces de maladies, & ajouté un grand nombre d'observations.* Edited and translated by Bosquillon. Paris: Théophile Barrois, le jeune et Méquignon, 1787.

———. *First Lines of the Practice of Physic, for the Use of Students.* Dublin: T. Armitage, 1777–84.

———. *First Lines of the Practice of Physic, for the Use of Students in the University of Edinburgh.* Edinburgh, 1783.

———. *Institutions de médecine pratique.* Vol. 2. Translated by P. Pinel. Paris: P. J. Duplain, 1785.

———. *Lectures on the Materia Medica.* Dublin: W. H. Whitestone, 1781.

———. *Nosology; or, An Arrangement of Diseases, by Classes, Orders, Genera and Species, with the Distinguished Characters of Each and the Outlines of the Systems, Of Sauvages, Linneaus, Vogel, Sagar, and Macbride.* Translated by William Cullen. Edinburgh: William Creech, 1800.

————. *Synopsis and Nosology Being an Arrangement and Definition of Diseases.* Boston, 1742.

————. *Synopsis and Nosology Being an Arrangement and Definition of Diseases.* Hartford, Conn.: Nathaniel Patten, 1792.

Dagoty, Gautier. *Exposition anatomique des organes des sens, jointe à la névrologie entière du Corps humain et conjecture sur l'électricité animale, avec des planches imprimées en couleurs naturelles.* Paris: Demonville, 1775.

Daignan, Guillaume. *Tableau des variétés de la vie humaine.* Paris: l'auteur, 1786.

Daquin, Antoine, Guy-Crescent Fagon, and Antoine Vallot. *Journal de santé de Louis XIV.* Edited by Stanis Perez. Grenoble: Jerôme Million, 2004.

D'Aquin, Antoine, Antoine Vallot, and Guy-Crescent Fagon. *Journal de Santé du Roi.* Paris: Auguste Durand, 1692.

Daquin, Joseph. *Analyse des eaux thermales d'Aix en Savoye.* Chambéry: F. Gorrin, 1773.

————. *Philosophie de la folie.* Paris: Née de la Rochelle, 1792.

Dartaize, H. *Le Cri de l'humanité pour les victimes egorgées sous Robespierre.* N.d.

Dazille, Jean Barthélemy. *Observations sur les maladies des negres, leurs causes, leurs traitements.* Paris: Didot, 1776.

De Bonnaire, Louis. *Examen critique, physique et théologique des convulsions et des caracteres divins qu'on croit voir dans les accidens des Convulsionnaires.* N.p., 1733.

Dechambre, Amédée. *Dictionnaire encyclopédique des sciences médicales.* Vol. 24, CRU–CYS. Paris: G. Masson and P. Asselin, 1880.

Decourcelle, Gilles-Josephand Étienne Chardon. *Elixir américain; ou, Le Salut des dames.* Chalons: Sombert, 1771.

Deduit. *Le Vendangeur aérostatique; ou, Les Adieux du baquet mesmérique.* London, 1786.

Delan, François-Hyacinthe. *Dissertation théologique contre les convulsions, adressees au laic auteur des reflexions sur la reponse au plan general.* N.p., 1735.

Delandine, Antoine François. *De la Philosophie corpusculaire et des procédés magnétiques chez les divers peuples.* Paris: Cuchet, 1785.

Delarive, Gaspard Charles. *Sur un établissement pour la guérison des aliénés.* Geneva: Bibliothèque Britannique, 1798.

Delorme, E. *Correspondance de Cl. Simon, 1792.* Extrait du Bulletin de l'Académie Delphine, 4th ser., vol. 12. Grenoble, 1899. Cited in Jean-Paul Bertrand, *Valmy* (Paris: Gallimard, 1992), 67.

Desbonnets. "Effets de la musique dans les maladies nerveuses." *Journal de médecine,* Notice vol. 59. St. Petersburg, 1784.

Desbout, Louis. *Dissertation sur l'effet de la musique dans les maladies nerveuses médecin de l'université de Pise, chirurgien à l'amirauté de S.M. imperiale de toutes les russies, de l'academi des ierogliphes de Florence, correspondant de la Société Royale.* St. Petersburg: Breitkopf, 1784.

Desessarts, Jean-Baptiste Poncet. *Recherche de la vérité; ou, Lettres sur l'œuvre des convulsions.* N.p., 1733.

D' Eslon, Charles. *Observations sur le magnétisme animal.* London and Paris: Didot, 1780.

D' Henin de Cuvillers, Étienne Félix Baron. *Le Magnétisme animal fantaxiéxoussique retrouvé dans l'antiquité; ou, Dissertation historique, etymologique et mythologique sur Esculape, Hippocrate et Galien, sur Apis, Sérapis, ou Osiris et sur Isis, suivi de recherches sur l'alchimie.* Paris, 1821.

Dictionnaire universel françois et latin vulgairement appelé Dictionnaire de 7 vols vulgairement appelé Dictionnaire de Trévoux. Paris: Compagnie des Libraires Associés, 1752.

Diderot, Denis. *Dialogues.* Washington, N.Y.: Kennikat Press, 1971.

————. *La Religieuse*. In *Œuvres*, vol. 2, edited by Laurent Versini. Paris: Robert Laffont, 1994. Translated by Leonard Tancock as *The Nun* (Harmondsworth, U.K.: Penguin Books, 1974).

————. "Les Bijoux indiscrets." In *Oeuvres romanesques*, edited by H. Benac. Paris: Garnier, 1965.

————. "Salon de 1767." In *Salons*, vol. 2. Paris: J. L. J. Brière, 1821.

————. "Sur les Femmes." In *Œuvres: Philosophie*, edited by Laurent Versini. Paris: Bouquins, 1994. First published 1772.

Diderot, Denis, Julien Busson, Marc-Antoine Eidous, and François-Vincent Toussaint, ed. and trans. *Dictionnaire universel de médecine, de chirurgie, de chymie, de botanique, d'anatomie, de pharmacie, d'histoire naturelle de James*. Paris: Briasson, 1746–48.

Digby, Kenelm. *Discours fait en une célèbre assemblée par le chevalier Digby, chancelier de la reine de Grande Bretagne, touchant la guerison des playes par la poudre de sympathie*. Paris: Augustin Courbe and Pierre Moet, 1658.

Dionis, Pierre. *Dionis dissertation sur la mort subite et sur la catalepsie, avec la relation de Plusieurs personnes qui en ont été attaquées*. Paris: Laurent d'Houry, 1718.

Dissertation sur les miracles et en particulier sur ceux qui ont été operés au tombeau de Mr de Pâris, en l'eglise de Saint Médard de Paris, avec la relation & les preuves de celui qui s'est fait le 3e Novembre en la personne d'Anne Le Franc de la Paroisse de Seulement Barthelemy. N.p., 1791.

Doppet, François-Amédée. *Aphrodisiaque externe; ou, Traité du fouet et de ses effets sur le physique de l'amour, ouvrage médico-philosophique, suivi d'une dissertation sur tous les moyens capables d'exciter aux plaisirs de l'amour*. Bordeaux: Ducros, 1970. First published 1788.

————. *La Mesmeriade; ou, Le Triomphe du magnétisme animal*. Geneva, 1784.

————. *Le Médecin de l'amour*. Paris: Leroy, 1787.

————. *Le Médecin philosophe: Ouvrage utile à tout citoyen, dans laquelle on trouve une nouvelle manière de guérir, puisée dans les affections de l'ame, & la gymnastique*. Paris and Turin: Leroy, 1787.

————. *Traité théorique et pratique du magnétisme animal*. Turin: Jean-Michel Briolo, 1784.

Dorat, Claude Joseph. *Les Malheurs de l'inconstance*, in *Romans Libertins du XVIIIe siècle*. Paris: Robert Laffont, 1993.

Doussin-Dubreuil, Jacques-Louis. *De l'Epilepsie en général, et particulièrement de celle déterminée par des causes morales*. Paris: chez l'Auteur, 1797.

Drake, Nathan. *Literary Hours; or, Sketches, Critical, Narrative, and Poetical*. London: Printed for Longman, Hurst, Rees, Orme, and Brown Paternoster-Row, 1820.

Duchêne, la Mère. *La mère Duchesne. Journal des femmes, etrennes de la mère Duchesne*. Paris: De l'imprimerie Crapart, 1792.

Du Deffand, Madame. *Lettres de Madame du Deffand*. Paris: Mercure de France, 2002.

Dufau, Pierre Armand. "Observation sur une hystérie vermineuse." *Journal de Médecine, Chirurgie, Pharmacie* 29 (1768).

Dufour, Jean-François. *Essai sur les opérations de l'entendement et sur les maladies qui le dérangent*. Amsterdam and Paris: Merlon, 1770.

Du Freny, Rivière. "La Malade sans maladie." In *Œuvres*, vol. 2. Paris: Briasson, 1781. First published 1699.

Dufresny, Charles. *Oeuvres*. Vol. 2. Paris: Barrois, 1779.

Du Laurens, André. *Discours de la conservation de la veue, des maladies mélancoliques, des cathares, & de la vieillesse*. Rouen: Claude Le Villain, 1600.

————. *Œuvres*. Edited and translated by G. Sauvageon. Paris, 1646.

Dumoulin, Jacques. *Nouveau Traité du rhumatisme et des vapeurs ou après avoir expliqué les causes, les symptômes, & les signes de ces maladies, l'on donne les remèdes propres et facultés pour les guérir.* Paris: Laurent d'Houry, 1703.

Duncan, Andrew. *Heads of Lectures on the Theory and Practice of Medicine.* 3rd ed. Edinburgh, 1781.

Dupont, Aimé-Marie-Joseph. *Recherches générales sur l'affection hystérique.* École de Médecine de Montpellier, 1801.

Dupont de Nemours, Pierre Samuel. *Idée sur les secours à donner.* Paris, 1786.

Duvernoy, Georges-Louis. *Dissertation sur l'hystérie.* Paris: Gabon et Cie, An IX, 1801.

Ettmüller, Michel. *Pratique speciale de medecine sur les maladies propres des hommes, des femmes & des petits enfants, avec des dissertations du même auteur, sur l'epilepsie, l'yvresse, le mal hypocondriaque, la douleur hypocondriaque, la corpulence, & la morsure de la vipère.* 2nd ed. enl. Lyon: Thomas Amaulry, 1698.

Fabre, Pierre. *Essai sur les facultés de l'ame, considérées dans leur rapport avec la sensibilité et l'irritabilité de nos organes.* Paris, 1785.

Falconer, William. *A Dissertation on the Influence of the Passions, upon the Disorders of the Body.* 3rd ed. London: Dilly, Poultr, 1796.

———. *An Essay on the Baths Waters in Four Parts, Containing a Prefatory Introduction on the Study of Mineral Waters in General.* London: Lowndes, 1770.

Falconet, Noel. *Système des fievres et des crises, selon la doctrine d'Hippocrate, des febrifuges, des vapeurs, de la goute, de la peste.* Paris: Antoine-Urbain Coustelier, 1723.

Falloppio, Gabriele. *Secreti diversi et miracolosi.* Edited by Borgaruccio Borgarucci. Venice: Francesco Camozio, 1578.

Fallowes, S. *The Best Method for the Cure of Lunatics.* London, 1705.

Falquet, Jean Claude. *Essai sur l'epilepsie.* Montpellier: Bonnariq, Germinal, 1804.

Féraud, Jean-François. *Dictionaire critique de la langue française.* 3 vols. Marseille: Mossy, 1787–88.

Fernel, Jean. *Les sept Livres de la physiologie.* Paris: Fayard, 2001. First published 1554.

Ferrand, Jacques. *De la Maladie d'amour; ou, Mélancholie erotique, discours curieux qui enseigne à cognoistre l'essence, les causes, les signes & les remedes de ce mal fantastique.* Paris: Denis Moreau, 1627.

Ferriar, John. *Medical Histories and Reflections.* Warrington, London, 1792.

Ficinus, Marsilius. "Expositio Prisciani & Marsilii in Theophrastum de sensu, phantasia, & intellectu." In *Opera omnia / Marsilio Ficino; Con una lettera introduttiva di Paul Oskar Kristeller; E una premessa di Mario Sancipriano.* Turin: Bottega d'Erasmo, 1962, 1983. First published 1497.

Finch, Anne. "A Pindaric Ode on the Spleen." In *Of the Spleen, its Description and History, Uses and Diseases . . . To Which Is Added Some Anatomical Observations in the Dissection of an Elephant. (A Pindaric Ode on the Spleen, by . . . the Countess of Winchilsea).* Edited by W. Stukeley. London, 1723. First published 1709.

Fitzgerald, Gérard. *Traités des maladies des femmes.* Translated by Fitzgerald. Paris: Du Chesne, 1758.

Flaubert, Gustave. *Dictionnaire des idées reçues.* Édition diplomatique des trois manuscrits de Rouen. Edited by Lea Caminiti. Paris: A. G. Nizet, 1966.

Forgues, V. "De l'Influence de la musique sur l'économie animale." Dissertation, l'École de Médecine de Montpellier, An VII.

Fosseyeux, Marcel. *L'hotel-Dieu de Paris, au XVIIe et au XVIIIe siècle.* Paris: Berger-Lervault, 1912.

Fothergill, D. *Conseils pour les femmes de 45 à 50 ans; ou, Conduite à tenir lors de la cessation des règles.* London: Briand, 1788.

Fouillou, Jacques. *Observations sur l'origine et le progrès des convulsions à l'occasion d'une lettre écrite au mois de janvier en faveur des convulsions.* N.p.: June 10, 1733.

Fournel, Jean-François. *Essai sur les probabilités du somnambulisme magnétique, pour servir à l'histoire du magnétisme animal.* Amsterdam, 1785.

———. *Remontrances des malades aux médecins de la Faculté de Paris.* Amsterdam, 1785.

Freind, John. *Emmenologie; ou, Traité médical d'evacuation ordinaire aux femmes où l'on explique les phénomènes, les retours, les vices, & la méthode curative, qui la concernent selon la loi de la méchanique.* Paris, 1730.

Frier, François. *Guide pour la conservation de l'homme.* Grenoble, 1789.

Fulgose, Messire Baptiste. *Contramours, l'anteros; ou, Contramour.* Paris: Matin Lejeune, 1581.

Fuller, Francis. *Medicina Gymnastica; or; A Treatise Concerning the Power of Exercise with Respect to the Animal Oeconomy; and the Great Necessity of It in the Cure of Several Distempers.* London: Robert Knaplock, 1705. Translated as *Medicina gymnastica; oder, Von der Leibesübung in Ansehung der Animalischen Oeconomie oder der zu Erhaltung der Gesundheit des menschlichen Lebens nöthigen Ordnung: und wie solche bey Curirung verschiedener Krankheiten unumgänglich nöthig sey* (Lemgo: Johannn Heinrich Meyer, 1750).

Furetière, Antoine. *Dictionaire universel contenant généralement tous les mots françois tant vieux que modernes, et les termes de toutes les sciences et des arts.* Rotterdam: A. et R. Leers, 1690.

Gachet, Louis Étienne. *Problème médico-politique pour ou contre les arcanes.* Paris, 1791.

Galen. *Oeuvres anatomiques, physiologiques et médicales.* Translated by Ch. Daremberg. Paris: J. B. Baillière, 1856.

———. *On Diseases and Symptoms.* Translated by Jan Johnston. Cambridge: Cambridge University Press, 2006.

———. *Selected Works.* Translated by P. N. Singer. New York: Oxford University Press, 1997.

———. *Systématisation de la medicine.* Translated and edited by Jacques Boulogne and Daniel Delattre. Lille: Presses Universitaires du Septentrion, 2003.

Galien, Claude. *Traité des passions de l'âme et de ses erreurs.* Translated by Rover van der Elst. Clichy: GREC, 1993.

Gamet, Jean-Marie. *Théorie nouvelle sur les maladies cancéreuses, nerveuses et autres.* Paris: Ruault, 1772.

Ganne, Ambroise. *De l'Homme physique et moral; ou, Recherches sur les moyens de rendre l'homme plus sage.* Strasbourg: J.-G. Treuttel, 1791.

Garat, Dominique-Joseph. *Mémoires historiques de la vie de M. Suard, sur ses ecrits et sur le six-huitième siècle.* Paris, 1820.

Gardanne, Charles-Pierre-Louis. *De la Ménopause; ou, De l'Age critique des femmes.* 2nd ed. Paris: Méquignon-Marvis, 1821.

Gardanne, Joseph-Jacques. *Des Maladies des créoles en Europe avec la manière de les traiter.* Paris: Valde, 1784.

Georget, Étienne-Jean. *De la Folie.* Paris, 1820.

Géraud, Pierre. *Aperçu sur l'hygiène des femmes enceintes, Présenté à l'école de médecine de Montpellier, le II Thermidor, an 7 de la République Française.* Montpellier: Jean Martel aîné, An VII.

Gilibert, Jean Emmanuel. *Aperçu sur le magnétisme animal; ou, Le Résultat des observations faites à Lyon sur ce nouvel agent.* Geneva, 1784.

———. *L'Anarchie médicinale; ou, La Médecine considérée comme nuisible à la société.* 3 vols. Neuchâtel, 1772.

Godwin, William. *Caleb Williams.* New York: Penguin, 1985. First published 1794.

Goldoni, Carlo. "La finta ammalata." In *Collezione completa delle commedie del signor Carlo Goldoni,* edited by Georg Schaz. Nabu Press, 2010. First published 1751.

———. *Mémoires De Goldoni: Pour Servir a L'histoire De Sa Vie Et a Cellede Son Théatre.* Paris: Colburn Libraire, 1814.

Goulin, Jean, and Labeyrie. *Dictionnaire raisonné de matière médicale.* Paris: F. Didot, 1773.

Goulin, le médecin des hommes. *Le véritable honneur est d'être utile aux hommes.* Paris: Vincent, 1772.

Gracián, Baltasar. *Art of Worldly Wisdom.* Boston: Shambhala Publications, 1993.

———. *The Courtiers Manual Oracle; or, The Art of Prudence.* London: M. Flesher, 1685.

Grafenberg, Johannes Schenck von. *Observationum medicarum rariorum. Libri VII.* Basilea, 1584.

Green, Monica., ed. and trans. *The Trotula: A Medieval Compendium of Women's Medicine.* Philadelphia: University of Pennsylvania, 2000.

Guibelet, Jourdain. *Trois Discours philosophiques.* Evreux: Antoine le Marié, 1603.

Guibert, Jacques-Antoine. *Eloge du maréchal de Catinat, du chancelier de l'Hospital, de Thomas de l'Académie Française. Suivi de l'éloge inédit de Claire-Françoise de Lespinasse. Publié par sa veuve, sur les manuscrits et d'après les corrections de l'auteur.* Paris: D'Hautel, 1806.

Guillon-Pastel, Marie-Nicolas-Silvestre. "Des maladies nerveuses." In *Magasin encyclopédique ou Journal des sciences, des lettres et des arts,* 6è année, tome second. Paris: Fuchs, 1800.

Guindant, Toussaint. *Exposition des variations de la nature dans l'espèce humaine.* Paris: Debure Père, 1771.

———. *La Nature opprimée par la médecine moderne.* Paris: Debure l'aîné, 1768.

Guyon, Louis. *Le Cours de médecine en françois contenant le miroir de beauté et santé corporelle.* 6th ed. Lyon: Jean Grégoire, 1673.

Guyot, Jean. *Dictionnaire médecinal portatif: contenant une méthode sure pour connoître & guérir les maladies critiques & chroniques.* Paris: D'Houry, 1763.

Hallaran, William Saunders. *Practical Observations on the Causes and Cure of Insanity.* 2nd ed. Cork: Edwards and Savage, 1818.

Haller, Albert de. *Dissertation sur les parties irritables et sensibles des animaux.* Translated by Samuel Tissot. Lausanne: Marc-Michel Bousquet et comp.,1755.

———. *Mémoires sur la nature sensible et irritable du corps animal.* Lausanne, 1756–60.

———. *Reflexions sur l'usage de l'opium, des calmants et des narcotiques pour la guérison des maladies. En forme de lettre.* Paris: Guillaume Cavelier Fils, 1726.

Haslam, John. *Observations on Insanity.* London, 1794.

Haygarth, John. *Of the Imagination, as a Cause and as a Cure of Disorders of the Body.* London: Rivington, 1800.

Haywood, Eliza. *The Injur'd Husband and Lasselia.* Lexington: University Press of Kentucky, 1999.

Heberden, William. *Medical Commentaries on the History and Cure of Diseases.* London: T. Payne, 1802.

Hecquet, Philippe. *La Suceuse convulsionnaire; ou, La Psylle miraculeuse.* N.p., 1736.

———. *Le Naturalisme des convulsions.* 2 vols. Soleure: Andrea Gymnicus, 1733.

————. *Le Naturalisme des convulsions dans les maladies de l'épidémie convulsionnaire.* Soleure: Andrea Gymnicus, 1733.

————. *Lettre sur la convulsionnaire en extase ou la vaporeuse en rêve.* N.p., 1736.

————. *Reflexions sur l'usage de l'opium, des calmants et des narcotiques pour la guérison des maladies. En forme de lettre.* Paris: Guillaume Cavelier fils, 1726.

Hélian. *Dictionnaire du diagnostic; ou, L'Art de connoître les maladies.* Paris: chez Vincent, 1771.

Henry, Peter A. *Collection of Some Extraordinary Nervous Cases, that Have Occured in the Course of Dr. Henry's Practice, Submitted to the Impartial Judgment of the Public in Order to Fully Demonstrate the Virtues and Efficacy of his Chymical Nervous Medicine in the Cure of Nervous Disroders.* London, 1780.

Herwig, Michael. *The Art of Curing Sympathetically, or Magnetically.* London: T. Newborough, 1700.

Hill, John. *The Virtues of Wild Valerian in Nervous Disorders; And the Manner of Taking it, Against Vapours and Melancholy as Tea; Against Fits and Hysteric Complaints, in the Powder and Against Confused Thoughts and Paralytic Complaints With Directions for Gathering and Preserving the Root; and for Chusing the Right Kind When it is Brought Dry Shewing that the Uncertainty of Effect in this Valuable Medicine is Owing to Adulteration or Ill Management.* 12th ed. London: Baldwin, 1772.

Hippocrates. *Airs, eaux, lieux.* Translated by Jacques Jouanna. 2 vols. Paris: Les Belles Lettres, 1996.

————. *Des Maladies des filles.* 2 vols. Paris: Rue et Hôtel Serpente, 1785.

————. *Des Vents—De l'Art.* Translated by Jacques Jouanna. Paris: Les Belles Lettres, 1986.

————. *La Maladie sacrée.* Translated by Jacques Jouanna. 2 vols. Paris: Les Belles Lettres, 2003.

————. *Maladies des femmes, Oeuvres complètes, livre VIII.* Translated by Littré. Paris: J.-B. Baillière, 1853.

Histoire de la Société Royale de Médecine année 1776 avec les mémoires de médecine & de physique médicale pour la même année tirés des registres de cette société. Paris: Philippe-Denys Pierres, 1779.

Histoire de la Société Royale de Médecine, année 1779, avec les mémoires de médecine et de physique expérimentale de la même année. Paris: Didot le jeune, 1782.

Hoffmann, Friedrich. *Fundamenta medicinae.* Translated by Lester S. King. London: Macdonald; New York: American Elsevier, 1971.

Holt, Edward. *The Public and Domestic Life of his Late . . . Majesty, George the Third: Comprising the Most Eventful and Important Period in the Annals of British History.* Vol.1. Sherwood, Neely, and Jones, 1820.

Hosack, David. *A System of Practical Nosology: To Which is Prefixed a Synopsis of the Systems of Sauvages, Linnaeus, Vogel, Sagar, Macbride, Cullen, Darwin, Crichton, Pinel, Parr, Swediar, Young and Good, with Reference to the Best Authors on Each Disease.* 2nd ed. corr. and enl. New York: C. S. Van Winkle, 1821.

Hufeland, Christoph Wilhelm. *Observations sur les fièvres nerveuses.* Berlin, 1807.

Huish, Robert. *The Public and Private Life of his Late . . . Majesty, George the Third: Embracing its Most Memorable Incidents . . . and Tending to Illustrate the Causes, Progress, and Effects, of the Principal Political Events of his Glorious Reign.* London: Thomas Kelly, 1821.

Hulshorff, M. "Discours sur les penchants, lu à l'Académie de Berlin." *La Gazette salutaire,* August 17, 1769.

Hunauld, Pierre. *Dissertation sur les vapeurs et les pertes de sang.* Paris: J.-Leloup, 1756; 2nd ed.

1771. Reprinted in Sabine Arnaud, ed., *La philosophie des vapeurs*. Paris: Mercure de France, 2009.

Huxham, John. *Essai sur les différentes espèces de fièvres*. Paris, 1746.

Important Facts and Opinions Relative to the King, Character of the Public Accounts at St James. London, 1789.

Jallabert, Jean. *Expériences sur l'electricité avec quelques conjectures sur la cause de ses effets*. Paris: Durand, Pissot, 1769.

James, Robert. *A Medicinal Dictionary, Including Physic, Surgery, Anatomy, Chymistry, and Botany, in all Their Branches Relative to Medicine*. 3 vols. London, 1743–45.

Jaquin, A.-P. *De la Santé*. Paris, 1762.

Joly, Henry. *Discours d'une estrange et cruelle maladie hypocondriaque venteuse, qui a duré onze ans: Accompagnée de l'hystérique passion, avec leurs noms, causes, signes, accidents terribles, & leurs remèdes*. Paris: Catherine Niverd, 1609.

Jonston, John. *Idea universae medicinae practicae*. Amsterdam: Ludovic Elzeviri, 1644.

Jorden, Edward. *A Briefe Discourse of a Disease Called the Suffocation of the Mother*. London: John Windet, 1603. Reprinted in Michael MacDonald, *Witchcraft and Hysteria in Elizabethan London: Edward Jorden and the Mary Glover Case* (London and New York: Tavistock/ Routledge, 1991.

Jouard, Gabriel. *Nouvel Essai sur la femme considérée comparativement à l'homme, principalement sous les rapports moral, physique, philosophique*. Paris: Imprimeur C. Hugelet, L'auteur, Ponthieu, libraire, Crohard, Libraire, 1804.

Kéralio, Marie Françoise Abeille de. *Les Visites*. Paris: Gattey, 1792.

Kinneir, David. *A New Essay on the Nerves and the Doctrine of the Animal Spirits Rationally Considered; Shewing the Great Benefit and the True Use of Bathing and Drinking the Bath Waters, in all Nervous Disorders, and Obstructions*. 2nd ed. enl. London: W. Innys and R. Mamby and Bath: J. Leake, 1739.

Kramer, Heinrich, and James Sprenger. *Malleus maleficarum*. Leiden: Mercurij Galli, 1487.

La Boissiere, Hervieu de. *Preservatif contre les faux principes et les maximes dangeureuses etablies par M de Montgeron pour justifier les seccours violents qu'on donne aux convulsionnaires*. Paris, 1787.

La Caze, Louis de. *Idée de l'homme physique et moral, pour servir d'introduction à un traité de médecine*. Paris: Guérin and Delatour, 1755.

La Chambre, Cureau Marin de. *L'Art de connoistre les hommes*. Amsterdam: Jacques Le Jeune, 1660.

———. *Les Charactères des passions*. Paris: Jacques d'Allin, 1662.

———. *Les Charactères des passions, ou il est traitté de la nature, des causes & des effects, des larmes, de la crainte, du desespoir*. Amsterdam: Antoine Michel, 1658.

Laclos, Choderlos de. *Dangerous Connections: A series of letters, selected from the correspondence of a private circle; and published for the instruction of society*. London: Printed for J. Ebers, 1812.

———. *Les Liaisons dangereuses*. Paris: Gallimard, 1989.

Lafage, Pierre. *Essai sur la nymphomanie; ou, Fureur utérine, présenté à l'Ecole de Montpellier et soutenu le 2 Floréal an VII*. Montpellier: Imprimerie de la Veuve de Jean Martel, an VIII.

La Framboisière, Nicolas Abraham de. *Œuvres*. Paris: Veusve Marc Orry, 1613.

Lallemant, Jean Baptiste Joseph. *Essai sur le mécanisme des passions en général*. Paris: Alexandre le Prieur, 1751.

La Mesnardière, Hippolyte Jules Pilet de. *Traitté de la mélancolie, sçavoir si elle est cause des effets que l'on remarque dans les possédées de Loudun.* La Flèche: M. Guyot and G. Laboe, 1635.

La Mettrie, Julien Offray de. *Histoire naturelle de l'ame, nouv. ed. revuë fort exactement, corrigé de quantité de fautes qui s'etoient glissées dans la premiere, & augmentée de la lettre critique de M. de la Mettrie à Madame la Marquise du Chattelet.* Oxford, 1747.

———. *Histoire naturelle de l'ame. Traduite de l'anglois par M. Charp par feu M. H*** de l'Académie des Sciences.* New ed. Oxford: Aux depens de l'auteur, 1747.

———. *Lettre critique de Mr. De La Mettrie sur l'histoire naturelle de l'ame.* N.p., 1747.

———. *L'homme Machine.* Leiden: E. Luzac fils, 1748.

———. *Man a Machine: French-English. Including Frederick the Great's 'Eulogy' on La Mettrie and Extracts from La Mettrie's 'The Natural History of the Soul.'* La Salle, Ill.: Open Court, 1912.

———. *Ouvrage de Pénélope; ou, Machiavel en médecine.* Paris: Payot, 2002.

———. *Ouvrage de Penelope; ou, Machiavel en medecine par Aletheius Demetrius.* 3 vols. Geneva: Her. de Cramer & Ph. Philibert, 1748, 1768.

———. *Traité du vertige: Avec la Description d'une catalepsie hystérique, & une lettre à Monsieur Astruc, dans laquelle on répond à critique qu'il a faite d'une dissertation de l'auteur sur les maladies vénériennes.* Paris: Prault, 1737.

La Molinière, Adrien Claude le Fort de. *Les Vapeurs.* Paris: Prault, 1753.

Lamothe, Bernard. "Dissertation sur l'hystéricie; ou, Vapeurs hystériques en général." Thèse de médecine, Montpellier, an X, 1802.

Lange, Martin. *Traité des vapeurs où leur origine, leurs effets, et leurs remèdes sont mécaniquement expliqués.* Paris: Veuve Denis Nion, 1689.

Lanthénas, François-Xavier. *De l'Influence de la liberté sur la santé, la morale et le bonheur.* Paris: Les Directeurs de l'Imprimerie du Cercle Social, 1792.

Lapon, J. B. *Philosophie médicale.* Paris, 1796.

La Quintine, Abbé de. *Dissertation sur le magnétisme des corps qui a remporté le pris proposé par l'Académie Royale de Belles-Lettres, sciences et arts pour l'année 1732.* Bordeaux: Pierre Brun, 1732.

La Roche, François de. *Analyse des fonctions du systême nerveux, pour servir d'introduction pratique à un examen des maux de nerfs.* 2 vols. Geneva: Du Villard Fils & Nouffer, 1778.

La Sarthe, Jacques L. Moreau de. *Histoire naturelle de la femme, suivie d'un traité d'Hygiène Appliquée à son régime physique et moral aux différentes époqes de la vie.* 3 vols. Paris: Duprat Lettelier et comp., 1803.

La Taste, Louis-Bernard de. *L'Esprit en convulsions; ou, Réponses aux lettres théologiques du R.P.P.D.B.M. du R.P. prieur des Blancs-Manteaux Dom la Taste.* N.p., 1734.

La Tour, Sousselier de. *L'Ami de la nature; ou, Manière de traiter les maladies par le prétendu magnétisme animal.* Dijon: J.-B. Capel, 1784.

La Tourette, Georges Gilles de. *Traité clinique et thérapeutique de l'hystérie d'après l'enseignement de la Salpêtrière.* Paris: E. Plon, Nourrit et Cie, 1891.

Laurent, P. *Introduction au magnétisme animal, suivie des principaux aphorismes du docteur Mesmer.* Paris, 1785.

Lautard, Jean-Baptiste. *La Maison des fous de Marseille.* Marseille: Imprimerie d'Achard, 1840.

Lavallée. "Première Observation sur l'hystéricisme et la cessation des règles." *Névropathie; ou, Mémoire sur les maladies nerveuses, proprement dites,* March 8, 1786.

Lavater, Johann Kaspar. *Essai sur la physiognomie destiné à faire connaître l'homme et le faire aimer*. 4 vols. The Hague: J. Van Karnebeek, 1781.

———. *Essays on Physiognomy for the Promotion of the Knowledge and the Love of Mankind*. Translated by Thomas Holcroft. 4 vols. London, 1804.

Lazare, Riviere. *Les Observations de medecine de Lazare Riviere*. Lyon: Jean Certe, 1688.

Lazerme, Jacobus. *Méthode pour guérir les maladies*. 2 vols. Paris: Veuve Cavelier. 1753.

Leake, John. *Practical Observations Towards the Prevention and Cure of Chronic Diseases Peculiar to Women*. 7th ed. London: Evans, Murray and Egerton, 1792.

Le Brun, L. S. D. *Traité théorique sur les maladies epidémiques*. Paris: Didot, 1776.

Le Brun, Réverend Père Pierre. *Histoire critique des pratiques superstitieuses qui ont séduit les peuples et embarassé les savants*. 2nd ed enl. Amsterdam: Jean-Frédéric Bernard, 1733.

Le Camus, Antoine. *Abdeker; ou, L'Art de conserver la beauté*. Edited by Alexandre Wenger. Grenoble: J. Millon, 2008. First published 1763.

———. *Médecine de l'esprit*. Paris: Ganeau, 1753.

Le Cat, Claude-Nicolas. *Œuvres physiologiques. Traité des sensations et des passions en général et des sens en particulier*. Paris: Vallat-la-Chapelle, 1767.

———. *Traité de l'existence, de la nature et des propriétés des fluides des nerfs, et principalement de son action dans le mouvement musculaire*. Berlin, 1753.

———. *Traité des sensations et des passions en général et des sens en particulier, Œuvres physiologiques*. Paris: Vallat-la-Chapelle, 1767.

Leclerc, C. *Traité des maladies morales qui ont affecté la nation française depuis plusieurs siècles*. Paris: Moutardier, 1794.

Le Clerc, Daniel. *Histoire de la médecine*. Amsterdam: George Gallet, 1702.

Leigh, John. *An Experimental Inquiry into the Properties of Opium and its Effects on Living Subjects*. Edinburgh: Elliot; London: Robinson, 1785.

Le Jeune, Robert. *Essai sur la mégalanthropogénèsie; ou, l'Art de faire des enfans d'esprit, qui deviennent des grands hommes*. Paris: Debray et Ant. Bailleul, 1801.

Le Journal des Sçavans (1786).

Le Loyer, Pierre. *Discours des spectres et des visions apparitions d'esprits*. Paris: Nicolas Buon, 1608.

———. *A Treatise of Specters or Straunge Sights, Visions and Apparitions Appearing Sensibly vnto Men: Newly Done out of French into English*. London: Val. Simmes, 1605.

Lepois, Charles. *Caroli Pisonis, selectionum observationum et consiliorum*. Paris: Boutestein and Langerak, 1714.

Le Pois, Nicolas. *De Cognoscendis et curandis, præcipue internis humani corporis morbis*. Lugduni Batavorum, 1736.

———. *Selectiorum observationum et consiliorum de praetervisis hactenus morbis affectibusque praeter naturam ab aqua, seu Serosa colluvie & diluvie ortis liber singularis*. Ponte ad Monticulum, 1618.

Les Convulsions du temps, attaquées dans leur principe et ruinees dans leur fondement par la consultation de 30 docteurs. Paris, 1737.

Les vieilles Lanternes, conte nouveau; ou, Allegorie faite pour ramener les uns et consoler les autres; Etrennes pour tout le monde: avec une clef pour rire et des notes pour pleurer. Paris: Lucrain, 1785.

Levacher de la Feutrie, A.-F. Thomas. *Dictionnaire de chirurgie*. Paris: Lacombe, 1767.

Lever, Maurice. *Anthologie érotique: Le XVIIIè siècle*. Paris: Robert Laffont, 2004.

————. *The Wicked Queen: The Origins of the Myth of Marie-Antoinette.* Translated by Julie Rose. New York: Zone Books, 1999.

Lewis, Matthew Gregory. *The Monk.* Orchard Park, N.Y.: Broadview, 1994.

Liébault, Jean. *Trois Livres appartenant aux infirmitez et maladies des femmes: Pris du Latin de M. Jean Liebaut Docteur Medecin à Paris, et faicts François.* Lyon: Jean Veyrat, 1598.

————. *Trois Livres des maladies et infirmitez des femmes.* Rouen: De l'Imprimerie de David du Petit val, 1651.

Lieutaud, Joseph. *Précis de la médecine pratique.* 3rd ed. Paris: Théophile Barrois, 1781.

Lignac, Louis François Luc de. *De l'Homme et de la femme considérés physiquement dans l'état du mariage.* Lille, 1773.

Linnaeus, Carl. *Linnaeus' Philosophia Botanica.* Translated by Stephen Freer. Oxford: Oxford University Press, 2003. First published 1788.

Lionet, Chalmers. *Journal de médecine* (Novembre 1759).

Lioult, P.-J. *Les Charlatans dévoilés.* Paris, 1800.

Louyer-Villermay, Jean-Baptiste. "Hypocondrie." In *Dictionnaire des sciences médicales, par une société de médecins et de chirurgiens.* Paris: Panckoucke, 1818.

————. "Hystérie." In *Dictionnaire des sciences médicales, par une société de médecins et de chirurgiens.* Paris: Panckoucke, 1818.

————. *Recherches historiques et médicales sur l'hypocondrie, isolée, par l'observation et l'analyse, de l'hystérie et de la mélancolie.* Paris: Méquignon, 1802.

————. *Traité des maladies nerveuses ou vapeurs et particulièrement de l'hystérie et de l'hypocondrie.* 2 vols. Paris, 1806, 1816.

Lysons, Daniel. *Practical Essays upon Intermitting Fevers, Dropsies, Diseases of the Liver, the Epilepsy, the Colic, Dysenteric Fluxes, and the Operation of Calomel.* Bath: S. Hazard, 1772.

M. J. F. *De la Passion de l'amour: De ses Causes, & des remedes qu'il faut y apporter, en la considérant comme une maladie.* Paris: Pichard, 1782.

Macquart, Louis Charles Henri. *Dictionnaire de la conservation de l'homme, ou d'hygiène et d'education physique et morale.* Vol. 1. Paris: Birault, 1798–99.

Makittrick Adair, James. *Essays on Fashionable Diseases.* N.p., 1786.

Mandeville, Bernard. *A Treatise of the Hypochondriack and Hysterick Passions, Vulgarly Call'd the Hypo in Men and Vapours in Women.* London, 1711.

Manningham, Richard. *The Symptoms, Nature, Causes and Cure of the Febricula; or, Little Fever, commonly called, The Nervous Hysteric Fever, the Fever of the Spirits; Vapours, Hypo and the Spleen.* London: Osborne, 1746.

Marat, Jean Paul. *Les aventures du comte Potowski.* Paris: Renaudot, 1989. First published 1847.

Marescot, M. *Discours véritable sur le faict de Marthe Brossier de Romorantin, pretendue démoniaque.* Paris: Mamert Patisson, 1599.

Maret, Hugues. *Encyclopédie méthodique.* Paris: Agasse, 1808.

————. *Mémoire dans lequel on cherche à déterminer quelle influence les mœurs des Français ont sur la santé.* Amiens: La Veuve Godard, Imprimeur du Roi et de l'Académie, 1772.

————. *Mémoire sur les moyens à employer pour rappeler à la vie les personnes que les vapeurs du charbon, ou quelques autres vapeurs méphitiques, ou le froid excessif, ou la submersion, ont réduites a l'état d'une mort apparente.* Dijon, 1776.

Marie de Saint-Ursin, P. J. *L'Ami des femmes; ou, Lettres d'un médecin concernant l'influence de l'habillement des femmes.* Paris: Barba, 1804.

Marivaux, Pierre de. *L'Ile aux esclaves.* Scène 3. Paris: Gallimard, 1993. First published 1725.

Marquet, François Nicolas. *Nouvelle Méthode facile pour connaître le pouls par les notes de musique*. 2nd ed. Amsterdam, 1769.

———. *Traité de l'apoplexie, paralysie, et autres affections soporeuses développées par l'expérience par feu M. Marquet*. Paris: J. P. Costard, 1770.

Marryat, Thomas. *The New Practices of Physick, Founded on Irrefragable Principles and Confirmed by Long and Painful Experience*. Dublin, 1764.

Matthey, André. *Nouvelles Recherches sur les maladies de l'esprit*. Paris: J. Paschou, 1816.

Mauduyt, Pierre-Jean-Claude. *Histoire de la société Royale de Médecine années 1777 et 1778 avec les mémoires de médecine & de physique médicale pour les mêmes années tirés des registres de cette société*. Paris: Philippe-Denys Pierres, 1780.

———. "Mémoire sur les différentes manières d'administrer l'électricité et observations sur les effets qu'elles ont produit." In *Extraits des mémoires de la société Royale de Médecine*. Paris, 1784.

Mauriceau, François. *Traité des maladies des femmes grosses et de celles qui sont accouchées*. Paris: Laurent d'Houry, 1694. First published 1668.

Mauriceau, François, and Hugh Chamberlen, trans. *The diseases of women with child, and in child-bed*. London: R. Clavel, W. Cooper, Benj. Billingsly, and W. Cadman, 1672.

Mead, Richard. "Of the Power and Influence of the Sun and Moon on Humane Bodies and of the Diseases that Rise from Thence." London: Richard Wellington, 1712.

Meilham, Senac de. *L'Emigré*. Paris: Gallimard, 1965. First published 1797.

Membres de la Société du Magnétisme. *Bibliothèque du magnétisme animal*. Paris: J.G. Dentu, 1817.

Mémoires de la Société médicale d'emulation, séante à l'Ecole de Médecine de Paris; Pour l'an V de la République (1797). Paris: Maradan, 1798, an VI.

Mercier, Louis-Sébastien. *Le nouveau Paris*. Paris: Mercure de France, 1994.

———. *Tableau de Paris*. Paris: Mercure de France, 1994.

Mesmer, Franz Anton. *Aphorismes de M. Mesmer, dictés à l'Assemblée de ses eleves, & dans lesquels on trouve ses principes, sa théorie, les moyens de magnétiser*. Edited by Caullet de Veaumorel. 2nd ed. Nice: Gabriel Flauteron, 1785.

———. *Les Miracles de Mesmer*. Paris, 1780.

———. *Mémoire sur la découverte du magnétisme animal*. Geneva and Paris: Didot le Jeune, 1779.

———. *Œuvres*. Paris: Payot, 1971.

———. *Précis Historique des faits relatifs au magnétisme animal*. London, 1781.

Michaud, L.-G. *Biographie universelle ancienne et moderne*. Vol. 8. Paris: Michaud Frères, 1813.

Misson, M. *Le Théâtre sacré des cévennes; ou, Récit des diverses merveilles opérées dans cette partie de la province du Languedoc*. London, 1707.

Monbron, Fougeret de. *Margot la ravaudeuse*. Cadeilhan: Calmann-Levy, 1992.

Montaigne. *Essais*. Edited by Pierre Villey. Paris: PUF, 1999. First published 1588.

Montenoy, Charles Palissot de. *Le Cercle ou Les Originaux*. Nancy, 1755.

More, Henry. *Enthusiasmus triumphatus*. Edited by M. V. De Porte. Los Angeles: University of California, William Andrews Clark Memorial Library, 1966. First published 1662.

Moreau De la Sarthe, Jacques Louis. *Histoire naturelle de la femme, suivie d'un traité d'hygiène apliquée à son régime physique et moral aux différentes époques de la vie*. Vol 1. Paris: Duprat Lettelier et comp., 1803.

Motherby, George. *A New medical Dictionary, or, General Repository of Physic*. London: J. Johnson, 1795.

Mourre, M. *Observations sur les insensés.* Toulon: Surre Fils, 1791.

Nicot, Jean. *Thresor de la langue francoyse, tant ancienne que moderne.* Paris: David Douceur, 1606.

Noel, Chomel, and J. Marret. *Dictionnaire oeconomique, contenant divers moyens d'augmenter son bien et de conserver sa santé.* Commercy: Henri Thomas, 1741.

Nollet, Jean Antoine Abbé. *Essai sur l'electricité des corps.* Paris: Frères Guérin, 1746.

Nynauld, Jean de. *De la Lycanthropie, transformation et extase des sorciers 1615.* Édition critique augmentée d'études, par l'équipe "Histoire critique de la sorcellerie." Edited by N. Jacques-Chaquin and M. Préaud. Paris: Frénésie éditions, coll. "Diaboliques, les Introuvables," 1990.

Nysten, Pierre-Hubert. *Dictionnaire de médecine et des sciences accessoires à la médecine, avec l'étmologie de chaque terme, suivi de deux volabulaires, l'un latin, l'autre grec.* Paris: J.-A Brosson, 1814.

Observations de medecine sur la maladie appelée convulsion par un médecin de la faculté de Paris. Paris: Saint Lambert, 1732.

Olivier, Jacques. *Alphabet de l'imperfection et malice des femmes.* Rouen: Robert Féon, 1630.

Panckoucke, Charles-Joseph, ed. *Encyclopédie méthodique.* Liège: Plomteux, 1782–1832.

Paracelse, Bombast Théophraste. *Operum medico-chimicorum.* Paris, 1603–5. Translated as *L'art d'alchimie, contenant un traité du mal caduc* (Paris: Presses littéraires de France, 1950).

Paré, Ambroise. "De la Génération." In *Œuvres,* vol. 9. Paris: Union Latine Editions, 1976. First published 1573.

———. *Les Œuvres d'Ambroise Paré, . . . divisées en 28 livres avec les figures et portraicts, tant de l'anatomie que des instruments de chirurgie, et de plusieurs monstres, reveues et augmentées.* 4th ed. Paris: Gabriel Buon, 1585.

Paré, Ambroise, and Thomas Johnson. *The Works of . . . Ambrose Parey: Translated out of Latine and Compared with the French.* London: Th. Cotes and R. Young, 1634.

Pargeter, William. *Observations on Maniacal Disorders.* London: Routledge, 1989. First published 1792.

Paulet, Jean-Jacques. *L'Antimagnétisme; ou, Origine, progrès, décadence, renouvellement et réfutation du magnétisme animal.* London, 1784.

Pechey, John. *The Whole Works of that Excellent Physician, Dr. Thomas Sydenham, 11th ed. Corrected from the Original Latin.* London: R. Ware and B. Wellington, 1740.

Perfect, William. *Methods of Cure in Some Particular Cases of Insanity: The Epilepsy Hypochondriacal Affection, Hysteric Passion, and Nervous Disorders, Prefixed with Some Accounts of Each of those Complaints.* Rochester: T. Fisher, 1780.

Perreau, J.-A. *Etudes de l'homme physique et moral, considéré dans ces différents ages.* Paris: Imprimerie des Annales d'Agriculture, 1797.

Perry, Charles. *A Mechanical Account and Explication of the Hysteric Passion, under all its Various Symptoms and Appearances and Likewise of all such other Diseases as Are Peculiarly Incident to the Sex . . . To Which Is Added, an Appendix. Being a Dissertation on Cancers in General; But more Especially such as Happen in the Breasts of Women.* London: printed for Mr. Shuckburgh, Mr. Osborn and Mess. Davis and Reymers, 1755.

Petetin, Jacques-Henri-Désiré. *Electricité animale, prouvée par la découverte des phénomènes physiques et moraux de la catalepsie hystérique, et de ses variétés, et par les bons effets de l'electricté artificielle dans le traitement de ces maladies.* Paris: Brunot-L'Abbe libraire et Gautier et Bretin; Lyon: Reyman, 1808.

———. *Mémoire sur la découverte des phénomènes qui présentent la catalepsie et le somnambu-*

lisme, symptômes de l'affection hystérique essentielle, avec des recherches sur la cause physique de ces phenomenes, 1787.

Petit, Marc-Antoine. *Discours sur la douleur*. Lyon: Reymann, an VII.

———. "Discours sur l'influence de la Révolution Française sur la santé publique prononcé le 30 septembre 1796." In *Essai sur la médecine du cœur; Auquel on a joint les principaux discours prononcés à l'ouverture des cours d'anatomie, d'opérations et de chirurgie clinique*. Lyon: Ballanche père et Fils, 1806.

———. *Traité des maladies des femmes enceintes, des femmes en couche et des enfans nouveaux nés, précédé du mécanisme des accouchemens, médecin de Paris, démonstrateur et professeur au jardin des plantes et membres de plusieurs académies, par les citoyens Baignères et Perral*. Paris: Baudoin, an VII.

Pinel, Philippe. *La Médecine clinique rendue plus précise et plus exacte par l'application de l'analyse ou recueil et résultat d'observations sur les maladies aigues faites à la Salpêtrière*. Paris: chez Brosson, Gabon, 1802.

———. *Nosographie philosophique*. 6th ed. 3 vols. Paris, 1816.

———. *Nosographie philosophique*. 6th ed. Paris, 1818.

———. *Nosographie philosophique; ou, La Méthode de l'analyse appliquée à la médecine*. Paris: Richard, Caille et Ravier, an VII (1798).

———. *Nosographie philosophique ou la méthode de l'analyse appliquée à la médecine, 3e éd. revue, corrigée et augmentée*. Paris: Brosson, 1807.

———. *Nosographie philosophique ou méthode de l'analyse appliquée à la médecine, seconde édition, très augmentée, et dans laquelle sont insérés les caractères spécifiques des maladies*. Paris: Brosson, an XI, 1803.

———. *Sur l'Aliénation mentale*. Paris: Théraplix, 1970. First published 1802.

———. *Traité médico-philosophique sur l'aliénation mentale, ou la manie*. Edited by François Azouzi. Geneva and Paris: Slatkine, 1980. First published 1800.

———. *Traité médico-philosophique sur l'aliénation mentale; ou, La Manie*. Paris: Richard, Caille et Ravier, 1801.

Piorry, P. A. *Tableau indiquant la manière d'examiner et d'interroger le malade*. Paris, 1832.

Planque, François. *Bibliothèque choisie de médecine, tirée des ouvrages périodiques, tant françois qu'étrangers, avec plusieurs pièces rares, & remarques utiles & curieuses. Tome troisième, avec figures*. Paris: d'Houry père, imprimeur-libraire, 1750.

Plato. *Ion*. Translated by E. Chambry. Paris: Garnier Flammarion, 1967.

———. *La République*. Translated by Pierre Pachet. Paris: Gallimard, 1993.

———. *Statesman, Philebus, Ion*. Translated by Harold North Fowler and W. R. M. Lamb. Cambridge, Mass.: Harvard University Press, 1925.

———. *Timaeus*. Translated by A. Rivaud. Paris: Gallimard, 1950.

Pomme, Pierre. *Abhandlung von den hysterisch und hypochondrischen Nerven-Krankheiten beider Geschlechter; oder, Von den vapeurs*. Breslau, 1775.

———. *Essai sur les affections vaporeuses des deux sexes*. Paris: Desaint and Saillant, 1760.

———. *Nouveau recueil des pièces publiées pour l'instruction du procès que le traitement des vapeurs a fait naître parmi les médecins*. Paris: Imprimerie J. Th. Hérissant Père, 1771.

———. *Nuevo Método para curar flatos, hypocondria, vapores y ataques hystericos de las mujeres de todos estados: Y en todo estado, con el qual los enfermos podrán por sí cuidar de su salud en falta de medico que les dirija*. Madrid: Miguel Escribano, 1776; Alfonso Lopez, 1786; Plácido Barco Lopez, 1794.

―――. *Réfutation de la doctrine médicale du docteur Brown, médecin écossais, suivie d'une notice sur l'électricité, le galvanisme et le magnétisme, sous le rapport des maladies nerveuses*, 3rd ed. corr. and enl. Arles: Imprimerie G. Mesnier, 1808.

―――. *Réponse de M. Pomme à Bret, Louis Bret, analyse comparée de quelques observations citées par M. Pomme.* Arles: Gespard Mesnier, 1806.

―――. *Sopra le Affezioni vaporese, Opera del Signore Pomme, figlio.* Naples: Raimondi Saggio, 1765.

―――. *Supplément au traité des affections vaporeuses des deux sexes, ou maladies nerveuses, dans lequel on trouve, 1°. une Nouvelle Edition, considérablement augmentée, du Mémoire et des observations cliniques sur l'abus du quinquina; 2°. la Réfutation de la doctrine médicale de Brown; 3°. Une Notice sur l'électricité, le galvanisme et le magnétisme.* Paris: Cussac, l'an XII, 1804.

―――. *Traité des affections vaporeuses des deux sexes: où l'on a tâché de joindre à une théorie solide une pratique sûre, fondée sur des observations.* Lyon: B. Duplain, 1763, 1765, 1767.

―――. *Traité des affections vaporeuses des deux sexes, ou maladies nerveuses, vulgairement appelées maux de nerfs.* Paris, 1760; Lyon 1763, 1765, 1767, 1769, 1771; Paris, 1782, 1798–99, 1804.

―――. *A Treatise on Hysterical and Hypochondriacal Diseases: In Which a New and Rational Theory Is Proposed, and a Cure Recommended.* London: P. Elmsky, 1777.

Pope, Alexander. *The Rape of the Lock.* Cambridge: Routledge English Texts, 1971. First published 1714.

Porée, Charles Gabriel. *Examen de la prétendue possession des filles de la paroisse de lands.* Antioche: Chez les héritiers de la Bonne foy, a la Vérité, 1738.

Portal, Antoine. *Précis des journaux tenus pour les malades qui ont été électrisés pendant l'année 1785, & des mémoires sur le même objet, adressés à la société Royale de Médecine pendant la même année.* Paris: de l'imprimerie Nationale, 1786.

Pressavin, Jean-Baptiste. *Les vrais Principes des vapeurs, ou Nouveau traité des maladies des nerfs, dans lequel on développe les vrais principes des vapeurs, & les moyens de guérir les maladies nerveuses.* Paris and Lyon: J. P. Costard, 1770.

―――. *Nouveau Traité des vapeurs; ou, Traité des maladies des nerfs dans lequel on développe les vrais principes des vapeurs.* Lyon: Veuve Réguilliat, Libraire Place Louis Le Grand, 1770.

Priscian. *On Theophrastus on Sense-Perception.* Translated by Pamela Huby. Ithaca, N.Y.: Cornell University Press, 1997.

Prost, Pierre-Antoine. *Essai physiologique sur la sensibilité.* Paris, 1805.

Purcell, John. *A Treatise of Vapours or Hysterick fits Containing an Analytical Proof of its Causes, Mechanical explanations of all its Symptoms and Accidents According to the Newest and Most Rational Principles: together with its Cure at Large.* London: Nicholas Cox, 1702.

Puységur, Armand Marie-Jacques de Chastenet. *Continuation du Journal du traitement magnétique du jeune Hébert.* Paris: J. G. Dentu, 1812.

―――. *Détail des cures opérées à Buzancy près de Soissons par le magnétisme animal.* Soisson, 1784.

―――. *Du Magnétisme animal, considéré dans ses rapports avec diverse branches de la physique générale.* Paris: Desenne, 1807.

―――. *Les Fous, les insensés, les maniaques et les frénétiques ne seraient-ils que des somnambules désordonnés.* Paris: J. G. Dentu, 1812.

―――. *Les Vérités cheminent tot ou tard elles arrivent.* Paris: J. G. Dentu, 1815.

―――. *Lettre dressée a Monsieur le Marquis de Puiségur, sur une observation faite à la Lune, précédée d'un système nouveau sur le méchanisme de la vue.* Amsterdam, 1787.

———. *Marques des mémoires pour servir à l'Histoire et à l'établissement du magnétisme animal.* Toulouse: Privat, 1986. First published 1786.

———. *Memoires pour servir à l'établissement du magnétisme animal.* London, 1784.

———. *Recherches expériences et observations physiologiques sur l'homme dans l'état de somnambulisme naturel provoqué par l'acte magnétique.* Paris: Dentu, 1811.

———. *Suite des memoires pour servir à l'établissement du magnétisme animal.* London, 1785.

Quaint, Richard. *A Dictionary of Medicine, Including a General Pathology, General Therapeutics, Hygiene, and the Diseases of Women and Children.* New ed. rev. New York: Appleton, 1883.

Quesnay, François. *Traité des effets et de l'usage de la saignée.* Paris: Veuve d'Houry, 1770.

Rabelais, François M. *Le tiers Livre des faictz et dictz du noble Pantagruel.* Paris: Livre de Poche, 1995. First published 1546.

Radcliffe, Ann. *The Italian.* New York: Oxford University Press, 1998. First published 1797.

———. *The Mysteries of Udolpho.* New York: Oxford University Press, 1998. First published 1794.

———. *The Romance of the Forest.* New York: Oxford University Press, 1999. First published 1791.

———. *A Sicilian Romance.* New York: Oxford University Press, 1993. First published 1790.

Radet, Jean-Baptiste. *Les Docteurs modernes, comédie parade suivie du banquet de santé.* Paris: Brunet, 1784.

Rapport des Commissaires chargés par le Roi de l'examen du magnétisme animal, imprimé par l'ordre du Roi. Nice: Gabriel Floteron, 1785.

Rast, Jean-Baptiste-Antoine. *Correspondance inédite de Albert de Haller, Barthiez. Tronchin, Tissot, avec le Dr Rast.* Lyon: A. Vingtrinier, 1856.

Raulin, Joseph. *Des Maladies occasionnées par les promptes et fréquentes variations de l'air, considére comme atmosphère terrestre.* Paris: Huart & Moreau, 1752.

———. "Lettre contenant des observations sur le Taenia autrement ver plat." In *Des Maladies occasionnées par les promptes et fréquentes variations de l'air considére comme atmosphère terrestre.* Paris: Huart & Moreau, 1752.

———. *Observations de médecine.* Paris: Moreau & Delaguete, 1754.

———. *Traité des affections vaporeuses du sexe, avec l'exposition de leurs symptômes . . . & la méthode de les guérir.* Paris: J.-T. Hérissant, 1758; 2nd ed, 1759.

———. *Traité des fleurs blanches avec la méthode de les guérir.* Paris: Hérissant, 1766.

Rausky, Franklin. *Mesmer; ou, La Révolution thérapeutique.* Paris: Payot, 1977.

Rayer, Pierre François Olive. *Mémoire sur le delirium tremens.* Paris: Baillière, 1819.

Raymond, Dominique. *Traité des maladies qu'il est dangereux de guérir.* Aug. ed. Paris: Léopold Colin, 1808.

"Recherches historiques sur le magnétisme animal, principalement dans l'ancienne Italie, sous les Empereurs, et dans les Gaules." In *Bibliothèque du magnétisme animal, MM. les Membres de la Société du Magnétisme,* vol. 2. Paris: J. G. Dentu, 1817.

Recueil général des pièces concernant le procez entre la Demoiselle Cadière, de la ville de Toulon, et le Père Girard, Jésuite, Recteur du Séminaire Royale de la Marine de ladite ville. Vol. 1. N.p., 1732.

Reflexions d'un eleve de M. Maris, professeur en affections vaporeuses, auteur & maître chirurgien-juré de Lyon. Avignon: Aux dépens de la compagnie, 1763.

Report from the Committee Appointed to Examine the Physicians who Have Attended his Majesty during his Illness, Touching the Present State of his Majesty's Health. London: Printed from the Cop. Published by authority, 1789.

Rétif De la Bretonne, Nicolas-Edme. *Les Nuits de Paris*. Edited by Jean Varlot and Michel Delon. Paris: Gallimard, 1987. First published 1788.

Retz, Noël. *Des Maladies de la peau et de celles de l'esprit, telles que les vapeurs, la mélancolie, la manie, & qui procèdent des affections du foie, leur origine, la description de celles qui sont le moins connues, les traitements qui leur conviennent*. Paris: Méquignon, 1786; 3rd ed. rev. 1790, 1810.

Révillon, Charles. *Recherches sur les causes des affections hypochondriaques, appelées communément vapeurs*. Paris, 1779; 2nd ed., Hérissant, 1786.

Richardson, Samuel. *Clarissa*. New York: Penguin, 1985. First published 1748.

———. *Pamela*. New York: Penguin, 1980. First published 1740.

Riviere, Lazare. *Les Observations de medecine, qui contient 4 centuries de guérisons très-remarquables*. 2nd ed. rev. Lyon: Jean Certe, 1688.

Robin, Abbé. *Du Traitement des insensés dans l'hôpital de Bethléem de Londres traduit de l'anglois; suivi d'observations sur les insensés de Bicêtre et de la Salpêtrière*. Amsterdam, 1787.

Robinson, Nicolas. *A New System of the Spleen and Vapours and Hypochondriack Melancholly*. London: Rivington, 1729.

Rochefort, Desbois de. *Cours elémentaire de matière médicale suivi d'un précis de l'art de formuler*. 2 vols. Avignon: Seguin, an XI.

Roger, Joseph-Louis. *Traité des effets de la musique sur le corps humain*. Translated by Étienne Sainte-Marie. Paris: Brunit, 1803.

Rogers, Timothy. *A Discourse Concerning Trouble of Mind and the Disease of Melancholy*. London: Thomas Pakhurst and Thomas Cockerill, 1691.

Rostaing. *Réflexions sur les affections vaporeuses; ou, Examen du traité des vapeurs des deux sexes, 3e éd. publiée en 1767, par Monsieur P****. Amsterdam, 1768; Paris, 1767; 2nd ed. Vincent 1768; 3rd ed. 1778.

Roubaud-Luce, Maurice. *Recherches médico-philosophiques sur la mélancolie*. Paris: Le Normand, 1817.Roucher-Deratte, Claude. *Leçons sur l'art d'observer*. Montpellier: Fontenay-Picot, 1807.

Rousseau, Jean-Baptiste. *L'Hypocondre; ou, La Femme qui ne parle point*. Amsterdam: Marc Michel Rey, 1751.

Rousseau, Jean-Jacques. *Emile*. Middlesex: Echo Library, 2007.

———. "Emile; ou, De l'Education." In *Œuvres complètes de J. J. Rousseau: La nouvelle Héloise, Emile, Lettre à M. de Beaumont*, vol. 2. Paris: Furne, 1835.

———. *Les Rêveries du promeneur solitaire*. Paris: Garnier, 1997. Translated by Russel Goulbourne as *Reveries of the Solitary Walker* (Oxford: Oxford University Press, 2011). First published 1782.

Roussel, Pierre. *Système physique et moral de la femme; ou, tableau philosophique de la Constitution, de l'état organique du tempérament, des mœurs et des fonctions propres au sexe*. Paris: Crapart, Caille et Ravier Libraires, 1803; Paris: Victor Masson, 1869.

Rowley, William. *The Rational Practice of Physic of William Rowley*. London: Newbery, 1793.

———. *A Treatise on Female, Nervous, Hysterical, Hypochondriacal, Bilious, Convulsive Diseases, Apoplexy and Palsy, with Thoughts on Madness, Suicide, &c. in Which Principal Disorders are Explained from Anatomical Facts, and the Treatment Formed on Several New Principles*. London: Printed for Nourse, 1788.

Rush, Benjamin. "An Account of the Influence of the Military and Political Events of the American Revolution upon the Human Body." In *Medical Inquiries and Observations*. 2nd ed. Philadelphia: printed; London: reprinted for C. Dilly in the Poultry, 1779.

————. *An Inquiry into the Influence of Physical Causes upon the Mental Faculty*. Philadelphia, 1786.

————. "An Inquiry into the Natural History of Medicine Among the Indians of North America." *Medical Inquiries and Observations* 1 (1774): 9–77.

————. *Inquiry into the Natural History of Medicine among the Indians of North-America and a Comparative View of their Diseases and Remedies with those of the Civilized Nations Read Before the American Philosophical Society, held at Philadelphia, on the 4th of February, 1774*, 36–38.

————. *Medical Inquiries and Observations* 2 (1797).

Sacombe, Jean-François. *La Luciniade du docteur Sacombe*. Nismes, 1815.

Saiffert, Andreas. *Beiträge zur übschäftlichen Artzneilehre der Suchen oder sogenannten langwierigen Krankheiten*. Paris: Druckerei der deutschen Sprachfreunde; Braunschweig and Leipzig: Friedrich Viewig, 1804.

Salverte, Eusèbe. *Des Rapports de la médecine avec la politique*. Paris, 1806.

Sans, Abbé. *Guérison de la paralysie par l'électricité, dans lequel on expose la méthode qu'il faut suivre pour guérir la paralysie par l'électricité*. Paris: Cailleau, 1778.

Santorio, Santorio. *Ars Sanctorii Sanctorii de statica medicina: Sectionibus octo comprehensa*. Leipzig, 1670.

Saurin, Bernard-Joseph. *Les Mœurs du temps*. Paris: Belin, 1788.

Sauvages, François Boissier de la Croix. *Dissertation sur les médicamens qui affectent certaines parties du corps humain plutôt que d'autres & sur les causes de cet effet*. Bordeaux: Pierre Brun, 1752.

————. *Nosologia methodica: Sistens morborum classes, genera et species juxta Sydenhami mentem et botanicorum ordinem*. Amsterdam: Tournes, 1763.

————. *Nosologie méthodique, dans laquelle les maladies sont rangées par classes, suivant le systeme de Sydenham, et l'ordre des botanistes*. Translated by Nicolas. 10 vols. Paris: Hérissant, 1771.

————. *Nosologie méthodique dans laquelle les maladies sont rangées par classe suivant le système de Sydenham et l'ordre des botanistes, ouvrage augmenté par quelques notes en forme de commentaire par M. Nicolas*. Paris: Hérissant, 1772.

————. *Nouvelles classes de maladies, qui dans un ordre semblable à celui des botanists, comprennent les genres & les espèces de toutes les maladies, avec leurs signes & leurs indications*. Avignon: B. D'Avanville, 1731.

Sénébier, Jean. *Traité sur l'art d'observer et de faire des expériences*. 3 vols. Geneva: Cl. Philibert & Bart Chirol, 1802.

Sennert, Daniel. *Institutiones medicinae*. Wittebergae: Schürer, 1611.

Servan, Joseph Michel Antoine. *Doutes d'un provincial, proposés à messieurs les médecins-commissaires*. Lyon, 1784.

Sévigné, Madame de. *Correspondance 1680–1696*. Vol. 3. Paris: Gallimard, 1978.

Sèze, Paul Victor de. *Recherches physiologiques et philosophiques sur la sensibilité ou la vie animale*. Paris: Prault, 1786.

Sibly, Ebenezer. *The Medical Mirror, or, A Treatise on Impregnation Showing the Origin of Diseases and the Principles of Life and Death*. London: printed for the proprietor, 1814.

Smith, William. *A Dissertation upon the Nerves*. London, 1768.

Smollett, Tobias. *The Adventures of Peregrine Pickle*. London: Oxford University Press, 1983.

————. *The Adventures of Roderick Random*. New York: Oxford University Press, 1999.

————. *The Critical Review, or, Annals of Literature by A Society of Gentlemen,* Volume the fifty-first. London: A. Hamilton, 1781.

————. *The Expedition of Humphry Clinker.* New York: The Century Co., 1902.

Société de Médecins et de Chirurgiens. *Dictionnaire des sciences médicales.* Paris: Panckoucke, 1824.

Société des Gens de Lettres. *L'Esprit des journaux Francois et étrangers.* Paris: Valade, 1786.

————. *Mercure de France,* 1786.

Société Royale de Médecine. *Projet d'instruction sur une maladie convulsive: fréquente dans les colonies de l'Amérique, connune sous le nom de tétanos.* Paris: de l'Imprimerie Royale, 1786.

Soranus of Ephesus. *Maladies des femmes.* 4 vols. Edited and translated by Paul Burguière, Danielle Gourevitch, and Yves Malinas. Paris: Les Belles Lettres, 1988.

Spatantigarude: vieux conte nouveau. London and Paris, 1785.

Spurzheim, J. G. *Observations on the Deranged Manifestations of the Mind; or, Insanity.* London: Baldwin Cradock and Joy, 1817.

Staël, Madame de. *Corinne, ou, l'Italie.* Paris: Gallimard, 1985.

Sterne, Laurence. *A Sentimental Journey.* New York: Penguin, 1986. First published 1768.

————. *Tristram Shandy.* York, 1759. New York: Modern Library, 1995.

Stukeley, William. *On the Spleen: its Description and History, Uses and Diseases, Particularly Vapors, with their Remedy; Being a Lecture Read at the Royal College of Physicians.* London, 1722.

Sue, Pierre. *Anecdotes historiques, littéraires et critiques sur la médecine, la chirurgie, & la pharmacie.* Brussels: Veuve Dujardin, 1789.

————. *Apperçu général, appuyé de quelques faits sur l'origine et le sujet de la médecine légale.* Paris: de l'Imprimerie de la Société de Médecine, 1799.

Sulaville, Jean Baptiste. *Le Moraliste mesmérien; ou, Lettres philosophiques sur l'influence du magnétisme.* London, 1784.

Sydenham, Thomas. *Abrégé de toute la médecine pratique, avec quelques augmentations dans la 2ième édition de cet ouvrage.* Paris: Huart l'aîné, 1728.

————. *Celeberrimi Opera omnia medica.* New ed. Geneva: Fractes de Tournes, 1696.

————. "An Epistle from Dr Tho. Sydenham, to Dr. Mr WM Cole; Treating of the Small Pox and Hysteric Disease." In *The Entire Works of Dr. Sydenham,* translated by John Swann. 1682. London: Edward Cave, 1742.

————. *Médecine pratique.* Translated by Augustin-François Jault. Paris: Didot le Jeune, 1774.

————. *Médecine pratique avec des notes, ouvrage traduit . . . sur la dernière édition anglaise, par feu M. A.-F. Jault, . . . Cette édition est augmentée d'une Notice sur la vie et les écrits de Sydenham, par M. Prunelle.* Montpellier: Vve Picot, 1816.

Sydenham, Thomas, and R. G. Latham. *The Works of Thomas Sydenham.* 2 vols. The Sydenham Society. London: Printed for the Sydenham Society, 1848.

Tagereau, Vincent. *Discours sur l'impuissance de l'homme et de la femme.* Paris, 1612.

Tallavignes, Jean-Antoine. *Dissertation sur la médecine où l'on prouve que l'homme civilisé est plus sujet aux maladies graves.* Carcassonne: B.-V. Gardel-Teissié, 1821.

Taxil, Jean. *Traicté de l'épilepsie, maladie vulgairement appellée au pays de Provence, la gouttete aux petits enfants.* Tournon: Claude Michel, Imprimeur de l'Université, 1602.

Thouret, Michel Augustin. *Recherches et doutes sur le magnétisme animal.* Paris: Prault, 1784.

Tissot, Samuel-Auguste André David. *Advice to the People in General, with Regard to Their Health: But more Particularly Calculated for Those, who, by Their Distance from Regular Phy-*

sicians, or Other Very Experienced Practitioners, Are the Most Unlikely to be Seasonably Provided with the Best Advice and Assistance . . . Translated from . . . Dr. Tissot's Avis au Peuple . . . by J. London: T. Becket and P. A. de Hondt, 1765.

———. *Avis au peuple sur sa santé.* Lausanne: F. Grasset, 1761.

———. *Avis au peuple sur sa santé; ou, Traité des maladies les plus fréquentes, seconde éd., augmentée sur la dernière de l'auteur, de la description et de la cure de plusieurs maladies, & principalement de celles qui demandent de prompts secours, ouvrage composé en faveur des habitans de la campagne, du peuple des viles & de tous ceux qui ne peuvent avoir facilement les conseils des médecins.* Paris: Didot le Jeune, 1763.

———. *De la Santé des gens de lettres.* Paris: Éditions la Différence, 1775.

———. *De l'Influence des passions de l'ame dans les maladies et des moyens d'en corriger les mauvais effets.* Besançon: Imprimerie de Briot, 1795.

———. Lettre a Monsieur Zimmerman, Dc. Med. des Academies de Basle, de Palerme, de la Société Oecon. de Berne &c. &c. &c. Sur l'épidemie courante par M. Tissot, D.M. de la Soc. Royale de Londres, de l'Acad. de Basle, de la Soc. Oec. De Berne.

———. *L'Onanisme.* Paris: Éd. La Différence, 1991. First published 1760.

———. *Observations sur la santé des gens du monde.* Lausanne, 1770.

———. *Traité de l'épilepsie.* Edited by Kazimierz Karbowski. Lausanne: Fondation Eben-Hezer, 1984. First published 1770.

———. *Traité des nerfs et de leurs maladies.* Paris, 1778–80.

Trembley, Abraham. *Mémoire pour servir à l'histoire d'un genre de polipes d'eau douce à bras en forme de cornes.* Leiden: J & H Verbeek, 1744.

Trotter, Thomas. *A View of the Nervous Tempérament Being a Practical Inquiry into the Increasing Prevalence, Prevention and Treatment of Those Diseases, Commonly Called Nervous, Bilious, Stomach and Liver Complaints; Indigestion, Low Spirits; Gout.* London, 1807.

Urban, Sylvanus. *The Gentleman's Magazine* 61, no. 1 (1787).

Vallemont, Abbé de. *La Physique occulte; ou, Traité de la baguette divinatoire.* The Hague: Adrien Montgens, 1722.

Vandermonde, Charles-Augustin. "Colique de Poitou." *Journal de Médecine, Chirurgie, Pharmacie, etc.* 8 (1758).

———. *Dictionnaire portatif de santé dans lequel tout le monde peut prendre une connoissance suffisante de toutes les maladies . . . instructions nécessaires pour être soi-même son propre médecin.* 2 vols. Paris: Vincent, 1770.

van Helmont, Jan Baptiste. *De Magnetica vulnerum curatione.* Paris: V. Le Roy, 1621.

———. *Oriatrike; or, Physik Refined, The Common Errors Therein Refuted and the Whole Art Reformed and Rectified Being a New Rise and Progress of Phylosophy and Medicine, for the Destruction of Diseases and Prolongation of Life.* London: Lodowick Loyd, 1662.

van Swieten, Gerard. *Commentaires des aphorismes de médecine d'Herman Boerhave sur la connaissance et la cure des maladies.* Translated by Moublet. 2 vols. Avignon: Roberty & Guilermont, 1766.

———. *The Commentaries upon the Aphorisms of Herman Boerhaave.* 2nd ed. London: Horsfield & Longman, 1765.

———. *His Life & Original Writings.* Berkeley: University of California Press, 1966.

van Swinden, Jan Henri. *Analogie de l'électricité et du magnetism; ou, Recueil de mémoires, couronnés par l'Académie de Bavière avec des notes & des dissertations nouvelles.* 2 vols. The Hague: Veuve Duchesne, 1785.

Varandée, Jean. *De Morbis & affectibus mulierum.* Leiden: sumptibus Bartholomiei Vicenti, 1619.

Translated, revised, and extended by I. Bonamour as *Traité des maladies des femmes* (Paris: Robert de Ninville, 1666).

———. *Traité des maladies des femmes.* Édition augmentée. Paris: Robert de Ninville, 1666.

Venel, Jean-André. *Essai sur la santé et l'éducation médecinale des filles destinées au mariage.* Yverdon, 1776.

Venette, Nicolas. *La Génération de l'homme; ou, Le Tableau de l'amour conjugal, divisé en quatre parties.* Cologne: Claude Joly, 1721. First published 1687.

———. *Tableau de l'amour humain considéré dans l'état du mariage.* Amsterdam, 1686.

Vernet, Jules. *La Magnétismomanie.* Paris, 1816.

Vesalius, Andreas. *La Fabrique du corps humain.* Edited by Louis Bakelants. Arles: Actes Sud, 1987.

Vicq d'Azyr, Félix, ed. *Encyclopédie méthodique, médecine, par une société de médecins.* Paris, 1819–30.

———, ed. *Remarques sur la médecine agissante.* Paris, 1786.

———. *Traité d'anatomie et de physiologie.* Paris: Didot, 1786.

Vieussens, Raymond de. *Histoire des maladies internes suivie de la névrographie.* N.p., 1674.

———. *Réponse du Sieur Vieussens, docteur en médecine, de la faculté de Montpellier à trois lettres imprimées du Sieur Chirac, professeur en médecine de l'Université de la même ville.* Montpellier: Honoré Pech, 1698.

———. *Traité nouveau des liqueurs du corps humain.* Toulouse, 1715.

Vigarous, Joseph Marie Joachim. *Cours élémentaire de maladies des femmes; ou, Essai sur une nouvelle méthode pour étudier et pour classer les maladies de ce sexe.* Paris: De l'Imprimerie de Crapelet, Déterville libraire, 1801.

Villemert, Boudier de. *L'ami des femmes; ou, Morale du sexe.* Paris: Royez, 1788.

Virey, Julien-Joseph. *De la Femme sous ses rapports physiologique, moral et littéraire.* Paris, 1823.

———. *De l'Influence des femmes sur le goût de la littérature et des beaux-arts.* Paris, 1810.

———. *Histoire naturelle du genre humain.* New ed. Paris: Crochard, 1824, 1825.

Viridet, Jean. *Dissertation sur les vapeurs.* Yverdon: J. Jacques Guenath, 1726.

Vitet, Louis. *La Médecine expectante.* Paris, 1806.

Voltaire. *Correspondance.* Vol. 10. Paris: Gallimard, 1986.

———. *Correspondance avec les Tronchin.* Edited by André Delattre. Paris: Mercure de France, 1950.

———. *Questions sur l'Encyclopédie, dictionnaire philosophique.* In *Œuvres complètes*, ed. Beuchot, vol. 18. Paris: Garnier Frères, Libraires Editeurs, 1878.

Von Haller, Albrecht. *Élémens de physiologie; ou, Traité de la structure et des usages des differentes parties du corps humain.* Paris: Prault, 1752.

V. Y. E. G. *Le Voyage du diable et de la folie, ou Causes des révolutions de France, Braband, Liége et autres.* Imprimé dans la lune, 1793.

Weiner, Dora B. *The Clinical Training of Doctors: An Essay of 1793.* Baltimore: Johns Hopkins University Press, 1980.

Wesley, John. *The Desideratum: or, Electricity Made Plain and Useful. By a Lover of Mankind, and of Common Sense, 1760, Hunter and Macalpine Three Hundred Years of Psychiatry, 1535–1860.* London, 1963.

Weyer, Johann. *De Ira morbo, eiusdem curatione philosophica, media & theologica, liber.* Basel, 1577.

———. *De Lamiis liber.* Basel, 1577.

———. *De l'Imposture des diables.* Paris: Théraplix, 1970. First published 1570.

————. De l'Imposture des diables, enchanteurs et sorciers. Translated by Guérin. Paris, 1557.

————. De Praestigiis daemonum, et incantationibus ac veneficiis, libri V. Basel, 1563.

————. Histoire, disputes et discours des illusions et impostures des diables, des magiciens infâmes, sorcières et emprisonneurs; des ensorcelez et démoniques et de la guérison d'iceux: item de la punition que méritent les magiciens, les empoisonneurs et les sorciers; Le tout compris en six livres (augmentez de moitié en ceste dernière édition. Deux dialogues touchant le pouvoir des sorcières et la punition qu'elles méritent par Thomas Erastus. 2 vols. Geneva: Jacques Chovet, 1579.

————. Medicarum observationum rararum liber. Basel, 1567.

————. Witches, Devils, and Doctors in the Renaissance; Johann Weyer, De praestigiis daemonum. Edited by George Mora and translated by John Shea. Binghamton, N.Y.: Medieval and Renaissance Texts and Studies, 1991.

Whiter, Walter. A Dissertation on the Disorder of Death. London, 1819.

Whytt, Robert. Les Vapeurs; ou, Maladies nerveuses et hypochondriaques ou hystériques; reconnus & traitées dans les deux sexes, nouvelle édition, à laquelle on a joint un extrait d'un ouvrage anglois du même auteur, sur les mouvements vitaux & involontaires des animaux, servant d'introduction à celui-ci. Translated by Alexander Monro. Vol. 1. Paris: P. Fr. Didot jeune, 1777.

————. Observations on the Nature, Causes, and Cure of those Disorders Which Have Been Commonly Called Nervous, Hypochondriac, or Hysteric: To Which Are Prefixed Some Remarks on the Sympathy of the Nerves. Edinburgh: Becket, and De Hondt, 1767.

Willis, Thomas. "An Essay of the Pathology of the Brain and Nervous Stock in which Convulsive Diseases are Treated of." Translated out of Latine into English by S. P. In The Remaining Medical Works of that Famous and Renowned Physician Dr Thomas Willis of Christ Church in Oxford, and Sidley, Professor of Natural Philosophy in that Famous University. London: T. Dring, C. Harper, J. Leigh and S. Martyn, 1681, 1684.

Wilson, Andrew. Medical Researches Being an An Enquiry into the Nature and Origin of Hysterics in the Female Constitution, and into the Distinction between that Disease and Hypochondriac or Nervous Disorder. London: S. Hooper, and G. Robinson, 1777.

Wollstonecraft, Mary. Maria. New York: Penguin, 1992. First published 1798.

Young, George. A Treatise on Opium Founded on Practical Observation. London: A. Millar, 1753.

Secondary Sources

Ackerknecht, Erwin. La médecine hospitalière à Paris (1794–1848). Paris: Payot, 1986.

Albala, Ken. Eating Right in the Renaissance. California Series in Food and Culture. Berkeley: University of California Press, 2002.

Anderson, Wilda C. Between the Library and the Laboratory: The Language of Chemistry in Eighteenth-Century France. Baltimore: Johns Hopkins University Press, 1984.

Andrews, Jonathan, and Andrew Scull. Undertaker of the Mind: John Monro and Mad-Doctoring in Eighteenth-Century England. Berkeley: University of California Press, 2001.

Apostolidès, Jean-Marie. Le roi-machine: Spectacle et politique au temps de Louis XIV. Paris: Éditions de Minuit, 1981.

Arnaud, Sabine. "Autobiographies of Illness as Medical Legitimacy in the Works of Francis Fuller, George Cheyne, Charles Révillon and the Count de Puységur." In Tensions/Spannung, edited by Christoph Holzhey. Berlin: Turia & Kant, 2010.

————. "Capturer l'indéfinissable: Métaphores et récits sur l'hystérie dans les écrits médicaux français et anglais entre 1650 et 1800." Annales: Histoire, Sciences Sociales 1 (2010): 63–85.

————. "De la dénomination d'une maladie à son assignation: L'hystérie et la différence sexuelle face à la pathologie entre 1750 et 1820." In *Les discours sur l'égalité/l'inégalité des sexes de 1750 aux lendemains de la Révolution*, edited by Nicole Pellegrin and Martine Sonnet. Saint-Étienne: Presses de l'Université de Saint-Étienne, 2012.

———. Introduction to *La philosophie des vapeurs*, by Pierre Hunauld. Paris: Mercure de France, 2009.

———. *La philosophie des vapeurs: Suivi d'une dissertation sur les vapeurs et les pertes de sang*. Paris: Mercure de France, 2009.

———. "L'art de vaporiser à propos, pourparlers entre un médecin et une marquise vaporeuse." *Dix-huitième siècle* 39 (2007): 505–19.

———. *L'invention de l'hystérie au temps des Lumières (1670–1820)*. Paris: Éditions EHESS, 2013.

———. "Medical Writing, Philosophical Encounters, and Professional Strategy: The Defiance of Pierre Pomme." *Gesnerus—Swiss Journal of the History of Medicine and Science* 66 (2009): 218–36.

———. "Ruse and Reappropriation in *Philosophy of Vapors; or, Reasoned Letters of an Attractive Woman* by C. J. de B. de Paumerelle." *French Studies* 65 (2011): 174–87.

———. "S/he Who Traps Last Traps Best: The Diagnosis of Vapours and the Writing of Observations in Late Eighteenth Century France." In *The Body and Its Images in Eighteenth-Century Europe/Le corps et ses images dans l'Europe du dix-huitième siècle*, edited by Sabine Arnaud and Helge Jordheim. Paris: Champion, 2012.

———. "'. . . Vous en guérirez et tout sera dit': De la circonspection dans l'énonciation de la maladie au XVIIIe siècle (Suisse, France, Angleterre)." In *Qu'est-ce qu'un bon patient? Qu'est-ce qu'un bon médecin?*, edited by Claire Crignon and Marie Gaille. Paris: Seli Arslan, 2010.

Aubert, François, ed. *Pierre Pomme, médecin: Récit familial*. Arles: Éd. des Amis du Vieil Arles, 1998.

Azouvi, François. "Sens et fonction épistémologiques de la critique du magnétisme animal par les Académies." *Revue d'histoire des sciences* 29, no. 2 (1976): 123–42.

Bachelard, Gaston. *La formation de l'esprit scientifique*. 1938. Paris: Vrin, 2004. Translated with an introduction and annotations by Mary McAllester Jones as *The Formation of the Scientific Mind: A Contribution to a Psychoanalysis of Objective Knowledge* (Manchester: Clinamen, 2002).

Barbé, Léo. "La bibliothèque d'un médecin gascon au milieu du XVIIIe siècle." *Bulletin de la Société Archéologique, Historique, Littéraire & Scientifique du Gers*, 1991, 200–213.

Barker-Benfield, Graham John. *The Culture of Sensibility: Sex and Society in Eighteenth-Century Britain*. Chicago: University of Chicago Press, 1992.

Barras, Vincent, Stefan Hächler, and Stéphane Pilloud. "Consulter par lettre au XVIIIe siècle." *Gesnerus—Swiss Journal of the History of Medicine and Science*, 61, nos. 3/4 (2004): 232–53.

Barras, Vincent, and Philippe Rieder. "Écrire sa maladie au siècle des Lumières." In *La médecine des lumières tout autour de Samuel-Auguste Tissot*, edited by Vincent Barras and Micheline Louis Couvoiser. Geneva: Georg Bibliothèque d'Histoire des Sciences, 2001.

Barroux, Gilles. *Philosophie, maladie et médecine au XVIIIe siècle*. Paris: Honoré Champion, 2008.

Bataille, Georges. "Informe." *Documents* 7 (1929): 382.

———. *Inner Experience*. Translated with an introduction by Leslie Anne Boldt. Albany: State University of New York Press, 1954.

———. *L'expérience intérieure: Œuvres complètes*. Vol. 4. Paris: Gallimard, 1992.

Baudrillard, Jean. *Simulacra and Simulation*. Translated by Sheila Glaser. Ann Arbor: University of Michigan Press, 1994.

——. *Simulacres et simulation*. Paris: Galilée, 1981.

Beard, George. *A Practical Treatise on Nervous Exhaustion*. New York: William Wood & Company, 1880.

Beauvoir, Simone de. *Le deuxième sexe*. Paris: Gallimard, 1985. First published 1949. Translated by H. M. Parshley as *The Second Sex* (New York: Knopf, 1953).

Beizer, Janet. *Ventriloquized Bodies: Narratives of Hysteria in Nineteenth Century France*. Ithaca, N.Y.: Cornell University Press, 1994.

Benrekassa, Georges. "La preuve et l'aveu, ce que parler veut dire." *Dix-huitième siècle* 39 (2007): 109–28.

Benveniste, Émile. "La communication." In *Problèmes de linguistique générale*. Paris: Gallimard, 1974.

——. *Problems in General Linguistics*. Translated by Mary Elizabeth Meek. Coral Gables, Fla.: University of Miami Press, 1971.

Bernadac, Christian, and Sylvain Fourcassié. *Les possédées de Chaillot*. Paris: J. C. Lattès, 1983.

Berriot-Salvadore, Évelyne. "Les médecins analystes de la passion érotique à la fin de la Renaissance." In *La peinture des passions de la Renaissance à l'âge classique*, edited by B. Yon. Saint-Étienne: Publications de l'Université de Saint-Étienne, 1995.

——. *Un corps, un destin: La Femme dans la médecine de la Renaissance*. Paris: H. Champion, 1993.

Bertucci, Paola. "Revealing Sparks: John Wesley and the Religious Utility of Electrical Healing." *British Society for the History of Science* 39, no. 3 (September 2006): 341–62.

Blake, John B. "George Rosen: From Medical Police to Social Medicine; Essays on the History of Health Care." *American Historical Review* 81, no. 3 (1976): 559.

Blanchot, Maurice. "L'oubli, la déraison." In *L'entretien infini*. Paris: Gallimard, 1969. Translated by Susan Hanson (with foreword) as "Forgetting, Unreason," in *The Infinite Conversation* (Minneapolis: University of Minnesota Press, 1993).

Bocchicchio, Rebecca P. "Blushing, Trembling and Incapable of Defense: The Hysterics of the British Recluse." In *The Passionate Fictions of Eliza Heywood: Essays on Her Life and Work*, edited by Kirsten Saxton. Lexington: University Press of Kentucky, 2000.

Bound Alberti, Fay, ed. *Medicine, Emotion, and Disease, 1750–1950*. New York: Palgrave Macmillan, 2006.

Brandt, Christina. "Genetic Code, Text, and Scripture: Metaphors and Narration in German Molecular Biology." *Science in Context* 18, no. 4 (2005): 629–48.

Brockliss, Laurence. *Calvet's Web: Enlightenment and the Republic of Letters in Eighteenth-Century France*. New York: Oxford University Press, 2002.

——. "Consultation by Letter in Early Eighteenth-Century Paris: The Medical Practice of Étienne-François Geoffroy." In *French Medical Culture in the Nineteenth Century*, edited by Ann La Berge and Mordechai Feingold, 79–117. Amsterdam: Rodopi, 1994.

Brockliss, Laurence, and Colin Jones. *The Medical World of Early Modern France*. Oxford and New York: Clarendon Press, 1997.

Bronfen, Elisabeth. "Hysteria, Phantasy, and the Family Romance: Ann Radcliffe's Romance of the Forest." *Women Writing* 1, no. 2 (1994): 171–80.

Burton, Robert. *The Anatomy of Melancholy, what it Is: with all the Kindes, Causes, Symptomes, Prognostickes, and Severall Cures of it: in Three Maine Partitions with their Seuerall Sections, Members, and Subsections: Philosophically, Medicinally, Historically, Opened and Cut*. Oxford: John Lichfield, James Short, 1621.

Butler, Judith. *Gender Trouble: Feminism and the Subversion of Identity.* New York: Routledge, 1990.

Cadden, Joan. *Meanings of Sex Difference in the Middle Ages: Medicine, Science and Culture.* Cambridge: Cambridge University Press, 2003.

Campbell, John Angus, and Keith Benson. "Review Essay: The Rhetorical Turn in Science Studies." *Quarterly Journal of Speech* 1, vol. 82 (1996): 74–91.

Canguilhem, Georges. *La formation du concept de réflexe au XVIIe et au XVIIIe siècle.* Paris: PUF, 1955.

———. *Le normal et le pathologique.* 1966. Paris: Presses Universitaires de France, 1991.

———. *The Normal and the Pathological.* Translated by Carolyn R. Fawcett in collaboration with Robert S. Cohen. New York: Zone Books, 1989.

Carroy-Thirard, Jacqueline. "Hystérie, théâtre, littérature au XIXe siècle." *Psychanalyse à l'Université* 7, no. 26 (1982): 299–317.

———. *Le mal de Morzine: De la possession a l'hystérie (1857–1877).* Paris: Solin, 1981.

———. "Possession, extase, hystérie au XIXe siècle." *Psychanalyse à l'Université* 5, no. 19 (1980): 499–515.

Carson, John. *The Measure of Merit: Talents, Intelligence, and Inequality in the French and American Republics, 1750–1940.* Princeton, N.J.: Princeton University Press, 2007.

Castle, Terry. *The Female Thermometer: 18th Century Culture and the Invention of the Uncanny.* Oxford: Oxford University Press, 1995.

Chamayou, Grégoire. *Les corps vils: Expérimenter sur les êtres humains aux XVIIIe et XIXe siècles.* Paris: Les Empêcheurs de penser en rond/La Découverte, 2008.

Chappey, Jean-Luc. *La Société des observateurs de l'homme, 1799–1804: Des anthropologues au temps de Bonaparte.* Paris: Société des études Robespierristes, 2002.

Chartier, Roger. *Lectures et lecteurs dans la France d'Ancien Régime.* Paris: Éditions du Seuil, 1987.

———. "Texts, Printings, Reading." In *The New Cultural History*, edited by Lynn Hunt. Berkeley: University of California Press, 1989.

Chartier, Roger, and Alain Paire. *Pratiques de la lecture.* Paris: Payot & Rivages, 2003.

Chaunu, Pierre. "Sur la fin des sorciers au XVIIe siècle." *Annales: Économies, Sociétés, Civilisations* 24, no. 4 (1969): 895–911.

Cixous, Hélène. *Portrait de Dora.* Paris: Édition des Femmes, 1976. Translated by Simone Benmussa and George Moore as *Benmussa Directs: Portrait of Dora* (London: J. Calder; Dallas: Riverrun Press, 1979).

Cixous, Hélène, and Catherine Clément. *La jeune née.* Paris: 10/18, 1975. Translated by Betsy Wing as *The Newly Born Woman* (Manchester: Manchester University Press, 1986).

Coleman, William. *Death Is a Social Disease, Public Health and Political Economy in Early Industrial France.* Madison: University of Wisconsin Press, 1982.

Coleman, William, and Frederic L. Holmes, eds. *The Investigative Enterprise.* Berkeley: University of California Press, 1988.

Compagnon, Antoine. *La seconde Main; ou, Le Travail de la citation.* Paris: Seuil, 1979.

Corsi, Pietro. "After the Revolution: Scientific Language and French Politics, 1795–1802." In *The Practice of Reform in Health, Medicine, and Science, 1500–2000: Essays for Charles Webster*, edited by Margaret Pelling and Scott Mandelbrote. Aldershot, U.K.: Ashgate, 2005.

Crane, R. S. "Suggestions toward a Genealogy of the 'Man of Feeling'." *EHL: A Journal of English Literary History* 1, no. 3 (1934): 205–30.

Crignon-De Oliveira, Claire. *De la mélancolie à l'enthousiasme: Robert Burton (1577–1640) et Antony Ashley Cooper, comte de Shaftesbury (1671–1713)*. Paris: Éditions Honoré Champion, 2006.

Cunningham, Andrew. "Thomas Sydenham: Epidemics, Experiment, and the Cause." In *The Medical Revolution of the Seventeenth Century*, edited by Roger French and Andrew Wear. Cambridge: Cambridge University Press, 1989.

Cunningham, Andrew, and Roger K. French, eds. *Medical Enlightenment of the Eighteenth Century*. Cambridge: Cambridge University Press, 1990.

Cutting-Gray, Joanne. *Woman as 'Nobody' and the Novels of Fanny Burney*. Gainesville: University Press of Florida, 1992.

Dacome, Lucia. "Balancing Acts: Picturing Perspiration in the Long Eighteenth Century." *Studies in History and Philosophy of Biological and Biomedical Sciences*, no. 43 (2012): 379–91.

———. "Useless and Pernicious Matter: Corpulence in Eighteenth-Century England." In *Cultures of the Abdomen: Diet, Digestion, and Fat in the Modern World*, edited by Christopher E. Forth and Ana Carden-Coyne. New York: Palgrave, 2005.

Darnton, Robert. "First Steps toward a History of Reading." In *The Kiss of Lamourette: Reflections in Cultural History*. New York: W. W. Norton, 1990.

———. *La fin des Lumières: Le Mesmérisme et la Révolution*. Translated by Marie-Alyx Revellat. Paris: Éditions Odile Jacob, 1995.

———. *Mesmerism and the End of the Enlightenment in France*. Cambridge, Mass.: Harvard University Press, 1968, 2009.

Daston, Lorraine. "The Ideal and Reality of the Republic of Letters in the Enlightenment." *Science in Context* 4 (1991): 367–86.

———. "The Language of Strange Facts." In *Inscribing Science: Scientific Texts and the Materiality of Communication*. Stanford, Calif.: Stanford University Press, 1998.

———. "Taking Notes." *Isis* 95 (2004): 443–48.

———. *Wunder, Beweise und Tatsachen: Zur Geschichte der Rationalität*. Frankfurt am Main: Fischer Taschenbuch Verlag, 2001.

Daston, Lorraine, and Peter Galison. *Objectivity*. New York: Zone Books, 2007.

———. "Objectivity and Its Critics." *Victorian Studies* 50 (2008): 666–77.

Daston, Lorraine, and Elizabeth Lunbeck, eds. *Histories of Scientific Observation*. Chicago: University of Chicago Press, 2011.

Daston, Lorraine, and Katherine Park. *Wonders and the Order of Nature*. New York: Zone Books, 1998.

Daston, Lorraine, and Gianna Pomata. *The Faces of Nature in Enlightenment Europe*. Concepts & Symboles du Dix-Huitième Siècle Européen = Concepts & Symbols of the Eighteenth Century in Europe. Berlin: BWV-Berliner Wissenschafts-Verlag, 2003.

Daston, Lorraine, and Fernando Vidal. *The Moral Authority of Nature*. Chicago: University of Chicago Press, 2003.

Davis, Natalie Zemon. *Fiction in the Archives: Pardon Tales and Their Tellers in Sixteenth-Century France*. Stanford, Calif.: Stanford University Press, 1987.

Dean-Jones, David Emil. "Galen, On the Constitution of the Art of Medicine: Introduction, Translation and Commentary." PhD diss., University of Texas at Austin, 1993.

Dear, Peter Robert. *The Literary Structure of Scientific Argument: Historical Studies*. Philadelphia: University of Pennsylvania Press, 1991.

De Baecque, Antoine. "La folie du jeu." In *L'État de la France pendant la Révolution 1789–1799*, edited by Michel Vovelle. Paris: Éditions la Découverte, 1988.

———. *La gloire et l'effroi: Sept morts sous la terreur.* Paris: B. Grasset, 1997. Translated by Charlotte Mandell as *Glory and Terror: Seven Deaths under the French Revolution* (New York: Routledge, 2001).

———. *La Révolution à travers la caricature.* Paris: Presses du CNRS, 1989.

———. *Le corps de l'histoire: Métaphores et politique 1770–1800.* Paris: Calmann-Lévy, 1993. Translated by Charlotte Mandell as *The Body Politic: Corporeal Metaphor in Revolutionary France, 1770–1800* (Stanford, Calif.: Stanford University Press, 1997).

———. *Les éclats du rire: La culture des rieurs au XVIIIe siècle.* Paris: Calmann-Levy, 2000.

De Baecque, Antoine, and Françoise Mélonio. *Lumières et liberté: Les dix-huitième et dix-neuvième siècles; Histoire culturelle de la France.* Vol. 3. Paris: Seuil, 2004.

Debus, Allen G. *The Chemical Promise: Experiment and Mysticism in the Chemical Philosophy, 1550–1800; Selected Essays of Allen G. Debus.* Canton, Mass.: Science History Publications, 2006.

———. *Chemistry and Medical Debate: Van Helmont to Boerhaave.* Canton, Mass.: Science History Publications, 2001.

De Certeau, Michel. *La fable mystique.* Paris: Gallimard, 1982.

———. *L'écriture de l'histoire.* Paris: Gallimard, 1975.

———. *Le lieu de l'autre.* Paris: Gallimard/Seuil, 2005.

———. *The Writing of History.* Translated by Tom Conley. New York: Columbia University Press, 1988.

Deleuze, Gilles. *Kafka: Pour une littérature mineure.* Paris: Éd. de Minuit, 1984. Translated by Dana Polan as *Kafka: Toward a Minor Literature* (Minneapolis: University of Minnesota Press, 1986).

———. *The Logic of Sense.* New York: Columbia University Press, 1990.

———. *Logique du sens.* Paris: Éditions de Minuit, 1969.

Delon, Michel. *L'invention du boudoir.* Cadeilhan: Zulma, 1999.

Deneys-Tunney, Anne. *L'écriture du corps: De Descartes à Laclos.* Paris: PUF, 1992.

Derrida, Jacques. *La dissémination.* Paris: Seuil, 1972. Translated by Barbara Johnson as *Dissemination* (Chicago: University of Chicago Press, 1981).

Dhombres, Nicole, and Jean Dhombres. *Naissance d'un pouvoir: Sciences et savants en France (1793–1824).* Bibliothèque Historique Payot. Paris: Payot, 1989.

Dieckmann, Herbert. *Cinq leçons sur Diderot.* Geneva: Droz, 1959.

Dixon, Laurinda. *Perilous Chastity: Women and Illness in Pre-Enlightenment Art and Medicine.* Ithaca, N.Y.: Cornell University Press, 1995.

Dixon, Thomas. "Patients and Passions: Languages of Medicine and Emotion, 1789–1850." In *Medicine, Emotion, and Disease, 1750–1950,* edited by Fay Bound Alberti. New York: Palgrave Macmillan, 2006.

Doerner, Klaus. *Madmen and the Bourgeoisie: A Social History of Insanity and Psychiatry.* Oxford: Basil Blackwell, 1981.

Dorat, Claude Joseph. *Les malheurs de l'inconstance. Romans libertins du XVIIIe siècle,* edited by Raymond Trousson. Paris: Robert Laffont, 1993.

Dorlin, Elsa. *La matrice de la race: Généalogie sexuelle et coloniale de la nation française.* Paris: Éditions la découverte, 2006.

Du Bled, Victor. *Les médecins avant et après 1789.* Paris: Perrin, 1908.

Duchesneau, François. *La physiologie des Lumières: Empirisme, modèles et théories.* The Hague: Martinus Nijhoff Publishers, 1982.

Duden, Barbara. *Geschichte unter der Haut: Ein Eisenacher Arzt und seine Patientinnen um 1730.*

Stuttgart: Klett-Cotta, 1987. Translated by Thomas Dunlap as *The Woman Beneath the Skin* (Cambridge, Mass.: Harvard University Press, 1991).

Edelman, Nicole. *Les métamorphoses de l'hystérique, du début de XIXe siècle à la grande guerre.* Paris: La Découverte, 2003.

Edelman, Nicole, Jean-Pierre Peter, and Luis Montiel, eds. *Histoire sommaire de la maladie et du somnambulisme de Lady Lincoln.* Paris: Tallandier, 2009.

Elias, Norbert. *La société de cour.* Paris: Flammarion, 2008.

Ellenberger, Henri F. *The Discovery of the Unconscious: The History and Evolution of Dynamic Psychiatry.* New York: Basic Books, 1970.

Emch-Deriaz, Antoinette. *Tissot: Physician of the Enlightenment.* New York: Peter Lang, 1992.

Ender, Evelyne. *Sexing the Mind: Nineteenth-Century Fictions of Hysteria.* Ithaca, N.Y.: Cornell University Press, 1995.

Evans, Martha Noel. *Fits and Starts: A Genealogy of Hysteria in Modern France.* Ithaca, N.Y.: Cornell University Press, 1991.

Fara, Patricia. *An Entertainment for Angels: Electricity in the Enlightenment.* New York: Columbia University Press, 2002.

Fissell, Mary E. "Gender and Generation: Representing Reproduction in Early Modern England." *Gender and History* 7 (1995): 433–56. Reprinted in *The Sexualities in History Reader,* edited by Kim Phillips and Barry Reay. London: Routledge, 2001.

———. "Hairy Women and Naked Truths: Gender and the Politics of Knowledge in Aristotle's Masterpiece." *William and Mary Quarterly* 60 (2003): 43–74.

———. "Imagining Vermin in Early Modern England." In *The Animal Human Boundary,* edited by William Chester Jordan and Angela N. H. Creager. Rochester: University of Rochester Press, 2002.

———. "Making a Masterpiece: The Aristotle Texts in Vernacular Medical Culture." In *Right Living: An Anglo-American Tradition of Self-Help Medicine,* edited by Charles E. Rosenberg. Baltimore: Johns Hopkins University Press, 2003.

———. *Vernacular Bodies: The Politics of Reproduction in Early Modern England.* Oxford: Oxford University Press, 2004.

Forge, Evelyn L. "Evocations of Sympathy: Sympathetic Imagery in Eighteenth-Century Social Theory and Physiology." *History of Political Economy* 35, Annual Supplement (2003): 282–308.

Foucault, Michel. *Dits et écrits.* 4 vols. Paris: Gallimard, 1994.

———. *The Foucault Reader.* Edited by Paul Rabinow. Pantheon, 1984.

———. *Histoire de la folie à l'âge classique suivi de mon corps, ce papier, ce feu et la folie, l'absence d'oeuvre.* Paris: Gallimard, 1972.

———. *Le pouvoir psychiatrique: Cours au Collège de France, 1973–1974.* Edited by Jacques Lagrange, under the direction of François Ewald and Alessandro Fontana. Paris: Gallimard/Le Seuil, 2003. Translated by Graham Burchell as *Psychiatric Power: Lectures at the Collège de France, 1973–74* (Basingstoke, U.K.; New York: Palgrave Macmillan, 2006).

———. *Les anormaux: Cours au Collège de France, 1974–1975.* Edited by Valerio Marchetti and Antonella Salomoni, under the direction of François Ewald and Alessandro Fontana. Paris: Gallimard/Le Seuil, 1999. Translated by Graham Burchell as *Abnormal: Lectures at the Collège de France, 1974–1975* (New York: Picador, 2003).

———. *Madness and Civilisation.* London: Routledge, 2001.

———. *Maladie mentale et psychologie.* Paris: PUF, 1995.

———. "Médecins, juges et sorciers au XVIIe siècle." In *Dits et écrits,* vol. 1. Paris: Gallimard, 1994.

————. *Naissance de la clinique.* Paris: PUF, 1963. Translated by A. M. Sheridan as *The Birth of the Clinic* (New York: Pantheon Books, 1973).

Garin, Eugenio. "Phantasiá e imaginatió fra Ficino e Pomponazzi." In *Umanisti, artisti, scienzati: Studi sul rinascimento italiano.* Rome: Editori Riuniti, 1989.

Garrabé, Jean, ed. *Philippe Pinel.* Les Empêcheurs de Penser en Rond, edited by Philipe Pignarre. Le Plessis-Robinson: Synthelabo, 1994.

Gelfand, Toby. *Professionalizing Modern Medicine: Paris Surgeons and Medical Science and Institutions in the 18th Century.* Contributions in Medical History. Westport, Conn.: Greenwood Press, 1980.

Gilman, Sander. *Difference and Pathology: Stereotypes of Sexuality, Race and Madness.* Ithaca, N.Y.: Cornell University Press, 1985.

Gilman, Sander, Helen King, Roy Porter, G. S. Rousseau, and Elaine Showalter, eds. *Hysteria beyond Freud.* Berkeley: University of California Press, 1993.

Goldstein, Jan. *Console and Classify: The French Psychiatric Profession in the Nineteenth Century.* 1987. Chicago: University of Chicago Press, 2001.

————. *Hysteria Complicated by Ecstasy.* Princeton, N.J.: Princeton University Press, 2010.

Golinski, Jan. "Barometers of Change: Meteorological Instruments as Machines of Enlightenment." In *The Science in Enlightened Europe,* edited by William Clark, Jan Golinski, and Simon Schaffer. Chicago: University of Chicago Press, 1999.

Gooding, David, T. J. Pinch, and Simon Schaffer, eds. *The Uses of Experiment: Studies in the Natural Sciences.* Cambridge: Cambridge University Press, 1989.

Goodman, Dena. *The Republic of Letters: A Cultural History of the French Enlightenment.* Ithaca, N.Y.: Cornell University Press, 1994.

Goring, Paul. *The Rhetoric of Sensibility in Eighteenth-Century Culture.* Cambridge: Cambridge University Press, 2005.

Goulemot, Jean-Marie. *Ces livres qu'on ne lit que d'une main: Lecture et lecteurs de livres pornographiques au XVIIIe siècle.* Paris: Minerve, 1994.

Gowlands, Angus. *The Worlds of Renaissance Melancholy.* Cambridge: Cambridge University Press, 2006.

Green, Monica. "Bodies, Gender, Health, Disease: Recent Work on Medieval Women's Medicine." In *Studies in Medieval and Renaissance History: Sexuality and Culture in Medieval and Renaissance Europe,* edited by Philip M. Soergel. New York: AMS Press, 2005.

Groneman, Carol. "Nymphomania: The Historical Construction of Female Sexuality." *Signs* 19, no. 2 (1994): 337–67.

Grosvenor Myer, Valerie, ed. *Against the Spleen.* In *Laurence Sterne: Riddles and Mysteries.* Totowa, N.J.: Barnes & Noble, 1984.

Guerrini, Anita. "Alexander Monro Primus and the Moral Theatre of Anatomy." *The Eighteenth Century: Theory and Interpretation* 47, no. 1 (Spring 2006): 1–18.

————. "Anatomists and Entrepreneurs in Early Eighteenth-Century London." *Journal of the History of Medicine and Allied Sciences* 59, no. 2 (April 2004): 219–39.

————. "Duverney's Skeletons." *Isis* 94, no. 4 (December 2003): 577–603.

————. "The Hungry Soul: George Cheyne and the Construction of Femininity." *Eighteenth-Century Studies* 32, no. 3 (1999): 279–91.

————. *Obesity and Depression in the Enlightenment: The Life and Times of George Cheyne.* Norman: University of Oklahoma Press, 2000.

————. "Theatrical Anatomy: Duverney in Paris, 1670–1720." *Endeavour* 33 (March 2009): 7–11.

Guthrie, Douglas. "The Patient: A Neglected Factor in the History of Medicine." *Proceedings of the Royal Society of Medicine* 38 (April 4, 1945), 490–94.

Hannaway, Caroline C. F. "Medicine, Public Welfare, and the State in 18th Century France: The Société Royale de Médecine of Paris (1776–1793)." PhD diss., Johns Hopkins University, 1974.

Harkness, Deborah. "Nosce Teipsum: Curiosity, the Humoural Body and the Culture of Therapeutics in Late Sixteenth- and Early Seventeenth-Century England." In *Curiosity and Wonder from the Renaissance to the Enlightenment*, edited by Robert John Weston Evans and Alexander Marr. Aldershot, U.K.: Ashgate, 2006.

Hawes, Clement. *Mania and Literary Style: The Rhetoric of Enthusiasm from the Ranters to Christopher Smart.* Cambridge Studies in Eighteenth-Century English Literature and Thought. Cambridge: Cambridge University Press, 1996.

Heilbron, John. *Electricity in the 17th and 18th Centuries: A Study of Early Modern Physics.* Berkeley: University of California Press, 1979.

Heyd, Michael. "Be Sober and Reasonable." In *The Critique of Enthusiasm in the 17th and Early 18th Centuries.* New York: E. J. Brill, 1995.

Howie, Gillian, and Andrew Shail, eds. *Menstruation: A Cultural History.* New York: Palgrave Macmillan, 2005.

Huet, Marie-Hélène. *Monstrous Imagination.* Cambridge, Mass.: Harvard University Press, 1993.

Imbroscio, Carmelina. *Le malattie dell'anima tra scienza e pregiudizio: Letteratura medica e letteratura parodica nel settecento francese.* Bologna: Clueb, 2002.

Irigaray, Luce. *Spéculum de l'autre femme.* Paris: Éd. de minuit, 1974. Translated by Gillian C. Gill as *Speculum of the Other Woman* (Ithaca, N.Y.: Cornell University Press, 1985).

Irlam, Shaun. *Relations: The Poetics of Enthusiasm in Eighteenth-Century Britain.* Stanford, Calif.: Stanford University Press, 1999.

Jacyna, L. S. "Somatic Theories of Mind and the Interests of Medicine in Britain, 1850–1879." *Medical History* 26, no. 3 (July 1982): 233–58.

Jaquier, Claire. *L'erreur des désirs: Romans sensibles au XVIIIe siècle.* Lausanne: Payot, 1998.

Jewson, Nicholas D. "The Disappearance of the Sick-Man from Medical Cosmology, 1770–1870." *Sociology* 10, no. 2 (1976): 225–44.

Jones, Chris. *Radical Sensibility: Literature and Ideas in the 1790s.* London: Routledge, 1993.

Jones, Colin. "Plague and Its Metaphors in Early Modern France." *Representations* 53 (1996): 97–127.

Jonsson, Fredrik Albritton. "The Physiology of Hypochondria in Eighteenth-Century Britain." In *Cultures of the Abdomen: Diet, Digestion, and Fat in the Modern World*, edited by Christopher E. Forth and Ana Carden-Coyne. New York: Palgrave Macmillan, 2005.

Jordanova, Ludmilla. *Defining Features: Scientific and Medical Portraits, 1660–2000.* London: Reaktion Books, 2000.

———. *Languages of Nature: Critical Essays on Science and Literature.* New Brunswick, N.J.: Rutgers University Press, 1986.

———. *Nature Displayed: Gender, Science and Medicine, 1760–1820.* London: Longman, 1999.

———. "The Popularization of Medicine: Tissot on Onanism." *Textual Practice* 1, no. 1 (1987): 68–79.

———. *Sexual Visions: Images of Gender in Science and Medicine between the Eighteenth and the Twentieth Centuries.* Madison: University of Wisconsin Press, 1989.

Jorden, Edward, and Mary Glover Case. *Witchcraft and Hysteria in Elizabethan London.* London and New York: Tavistock/Routledge, 1991.

Jordlan, Gérard. *Une société à soigner, hygiène et salubrité publique en France au XIXe siècle.* Paris: Gallimard, 2010.

Kafka, Franz. *In der Strafkolonie.* Leipzig: Kurt Wolff Verlag, 1919. Translated by Willa Muir and Edwin Muir as *The Penal Colony* (New York: Schocken Books, 1948).

Kant, Immanuel. *An Answer to the Question: 'What Is Enlightenment?'.* Translated by H. B. Nisbet. London and New York: Penguin Books, 2009.

Kantorowicz, Ernst Hartwig. *Les deux corps du roi.* Paris: Gallimard, 1989.

Keel, Othmar. *L'avènement de la médecine clinique et moderne en Europe, 1750–1815: Politiques, institutions, savoirs.* Geneva/Montreal: Georg/Presses Universitaires de Montréal, 2002.

King, Helen. *The Disease of Virgins: Green Sickness, Chlorosis and the Problem of Puberty.* London: Routledge, 2004.

———. *Hippocrates' Woman: Reading the Female Body in Ancient Greece.* New York: Routledge, 1998.

———. "The Mathematics of Sex: One to Two or Two to One?" In *Studies in Medieval and Renaissance History: Sexuality and Culture in Medieval and Renaissance Europe,* edited by Philip M. Soergel. New York: AMS Press, 2005.

———. *Midwifery, Obstetrics and the Rise of Gynaecology: The Uses of a Sixteenth-Century Compendium.* Aldershot, U.K.: Ashgate, 2007.

King, Lester S. "Boissier de Sauvages and the Eighteenth-Century Nosology." *Bulletin of the History of Medicine* 40 (1966): 43–51.

———. *The Medical World of the Eighteenth Century.* Huntington: R. E. Krieger, 1958.

Klein, Lawrence Eliot, and Anthony J. LaVopa, eds. *Enthusiasm and Enlightenment in Europe, 1650–1850.* San Marino: Huntington Library, 1998.

Klein, Robert. "L'enfer de Ficin." In *Umanesimo e esoterismo,* edited by E. Castelli. Padova: cedam, 1960.

Klibansky, Raymond, Erwin Panofsky, and Fritz Saxl. *Saturn and Melancholy: Studies in the History of Natural Philosophy, Religion, and Art.* New York: Basic Books, 1964.

Kniebiehler, Yvonne, and Catherine Fouquet, eds. *La femme et les médecins.* Paris: Hachette, 1983.

Koselleck, Reinhart. *Historische Semantik und Begriffsgeschichte.* Stuttgart: Klett-Cotta, 1979.

Kreiser, Robert. *Miracles, Convulsions and Ecclesiastical Politics in Early 18th Century Paris.* Princeton, N.J.: Princeton University Press, 1978.

La Berge, Ann. *Mission and Method: The Early Nineteenth-Century French Public Health Movement.* Cambridge: Cambridge University Press, 1992.

Labrude-Estenne, Jacqueline. *Médecins et médecine dans l'univers romanesque de T. Smollet et de L. Sterne, 1748–1771.* Paris: Presse de la Sorbonne, 1995.

Lacan, Jacques. "Intervention sur le transfert." In *Écrits.* Paris: Seuil, 1996. Translated by Bruce Fink as "Presentation on Transference," in *Écrits: The First Complete Edition in English* (New York and London: W. W. Norton, 2006).

Laqueur, Thomas. *Le sexe en solitaire: Contribution à l'histoire culturelle de la sexualité.* 2003. Paris: Gallimard, 2005.

———. *Making Sex: Body and Gender from the Greeks to Freud.* Cambridge, Mass.: Harvard University Press, 1990.

———. "Orgasm, Generation, and the Politics of Reproductive Biology." In *The Making of the Modern Body: Sexuality and Society in the Nineteenth Century,* edited by Catherine Gallagher and Thomas Laqueur. Berkeley: University of California Press, 1987.

Lawrence, Christopher. "The Nervous System and Society in the Scottish Enlightenment." In

Natural Order: Historical Studies of Scientific Culture, edited by Barry Barnes and Steven Shapin. Beverly Hills: Sage, 1979.

Lawrence, Christopher, and Steven Shapin, eds. "Medical Minds, Surgical Bodies." In *Science Incarnate: Historical Embodiments of Natural Knowledge*. Chicago: University of Chicago Press, 1998.

Lennox, Charlotte. *The Female Quixote; or, The Adventures of Arabella*. New York: Oxford University Press, 1970.

Léonard, Jacques. "À propos de l'histoire de la saignée (1600-1900)." In *Affaires de sang*, edited by Arlette Farge. Paris: Imago, 1988.

———. *La médecine entre les pouvoirs et les savoirs*. Paris: Aubier, 1981.

———. *Médecins, maladies et société dans la France du XIXe siècle*. Paris: Sciences en situation, 1992.

Licoppe, Christian. *La formation de la pratique scientifique: Le discours de l'expérience en France et en Angleterre, 1630–1820*. Paris: La Découverte, 1996.

———. "La formation de la pratique scientifique: Le discours de l'expérience en France et en Angleterre (1630–1820)." *Annales: Histoire, Sciences Sociales* 53, nos. 4–5 (1998): 995–1000.

Livi, Jocelyne. *Vapeurs de femmes: Essai historique sur quelques fantasmes médicaux et philosophiques*. Paris: Navarin Editeur, 1984.

Lockwood, Dean Putnam. *Ugo Benzi: Medieval Philosopher and Physician (1376–1439)*. Chicago: University of Chicago Press, 1951.

Logan, Peter Melville. *Nerves and Narrative: A Cultural History of Hysteria in 19th Century Prose*. Berkeley: University of California Press, 1997.

Lord, Alexandra. "'The Great Arcana of Deiti': Menstruation and Menstrual Disorders in Eighteenth Century British Medical Thought." *Bulletin of the History of Medicine*, 1999, 71–99.

Louis-Courvoisier, Micheline. *Soigner et consoler: La vie quotidienne dans un hôpital à la fin de l'ancien régime (Genève, 1750–1820)*. Geneva: Georg, 2000.

Lyon-Caen, Nicolas. "La minute du miracle: Engagement janséniste et appartenance sociale des notaires parisiens au XVIIIe siècle." *Revue Historique*, no. 645 (2008): 61–83.

Macalpine, Ida, and Richard Alfred Hunter. *George III and the Mad-Business*. London: Allen Lane, 1969.

Mackay, Christopher S., ed. and trans. *Malleus maleficarum*. 2 vols. Cambridge: Cambridge University Press, 2006.

Maclean, Ian. *The Renaissance Notion of Woman: A Study in the Fortunes of Scholasticism and Medical Science in European Intellectual Life*. Cambridge: Cambridge University Press, 1980.

Maire, Catherine. *De la cause de Dieu à la cause de la nation: Le Jansénisme au XVIIIe siècle*. Paris: Gallimard, 1998.

———. *Les Convulsionnaires de Saint-Médard*. Paris: Gallimard, 1985.

Mandressi, Rafael. *Le regard de l'anatomiste: Dissections et invention du corps en Occident*. Paris: Éditions du Seuil, 2003.

Mandrou, Robert. *Magistrats et sorciers en France au XVIIe siècle: Une analyse de psychologie historique*. Paris: Seuil, 1980.

Marin, Louis. *Food for Thought*. Baltimore: Johns Hopkins University Press, 1997.

———. *Le récit est un piège*. Paris: Éditions de Minuit, 1978.

Markman, Ellis. *The Politics of Sensibility: Race, Gender and Commerce in the Sentimental Novel*. Cambridge: Cambridge University Press, 1996.

Marland, Hilary. "Languages and Landscapes of Emotion: Motherhood and Puerperal Insanity in the Nineteenth Century." In *Medicine, Emotion and Disease, 1700–1950*, edited by Fay Bound Alberti. Basingstoke, U.K.: Palgrave Macmillan, 2006.

Martin, Julian. "Sauvages's Nosology: Medical Enlightenment in Montpellier." In *The Medical Enlightenment in the Eighteenth Century*, edited by Andrew Cunningham and Roger French. Cambridge: Cambridge University Press, 1990.

Matlock, Jann. *Scenes of Seduction, Prostitution, Hysteria and Reading Difference in the Nineteenth Century France*. New York: Columbia University Press, 1994.

Mazauric, Simone. *Fontenelle et l'invention de l'histoire des sciences à l'aube des Lumières*. Paris: Fayard, 2007.

Mazzoni, Cristina. *Saint Hysteria: Neurosis, Mysticism, and Gender in European Culture*. Ithaca, N.Y.: Cornell University Press, 1996.

McClive, Cathy. "Menstrual Knowledge and Medical Practice in France, c. 1555–1761." In *Menstruation: A Cultural History*, edited by Gillian Howie and Andrew Shail. New York: Palgrave Macmillan, 2005.

McGann, Jerome. *The Poetics of Sensibility: A Revolution in Literary Style*. Oxford: Clarendon Press, 1996.

Mee, Jon. *Romanticism, Enthusiasm, and Regulation: Poetics and the Policing of Culture in the Romantic Period*. New York: Oxford University Press, 2003.

Micale, Mark S. *Approaching Hysteria: Disease and Its Interpretation*. Princeton, N.J.: Princeton University Press, 1995.

———. *Hysterical Men: The Hidden History of Male Nervous Illness*. Cambridge, Mass.: Harvard University Press, 2008.

Miles, Robert. *Ann Radcliffe, the Great Enchantress*. Manchester: Manchester University Press, 1995.

———. Introduction to "Female Gothic Writing." Special issue, *Women's Writing: The Elizabethan to Victorian Period* 1, no. 2 (1994): 131–42.

Montrelay, Michele. *L'ombre et le nom: Sur la féminité*. Paris: Éditions de Minuit, 1977.

Mora, George, ed. *Witches, Devils, and Doctors in the Renaissance: Johann Weyer, De praestigiis daemonum*. Medieval and Renaissance Texts and Studies 73. Associate editor Benjamin Kohl. Translated by John Shea. Binghamton: State University of New York Center for Medieval and Early Renaissance Studies, 1991.

Moravia, Sergio. *Il tramonto dell'illuminismo: Filosofia e politica nella società francese (1770–1810)*. Bari: Laterza, 1968.

Muchembled, Robert. *Une histoire du diable XIIe-XXe siècle*. Paris: Seuil, 2000.

Mullan, John. *Sentiment and Sociability: The Language of Feeling in the Eighteenth Century*. 1988. New York: Oxford University Press, 2002.

Nagle, Christopher. *Sexuality and the Culture of Sensibility in the British Romantic Era*. New York: Palgrave Macmillan, 2007.

Nance, Brian. *Turquet de Mayerne as Baroque Physician: The Art of Medical Portraiture*. Amsterdam and New York: Rodopi, 2001.

Noyes, Russell. "The Transformation of Hypochondriasis in British Medicine, 1680–1830." *Social History of Medicine* 24, no. 2 (August 2011): 281–98.

Otis, Laura. *Membranes: Metaphors of Invasion in Nineteenth-Century Literature, Science, and Politics*. Baltimore: Johns Hopkins University Press, 1999.

Packard, Randall M., and Mary Fissell, eds. "Women, Health, and Healing in Early Modern Europe." Special issue, *Bulletin of the History of Medicine* 82 (Spring 2008).

Pancaldi, Giuliano. *Volta: Science and Culture in the Age of Enlightenment*. Princeton, N.J.: Princeton University Press, 2003.

Parent, Françoise. "De nouvelles pratiques de lecture." In *Histoire de l'édition française*, vol. 2, *Le livre triomphant, 1660–1830*, edited by Henri-Jean Martin. Paris: Fayard, 1984.

Park, Katharine. *Secrets of Women: Gender, Generation, and the Origin of Human Dissection*. New York: Zone Books, 2006.

Pauthier, Céline. *L'exercice illégal de la médecine (1673–1793): Entre défaut de droit et manière de soigner*. Paris: Glyphe & Biotem Editions, 2002.

Peter, Jean-Pierre. *De la douleur*. Paris: Quai Voltaire, 1993.

———. "Entre femmes et médecins: Violence et singularités dans les discours du corps d'après les manuscrits médicaux de la fin du XVIIIe siècle." *Ethnologie Française* 6 (1976): 341–48.

———, ed. "Puységur et l'enfant fou ou la raison originelle." In A. M. J. de Chastenet, marquis de Puységur, *Un somnambule désordonné? Journal du traitement magnétique du jeune Hébert*. Le Plessis-Robinson: Synthélabo, 1999.

———. "Une enquête de la Société royale de medicine: Maladies et maladies à la fin du XVIIIe siècle." *Annales: Économies, Sociétés, Civilisations* 22, no. 4 (1967): 711–51.

Peterson, Kaara L. "Historica Passio: Early Modern Medicine, King Lear, and Editorial Practice." *Shakespeare Quarterly* 57, no. 1 (Spring 2006): 1–22.

Peyre, Evelyne, and Joëlle Peyre. "De la 'Nature des Femmes' & de son incompatibilité avec l'exercice du pouvoir: Le poids des discours scientifiques depuis le XVIIIe siècle." In *La démocratie 'à la française'; ou, Les femmes indésirables*, edited by Eliane Viennot. Paris: Publications de l'Université de Paris VII, 2002.

Pigeaud, Jackie. *Aux portes de la psychiatrie: Pinel, l'ancien et le moderne*. Paris: Aubier, 2001.

———. *La maladie de l'âme: Étude sur la relation de l'âme et du corps dans la tradition médico-philosophique antique*. Paris: Les Belles Lettres, 1981.

Pinch, Adela. *Strange Fits of Passion: Epistemologies of Emotion, Hume to Austen*. Stanford, Calif.: Stanford University Press, 1996.

Pomata, Gianna. "Observation Rising: Birth of an Epistemic Genre, ca. 1500–1650." In *Histories of Scientific Observation*, edited by Lorraine Daston and Elizabeth Lunbeck. Chicago: University of Chicago Press, 2011.

———. "Sharing Cases: The Observationes in Early Modern Medicine." *Early Science and Medicine* 15, no. 3 (2010): 193–236.

———. "The Uses of Historia in Early Modern Medicine." In *Historia: Empiricism and Erudition in Early Modern Europe*, edited by Nancy Siraisi and Gianna Pomata. Cambridge, Mass.: MIT Press, 2005.

Pomata, Gianna, and Nancy Siraisi, eds. *Historia: Empiricism and Erudition in Early Modern Europe*. Cambridge, Mass.: MIT Press, 2005.

Porter, Dorothy, and Roy Porter. *In Sickness and in Health: The British Experience, 1650–1850*. London: Fourth Estate, 1988.

Porter, Roy. "'Expressing Yourself Ill': The Language of Sickness in Georgian England." In *Language, Self and Society: A Social History of Language*, edited by Peter Burke and Roy Porter. Cambridge: Polity Press, 1991.

———. *Patients and Practitioners: Lay Perceptions of Medicine in Pre-Industrial Society*. Cambridge: Cambridge University Press, 1985.

Porter, Roy, and Marie Mulvey Roberts. *Literature and Medicine during the Eighteenth Century*. New York: Routledge, 1993.

Porter, Roy, and George Sebastian Rousseau, eds. *The Ferment of Knowledge: Studies in the Historiography of Eighteenth Century Science*. Cambridge: Cambridge University Press, 1980.

Quinlan, Sean M. *The Great Nation in Decline. Sex, Modernity and Health Crises in Revolutionary France c. 1750–1850*. Aldershot, U.K.: Ashgate, 2007.

Ramsay, Matthew. *Professional and Popular Medicine in France, 1770–1830: The Social World of Medical Practice*. Cambridge: Cambridge University Press, 1988.

Rancière, Jacques. *The Flesh of Words: The Politics of Writing*. Translated by Charlotte Mandell. 1998. Stanford, Calif.: Stanford University Press, 2004.

Rausky, Franklin. *Mesmer; ou, La révolution thérapeutique*. Paris: Payot, 1977.

Rey, Roselyne. *Naissance et développement du vitalisme en France de la deuxième moitié du 18e siècle à la fin du premier Empire*. Oxford: Voltaire Foundation, 2000.

Rheinberger, Hans-Jörg. *Experiment, Differenz, Schrift: Zur Geschichte epistemischer Dinge*. Marburg/Lahn: Basilisken-Presse, 1985.

Rheinberger, Hans-Jörg, and Staffan Müller-Wille. *Vererbung: Geschichte und Kultur eines biologischen Konzepts*. Frankfurt am Main: Fischer, 2009.

Ricoeur, Paul. *Temps et récit*. Paris: Éd. du Seuil, 1983. Translated by Kathleen McLaughlin as *Time and Narrative*, vol. 1 (Chicago: University of Chicago Press, 2007).

Rieder, Philip. *La figure du patient au XVIIIe siècle*. Geneva: Droz, 2010.

Rigoli, Juan. *Lire le délire: Aliénisme, rhétorique et littérature en France au XIXe siècle*. Paris: Arthème Fayard, 2001.

Riskin, Jessica. *Science in the Age of Sensibility: The Sentimental Empiricists of the French Enlightenment*. Chicago: University of Chicago Press, 2002.

Risse, Guenter S. "Hysteria at the Edinburgh Infirmary: The Construction and Treatment of a Disease, 1770–1800." *Medical History* 32 (1988): 1–22.

Riva, Massimo. *Saturno e le grazie: Malinconici e ipocondriaci nella letteratura italiana del settecento*. Palermo: Sellerio Editore, 1992.

Robertson, Étienne-Gaspard. *Mémoires récréatifs, scientifiques et anecdotiques d'un physicien-aéronaute*. Vol. 1, *"La Fantasmagorie."* Langres: Clima, 2000.

Rocca, Julius. "Wiliam Cullen and Robert Whytt on the Nervous System." In *Brain, Mind and Medicine: Essays in Eighteenth-Century Neuroscience*, edited by Harry Whitaker, C. U. M. Smith, and Stanley Finger. New York: Springer, 2007.

Roche, Daniel. *A History of Everyday Things*. Cambridge: Cambridge University Press, 2000.

———. *Le siècle des Lumières en province: Académies et académiciens provinciaux, 1680–1789*. Paris: Mouton, 1978.

Roger, Jacques. *Les sciences de la vie dans la pensée française du XVIIIe siècle: La génération des animaux de Descartes à l'encyclopédie*. Paris: A. Colin, 1963. Translated by Robert Ellrich as *The Life Sciences in Eighteenth-Century French Thought* (Stanford, Calif.: Stanford University Press, 1997).

Rommevaux, Sabine. "Magnetism and Bradwardine's Rule of Motion in Fourteenth- and Fifteenth-Century Treatises." *Early Science and Medicine: A Journal for the Study of Science, Technology and Medicine in the Pre-Modern Period* 14, no. 6 (2010): 618–47.

Rosen, George. "Cameralism and the Concept of Medical Police." *Bulletin of the Medicine of History* 27, no. 1 (1953): 21–42.

———. "The Fate of the Concept of Medical Police, 1780–1890." *Centaurus* 50, nos. 1–2 (2008): 46–62. Originally published in Centaurus 5, no. 2 (June 1957): 97–113.

———. *From Medical Police to Social Medicine: Essays on the History of Health Care*. New York: Science History Publications, 1974.

———. *Madness in Society. Chapters in the Historical Sociology.* Chicago: University of Chicago Press, 1980.

Rosenberg, Charles, and Janet Golden, eds. *Framing Disease: Studies in Cultural History.* New Brunswick, N.J.: Rutgers University Press, 1992.

Rousseau, George Sebastian. *Enlightenment Borders: Pre- and Post-Modern Discourses; Medical, Scientific.* Manchester: Manchester University Press, 1991.

———. "Mysticism and Millenarianism: Immortal Doctor Cheyne." In *Hermeticism and the Renaissance,* edited by Ingrid Merkel and Allen G. Debus. Washington, D.C.: The Folger Shakespeare Library, 1988.

———. "Nerves, Spirits, and Fibres: Towards Defining the Origins of Sensibility." *Studies in the Eighteenth Century* 3 (1976): 137–57.

———. *Nervous Acts: Essays on Literature, Culture and Sensibility.* New York: Palgrave, 2004.

———. "Science and Discovery of the Imagination in the Enlightenment England." *Eighteenth Century Studies* 3 (1969): 108–55.

———. "A Strange Pathology: Hysteria in the Early Modern World, 1500–1800." In *Hysteria beyond Freud,* edited by Sander L. Gilman et al. Berkeley: University of California Press, 1993.

———. "Temperament and the Long Shadow of Nerves in the Eighteenth Century." In *Brain, Mind and Medicine: Essays in Eighteenth-Century Neuroscience,* edited by Harry A. Whitaker, Christopher Smith, and Stanley Finger. New York: Springer, 2007.

Saad, Mariana, ed. "La mélancolie et l'unité matérielle de l'homme—XVIIe et XVIIIe siècles: Introduction." *Generus* 63 (2006): 6–11.

Salomon-Bayet, Claire. *L'institution de la science et l'expérience du vivant: Méthode et expérience à l'Académie Royale des Sciences, 1777–1793.* Paris: Flammarion, 1978.

Schaffer, Simon. "Natural Philosophy and Public Spectacle in the Eighteenth Century." *History of Science* 21 (1983): 1–43.

Schefer, Jean Louis. *L'espèce de chose mélancholie.* Paris: Flammarion, 1978.

Schiebinger, Londa. *The Mind Has No Sex? Women in the Origins of Modern Science.* Cambridge, Mass.: Harvard University Press, 1989.

———. *Nature's Body: Gender in the Making of Modern Science.* Boston: Beacon Press, 1993.

Schmid, Thomas H. "My Authority: Hyper Mimesis and the Discourse of Hysteria in The Female Quixote." *Rocky Mountain Review of Language and Literature* 51, no. 1 (1997): 21–35.

Scott, Joan Wallach. *Gender and the Politics of History.* New York: Columbia University Press, 1988.

Scull, Andrew. *Hysteria: The Biography.* Oxford: Oxford University Press, 2009.

———. *Madhouse: A Tragic Tale of Megalomania and Modern Medicine.* New Haven, Conn.: Yale University Press, 2005.

———. *Social Order/Mental Disorder: Anglo-American Psychiatry in Historical Perspective.* Berkeley: University of California Press, 1989.

Seth, Catriona, ed. *Les rois aussi en mouraient: Les Lumières en lutte contre la petite vérole.* Paris: Desjonquères, 2008.

———. *Marie-Antoinette: Anthologie et dictionnaire.* Paris: Laffont, 2006.

Shapin, Steven. *Never Pure: Historical Studies of Science as If It Was Produced by People with Bodies, Situated in Time, Space, Culture, and Society, and Struggling for Credibility and Authority.* Baltimore: Johns Hopkins University Press, 2010.

————. "The Philosopher and the Chicken: On the Dietetics of Disembodied Knowledge." In *Science Incarnate: Historical Embodiments of Natural Knowledge*, edited by Christopher Lawrence and Steven Shapin. Chicago: University of Chicago Press, 1998.

Shapin, Steven, and Simon Schaffer. *Leviathan and the Air-Pump, Hobbes, Boyle and the Experimental Life*. Princeton, N.J.: Princeton University Press, 1985.

Sheriff, Mary. "Passionate Spectators: On Enthusiasm, Nymphomania and the Imagined Tableau." In *Enthusiasm and Enlightenment in Europe, 1650–1850*, edited by L. E. Klein and A. J. La Vopa. San Marino: Huntington Library, 1998.

Shin, Yang-Sook. "The Hysterical Structure of 'The Rape of the Lock'." *Journal of English Literature* 42, no. 3 (1996): 523–47.

Showalter, Elaine. "Hysteria, Feminism, and Gender." In *Hysteria beyond Freud*, edited by Sander L. Gilman et al. Berkeley: University of California Press, 1998.

Shuttleton, David E. "Methodism and George Cheyne's 'More Enlightening Principles'." In *Medicine and the Enlightenment*, edited by Roy Porter. Amsterdam: Rodophi, 1995.

————. "'Pamela's Library': Samuel Richardson and Dr. Cheyne's 'Universal Cure'." *Eighteenth-Century Life* 23.1 (1999): 59–79.

Siraisi, Nancy G. "Girolamo Cardano and the Art of Medical Narrative." *Journal of the History of Ideas* 52, no. 4 (1991): 581–602.

Sontag, Susan. *Illness as Metaphor*. New York: Farrar, Straus and Giroux, 1978.

Sournia, Jean-Charles. "La médecine révolutionnaire (1789–1799)." In *Médecine et sociétés*. Paris: Payot, 1989.

Starobinski, Jean. *Action and Reaction: The Life and Adventures of a Couple*. New York: Zone Books, 2003.

————. *Histoire de la médecine*. Lausanne: Éditions Rencontre, 1963.

————. *Histoire du traitement de la mélancolie des origines à 1900*. Bâle: Geigy, 1960.

————. *La Transparence et l'obstacle*. Paris: Gallimard, 2004. Translated by Arthur Goldhammer as *Transparency and Obstruction* (Chicago: University of Chicago Press, 1988).

Staum, Martin Sheldon. *Cabanis: Enlightenment and Medical Philosophy in the French Revolution*. Princeton, N.J.: Princeton University Press, 1980.

Staum, Martin S., and Donald E. Larsen, eds. *Doctors, Patients, and Society: Power and Authority in Medical Care*. Waterloo: Wilfrid Laurier University Press, 1981.

Steinke, Hubert. *Irritating Experiments: Haller's Concept and the European Controversy on Irritability and Sensibility, 1750–90*. Amsterdam and New York: Clio Medica, 2005.

Stengers, Jean, and Anne Van Neck. *Histoire d'une grande peur: La masturbation*. Brussels: Éd. de l'Université de Bruxelles, 1984.

Stolberg, Michael. "Medical Popularization and the Patient in the Eighteenth Century." In *Cultural Approaches to the History of Medicine: Mediating Medicine in Early Modern and Modern Europe*, edited by Willem de Blécourt and Cornelie Usborne. New York: Palgrave Macmillan, 2004.

Tallent, Alistaire. "The Hysterical Lesbian and the Unspeakable in Diderot's 'La Religieuse'." *Romance Review* (Fall 2002): 117–25.

Thomas, Chantal. *La reine scélerate*. Paris: Seuil, 1989.

Tort, Patrick. *La raison classificatoire*. Paris: Aubier, 1989.

Trevor-Roper, Hugh Redwald. *The European Witch-Craze: Of the Sixteenth and Seventeenth Centuries and Other Essays*. New York: Harper Collins, 1967.

Trillat, Étienne. *Histoire critique de l'hystérie*. Paris: Seghers, 1986.

Tronchin, Henri. *Un médecin du XVIIIe siècle, Théodore Tronchin. 1709–1781, d'après des documents inédits.* Paris: Plon Nourrit et Cie, 1906.

Vallot, Antoine, Antoine Daquin, and Guy-Crescent Fagon. *Journal de santé de Louis XIV.* Edited by Stanis Perez. Grenoble: Jerôme Million, 2004.

van Sant, Ann Jessie. *Eighteenth-Century Sensibility and the Novel: The Senses in Social Context.* Cambridge: Cambridge University Press, 1993.

Veith, Ilza. *Hysteria: The History of a Disease.* Chicago: University of Chicago Press, 1965, 1973.

Vidal, Daniel. *Miracles et convulsions jansénistes au XVIIIe siècle.* Paris: PUF, 1987.

Vidal, Fernando. *Les sciences de l'âme, XVIe-XVIIIe siècle.* Paris: Champion, 2006. Translated by Saskia Brown as *The Sciences of the Soul: The Early Modern Origins of Psychology* (Chicago: University of Chicago Press, 2011).

Vila, Anne C. *Enlightenment and Pathology: Sensibility in the Literature and Medicine of 18th Century France.* Baltimore: Johns Hopkins University Press, 1998.

———. "Sex and Sensibility: Pierre Roussel's *Système physique et moral de la femme.*" *Representations* 52 (1995): 76–93.

Vrettos, Athena. *Somatic Fictions: Imagining Illness in Victorian Culture.* Stanford, Calif.: Stanford University Press, 1995.

Wagner, Peter. "The Discourse on Sex—or Sex as a Discourse: Eighteenth Century Medical and Paramedical Erotica." In *Sexual Underworlds of the Enlightenment,* edited by G. S. Rousseau and R. Porter. Chapel Hill: University of North Carolina Press, 1988.

Weiner, Dora B. *The Citizen-Patient in Revolutionary and Imperial Paris.* 1993. Baltimore: Johns Hopkins University Press, 2002.

Wellman, Kathleen. *La Mettrie: Medicine, Philosophy, and Enlightenment.* Durham, N.C.: Duke University Press, 1992.

Wenger, Alexandre. *La fibre littéraire: Le discours médical sur la lecture au XVIIIe siècle.* Geneva: Droz, 2007.

———. "Le 'Procès Pierre Pomme': Légitimité professionnelle et violence polémique dans une querelle médicale du XVIIIe siècle." In *Le mot qui tue: Les violences intellectuelles de l'antiquité à nos jours,* edited by Vincent Azoulay and Patrick Boucheron. Paris: Champ Vallon, 2009.

Wild, Wayne. *Medicine-By-Post: The Changing Voice of Illness in Eighteenth-Century British Consultation Letters and Literature.* Amsterdam and New York: Rodopi, 2006.

Williams, Elizabeth A. *A Cultural History of Medical Vitalism in Enlightenment Montpellier.* Aldershot, U.K.: Ashgate, 2003.

———. "The French Revolution, Anthropological Medicine and the Creation of Medical Authority." In *Re-Creating Authority in Revolutionary France,* edited by Bryant T. Ragan and Elizabeth A. Williams. New Brunswick, N.J.: Rutgers University Press, 1992.

———. "Hysteria and the Court Physician in Enlightenment France." *Eighteenth-Century Studies* 35, no. 2 (Winter 2002): 247–55.

———. *The Physical and the Moral: Anthropology, Physiology, and Philosophical Medicine in France, 1750–1850.* Cambridge: Cambridge University Press, 1994.

Williams, Katherine. "Hysteria in Seventeenth-Century Case Records and Unpublished Manuscripts." *History of Psychiatry* 1, no. 4 (1990): 383–401.

Wilson, Lindsay. *Women and Medicine in the French Enlightenment: The Debate over Maladies des Femmes.* Baltimore: Johns Hopkins University Press, 1993.

Yeo, Richard. "Classifying the Sciences." In *The Cambridge History of Science,* vol. 4, edited by Roy Porter. Cambridge: Cambridge University Press, 2003.

Index

The letter *f* following a page number denotes a figure.